T0360544

A Cybernetic View of Biological Growth: The Maia Hypothesis

Maia is the story of an idea, and its development into a working hypothesis, that provides a cybernetic interpretation of how growth is controlled. Growth at the lowest level is controlled by regulating the rate of growth. Access to the output of control mechanisms is provided by perturbing the growing organism, and then filtering out the consequences to growth rate. The output of the growth control mechanism is then accessible for interpretation and modelling. Perturbation experiments have been used to provide interpretations of hormesis, the neutralization of inhibitory load and acquired tolerance to toxic inhibition, and catch-up growth. The account begins with an introduction to cybernetics, covering the regulation of growth and population increase in animals and man, and describes this new approach to access the control of growth processes. This book is suitable for postgraduate students of biological cybernetics and researchers of biological growth, endocrinology, population ecology and toxicology.

TONY STEBBING has been at the Plymouth Marine Laboratory (PML) since its inception in 1971. He worked initially to develop bioassay techniques for pollution studies. His later discovery of the stimulatory effect of low concentrations of toxic substances ('hormesis') led him to develop a novel method of accessing the output of growth control mechanisms. The coordination of scientific programmes occupied the later years of his career. He was awarded an Honorary Fellowship from PML, which has given him the opportunity to write this account of his research and its implications.

The marine colonial hydroid *Laomedea flexuosa*, used as a clone in the early experiments described in this book. I thank Peter Stebbing for the drawing; permission to use the image was given by The Royal Society.

A Cybernetic View of Biological Growth

The Maia Hypothesis

TONY STEBBING
Plymouth Marine Laboratory
(Honorary Fellow)

CAMBRIDGE
UNIVERSITY PRESS

University Printing House, Cambridge CB2 8BS, United Kingdom

One Liberty Plaza, 20th Floor, New York, NY 10006, USA

477 Williamstown Road, Port Melbourne, VIC 3207, Australia

314-321, 3rd Floor, Plot 3, Splendor Forum, Jasola District Centre, New Delhi - 110025, India

103 Penang Road, #05-06/07, Visioncrest Commercial, Singapore 238467

Cambridge University Press is part of the University of Cambridge.

It furthers the University's mission by disseminating knowledge in the pursuit of education, learning and research at the highest international levels of excellence.

www.cambridge.org
Information on this title: www.cambridge.org/9780521199636

© Tony Stebbing 2011

This publication is in copyright. Subject to statutory exception and to the provisions of relevant collective licensing agreements, no reproduction of any part may take place without the written permission of Cambridge University Press.

First published 2011

A catalogue record for this publication is available from the British Library

Library of Congress Cataloging in Publication data
Stebbing, Tony.
A cybernetic view of biological growth : the Maia hypothesis / Tony Stebbing.
p. cm.
Includes bibliographical references and index.
ISBN 978-0-521-19963-6 (hardback)
1. Growth–Regulation. 2. Biological control systems. 3. Cybernetics.
I. Title.
QP84.S727 2011
612.6–dc22
2010035500

ISBN 978-0-521-19963-6 Hardback

Cambridge University Press has no responsibility for the persistence or accuracy of URLs for external or third-party internet websites referred to in this publication, and does not guarantee that any content on such websites is, or will remain, accurate or appropriate.

Contents

Foreword

Compared with the high-profile discoveries of molecular genetics concerning individual genes and their specific expression at the tissue and organ level, studies of general patterns of growth in living organisms often receive less public attention in the media. Yet an understanding of the processes of biological growth at the level of the individual and in populations of organisms is key to human survival. Uncontrolled growth is a feature of human cancer cells which prejudices survival at the level of the individual, while unchecked growth of global human population density has to be considered against a background of finite limitations of planetary living space and natural resources. Indeed, this book begins with the premise that the basis of all biological growth is exponential, with an inherent tendency for constant doubling with time. A priori, therefore, it is argued that there would be evolutionary adaptive advantage to be gained by the acquisition of mechanisms which induce self-limitation of growth processes and therefore enhance the chances of survival of individuals and populations of organisms, including humans.

The author's interest in these phenomena started when he was studying the biological effects of toxic chemicals discharged into the sea. In laboratory experiments, the growth of simple marine organisms such as hydroid coelenterates was tested in response to the application of various toxins. As expected, growth of the test organisms was inhibited following exposure to high concentrations of chemical toxins such as copper. However, at low concentrations of some normally toxic agents it was found that growth rate was in fact enhanced. This phenomenon, called 'hormesis', was presumed to occur when a naturally occurring growth inhibitory mechanism, which normally modulates a growth promoter, was partially disinhibited. In a practical sense the phenomenon of hormesis demonstrates that the test organisms, and therefore naturally occurring species, may acquire some tolerance to toxic agents. However, in a theoretical sense it permits the development of ideas throughout this book concerning a wide range of biological growth processes which can be analysed from a cybernetic point of view and which are shown to be self-regulatory. Experimental evidence for self-limiting growth, generated by interacting promoter and inhibitory processes, is provided for unicellular and multicellular organisms, leading to the notion of the 'Maia hypothesis', deriving analogously from James Lovelock's grand-scale Gaia hypothesis but named after the Roman goddess of growth, who gave her name to the month of May.

The basic hypothesis is then extended from the physiological level in individual organisms to animal populations which, by analogy, show reduced rates of reproduction induced by negative feedback mechanisms when an animal population threatens to overburden the carrying capacity of its habitat. Finally, developing the concept of population-density control achieved by animal populations in nature, the book turns to the implications for humankind on Earth. It is argued that human overpopulation has been recognised for over a hundred years and is set to continue to increase, notwithstanding a fall in the rate of increase in recent decades. In a naturally occurring situation it is inferred that self-regulatory population density mechanisms would have come into play by now, but it is argued that the extent to which humans have manipulated their environment in recent centuries has overridden the onset of inhibitory population control. Political recognition of the Earth's carrying capacity for the human race is not possible at the present time, but scientific estimates suggest that sustainable population density had already been exceeded by the end of the twentieth century, as evidenced by the critical manifestation of anthropogenic global climate change. On this basis the book raises the challenge for humanity to move towards a stable global population and a steady-state world economy. The challenge is immense and provocative. However, it is argued that on a planet with a seemingly continuously increasing human population and finite global resources, the challenge for human ingenuity is to devise feedback mechanisms which would bring the world population into something like a steady state, or to live with the consequences of uncontrolled growth.

The book has much to offer as a rational exposition in current debates concerning human population density on Earth and the 'belief' that anthropogenic global climate change is a myth.

Emeritus Professor Ernest Naylor
School of Ocean Sciences
Bangor University
Wales

*Preface: 'A fragment of a possible world'**

Anyone who is practically acquainted with scientific work is aware that those who refuse to go beyond the facts, rarely get as far.

Thomas H. Huxley

To give priority to invisible process, rather than tangible structure.

Douglas Hofstadter

This book is the narrative of an idea about the control of biological growth that has developed over the course of my career. At an early stage it was given the name Maia, after the Roman goddess of spring, growth and fertility. The origin of the idea goes back to attempts to explain how it could be that toxic agents at low concentrations have the improbable consequence of stimulating growth. Later, perturbation experiments revealed the output of a control mechanism responsible for regulating growth. This breviary of the Maia hypothesis explains the central idea of the book. All that follows derives from the key finding of how to access the output of growth control mechanisms, as ultimately growth must be controlled and constrained. It is one of many homeodynamic processes, all controlled by cybernetic mechanisms. This account is pitched to make the biology accessible to cyberneticists, and the cybernetics accessible to biologists. These preliminary remarks therefore give the crux of the idea, summarising its development, in a few paragraphs, from growth experiments with simple organisms to the point where Maia bears on the two engines of growth on Earth.

During early sensitivity tests of a seawater bioassay with a clonal hydroid (see Frontispiece), it was discovered that low concentrations of toxic agents stimulate biological growth ('hormesis'). It was suspected

* Quoted by P.B. Medawar (1969) from the German philosopher and psychologist Oswald Kulpe (1895).

that such paradoxical stimulation was not an effect of the agent itself, but was due to an adaptive and generalised response to toxic inhibition. This led to the finding that the specific growth rate is maintained at a constant velocity optimal for the physiological processes involved in biosynthesis. There emerged a method that could be used to access the output of growth control mechanisms, thus revealing the alluring oscillatory behaviour of feedback mechanisms. An understanding of acquired resistance to toxicity followed, with hormesis as a byproduct. Data from perturbation experiments made it possible to establish the limits of the counter-response, and its capacity to neutralise toxic inhibition. The capacity to neutralise toxic inhibition led to a re-interpretation of the toxicologists' 'dose–response' relationship. Growth is considered here in its broadest sense, from physiological growth of tissues to the growth of populations of organisms. Although these fields belong to different branches of biology, they are related mathematically and by descent.

A key part of this story is that early in evolution there must have been a significant advance in the evolution of growth control from plants to animals. Autotrophs (plants) make their own organic constituents from inorganic materials, typically by photosynthesis, and gave rise to organotrophs (animals) that consume organic carbon created by plants and other animals. A necessary corollary was a change in the way that growth and reproduction were controlled.

In the ocean, the autotrophs responsible for photosynthesis are algae, which generate half of all the oxygen created on Earth. Within the space of a few weeks, offshore microalgae grow to form vast blooms, before they die back as nutrients are used up. Their life strategy is to grow rapidly and strip all the available nutrients from the seawater; as the nutrients are used up, the microalgae form spores capable of remaining dormant for many months until the nutrients have been recycled and favourable conditions for growth return. Autotrophs control their internal processes of biosynthesis and replication, kept constant by means of a specific rate-sensitive feedback mechanism, which has the consequence of generating exponential population growth. Such growth may be adaptive at first, but becomes unsustainable later. Survival of this 'boom and bust' lifestyle is made possible by spore formation and dormancy, providing an opportunity to literally 'drop out' of the photic zone. A discontinuous life inevitably limits the scope of such organisms, but a continuous life can only become sustainable with the self-limitation of exponential growth.

Organotrophs evolved from autotrophs and depend on consuming organic compounds, typically as microscopic algae, planktonic invertebrates and their products. If organotrophs were to behave as autotrophs do, they would rapidly extinguish the plants and animals on which they depend, and ultimately bring about their own demise. So to ensure survival, organotrophs must avoid extinguishing food species, and exploit them at a level that allows the food species a sustainable existence, and thus ensure their own survival. The evolutionary solution to this problem was the superimposition of a population self-limitation loop on the rate-control mechanism inherited from autotrophs. This has the advantage of allowing first the maximum growth rate made possible by positive feedback, leading later to self-limitation by negative feedback. For the organotroph, it is considered probable that these requirements are met in the logistic equation as a nested, dual loop control mechanism, in which the inner loop controls the specific rate of biosynthesis, while the cumulative products of growth are monitored by the outer loop, which imposes a limit on population size.

The logistic control mechanism allows positive feedback, when a 'virtuous' circle and exponential growth has survival value, but prevents a 'vicious' circle when growth threatens 'runaway' and instability. So when exponential growth becomes maladaptive, self-limitation by the outer loop cuts in and arrests further growth. In organotrophs therefore, it is proposed that rate control is coupled to the self-limitation of population growth. A key property of population self-limitation is that the maximum sustainable carrying capacity becomes the goal setting for the growth control mechanism, such that numbers can be maintained within the limits of the habitat to support them. So with exponential increase constrained, there is advantage both for the consumer and the consumed, and so organotrophs evolved to lead a continuous and sustainable life, when conditions for growth allow. The logistic equation viewed as a control mechanism provides a means of maintaining sustainable populations. Such control mechanisms are conjectural, yet provide a minimal mechanism that can reproduce the population growth behaviour of such organisms.

Multiplication is not reserved for individual organisms, as there are modular replicators in all organisms. By a process termed 'endosymbiosis', some micro-organisms became internal modules of higher organisms. J.B.S. Haldane (1892–1964) suggested that inhibitory exocrines, which limit population numbers, may have become the endocrines that constrain the multiplication of self-replicating modules within metazoa, anticipating the development of the idea of endosymbiosis. The limits to the number of cells within a tissue are genetically prescribed, and are imposed by such specific inhibitors as endocrines, apparently descended from specific inhibitors known as population limiters in many aquatic species from protozoa to amphibians. So it seems likely that a mechanism that evolved in free-living unicells and their populations later became internalised by more complex life forms, providing control and limitation to those replicating modules that make up the bodies of metazoa. Here the logistic control mechanism is adopted once more as a minimal control mechanism for replicators, as their requirements for growth control are similar. The control mechanism is used as a conceptual model to interpret failure of control in neoplastic growth. Cancer is understood as a failure by some cells to respond to specific inhibitors that, within tissues, limit cell numbers. Not only does limitation cease, but regulation reverts to the inner loop, which continues to maintain growth at a constant specific rate, resulting in a positive feedback. Consequently, unlimited exponential increase ensues, and the growth of cell numbers proceeds without limit.

It can then be asked, if self-control and self-limitation are such important attributes of all biological systems, what of human populations? Early man was a hunter-gatherer, and so sparsely distributed that density-dependent limitation was ineffective, because space was not limiting. We can therefore not expect any intrinsic mechanism to provide human population regulation of the kind found in animals. In the absence of the ability to limit population growth, it reverts to control by the inner loop alone, and unlimited population growth. Self-limitation characterises biological populations from unicells to mammals, yet ironically, it appears that our species lost the innate capacity to limit population numbers. As a result, our numbers have already exceeded the carrying capacity of the Earth for humanity, which has been estimated at 5 billion using ecological footprint analysis.

In 1992, D.H. Meadows, D.L. Meadows and J. Randers in *Beyond the Limits* expressed the fear that positive socio-economic feedback loops induce the exponential growth of the human population and economic capital. Biological growth and the growth of economic capital have the same structure and systems, in that they both create growth by positive feedback due to birth or investment, and slow down due to negative feedback by death or the depreciation of capital assets. Not unlike biological growth, the investment of capital earns interest, leading to more capital and so more interest. Both biological and economic increase is exponential, which without constraint ultimately becomes destabilising 'runaway'. Self-limitation of the kind found in animal populations is apparently due to the limitation of population size in relation to the maximum sustainable carrying capacity of its habitat. If we take biological systems as exemplary of sustainability, this tells us that the size of the human population, and of the economy, both need to be limited in relation to the ecological capacity of the Earth to assimilate the consequences of human activity. Presently this is not happening.

Power is required to drive the economy, which continues to be provided primarily by fossil fuels, resulting in carbon dioxide emissions that are increasing beyond the capacity of natural processes to return them to the earth. The result is global warming and the destabilisation of the global climate system, which will itself ultimately impose its own limit on growth. The point is that the Earth is finite, and that human populations and economic growth must ultimately be constrained by the capacity of the Earth to assimilate the polluting byproducts of man and his economic activity. Climate change shows us that population and economic growth have already exceeded the capacity of the Earth to draw down atmospheric carbon dioxide. This is not only due to the release of carbon dioxide, but also to the clearing of the forests that provide the natural means of drawing down carbon dioxide. In the absence of humanity choosing to limit the growth of the economy at a sustainable level, the effect of carbon dioxide as the principal metabolite of economic growth will impose its own constraints. Such effects can already be recognised in the various facets of man-made global change.

I read James Lovelock's book *Gaia* in 1979, when it was first published, and realised that I was grappling with a similar problem. Like Lovelock, I had strong evidence of a cybernetic mechanism at work, but there was, it seemed, no known control mechanism on which the evidence could be hung. Maia is a much younger, fraternal hypothesis that still needs to establish itself as a factor in the way in which the biosphere regulates growth processes. This book provides evidence that for growth at all levels, from cells within organisms to populations of free-living populations of organisms, growth rate is controlled and numbers are limited in relation to the capacity of their interior or exterior environment to sustain them. I dedicate this volume to James Lovelock, in appreciation of his achievement in demonstrating Gaian control, and in gratitude for his inspiration to a young scientist over 30 years ago when this hypothesis was in its formative stages.

Acknowledgements

A list of sources is given for further reading, and scientific papers are cited for those readers who wish to explore more deeply the ground I cover. At the end of each chapter there is a short summary, which may be read either as an introduction, or in retrospect. In some cases, findings that are only now coming to fruition had their origin in research carried out 50 years ago or more, and some have been traced from their origins to the present time.

I am indebted and thankful to Professor Ernest Naylor, my tutor at university, and mentor since, for writing the Foreword to this book. I have held an Honorary Fellowship at Plymouth Marine Laboratory for the 12 years since my retirement in 1998. All of the Directors – Fauzi Mantoura, Nick Owens and Steve de Mora – have supported and encouraged the writing of this book.

The book would not have been published without the contribution of my good friend Geoff Smaldon, whose experience in science publishing has proved invaluable in providing guidance and understanding. I am grateful to have the encouragement of Gordon Heath, and his collaboration in the development of ideas on hierarchy and the economy, particularly during the early stages in the development of Maia. I am also thankful to my friends and colleagues, Arnold Taylor, John Widdows and Les Wyatt, who have been prepared to listen to and provide helpful criticism of my ideas. I was fortunate at an early stage to have as collaborators Lex Hiby and John Norton, whose control models helped provide conviction that the experimental data really did reflect the output of growth control mechanisms operating within biological systems. Through most of the intensive period of experimental work on hydroids and yeast I was also fortunate to have the help of Mary Brinsley. Much later, in retirement, she also transformed my rough diagrams into the electronic figures used here.

No writer can do without critical readers. I am grateful to Tim Stebbing, who read the whole book (and some chapters twice), making a substantial contribution to the book. Arnold Taylor read the early chapters and Chapter 8, and Dick Woodley read over half of the book. The following people read and provided criticism on certain chapters: David Livingstone (Chapter 6), Mike Moore (Chapters 8 and 9), Ed Calabrese (Chapters 9 and 10), Clive Peedell (Chapter 12) and Renxiang Ding (part of Chapter 13). To each I give my grateful thanks for their contributions.

I also thank John Bury for his help in understanding the management practices that allow the commons of the New Forest to remain so after many centuries. He was a past Official Verderer of the New Forest, a position dating back to the thirteenth century. I am most grateful to my dear friend Stella Turk, who gave me free range of her husband Frank's library, and allowed me to take whatever I needed. Last, but first in my thoughts, I am ever-thankful to my wife Valerie for her love, understanding and encouragement throughout the writing. The book could not have been written without her.

Throughout the writing I have had the benefit of the Library of the Marine Biological Association, and the help of the Librarian and her staff. I must also acknowledge the debt owed to Wikipedia, as an invaluable avenue of enquiry.

I am indebted to my commissioning editor, Dominic Lewis, who had faith in the idea when others did not. I am grateful also to Rachel Eley and Laura Clark at Cambridge University Press, who worked hard to get the book to press. I would also like to express my appreciation and thanks to Nancy Boston for her care and thoroughness in editing the book.

Any errors, lapses in citation, or oversights in seeking permissions are my own responsibility.

1

Introduction

The broadest and most complete definition of Life will be – The continuous adjustment of internal relations to external relations.

Herbert Spencer

The truths of cybernetics are not conditional on their being derived from some other branch of science. Cybernetics has its own foundations.

W. Ross Ashby

Among the most fertile ideas introduced into biology in recent years are those of cybernetics ... control theory obtrudes everywhere into biology.

Peter B. Medawar

THE DEVELOPMENT OF BIOLOGICAL CYBERNETICS

I watched a kestrel from a building high on the Citadel overlooking Plymouth Sound. It hovered at arms' length from the window in a stiff breeze, holding its position for long periods so perfectly that one could not detect the slightest movement of its head. To maintain this static hover, its wings beat quickly, varying a little in frequency and angle to adjust to the buffeting breeze, as it watched intently for any movement of prey on the ground below. It reminded me of Gerard Manley Hopkins' lines from his poem The Windhover: 'how he rung upon the rein of a wimpling wing'. Occasionally the bird sheared away: 'then off, off forth on swing'; to return again and take up its position as before, as if to dispel any doubts I might still have of its control in the air. 'The achieve of, the mastery of the thing!'

In the same way our control systems master the continuous variation in the environment in which we live, holding steady a host of different internal processes against the continuous fluctuation of external change. In warm-blooded creatures, control over temperature is within a degree of 37°C. The precise control over our internal temperature is just as astonishing as the kestrel's control of flight but, unlike the kestrel in flight, the continuous adjustment of temperature is invisible to us, and is therefore often overlooked. The kestrel hovers, its wings beating rapidly, its tail making continuous small adjustments, working hard to keep its eye steady for prey. But we are all but oblivious of the subtle and continuous adjustments of our own thermal regulation, through the many changes in temperature of a typical day; leaving a warm house on a winter's morning, or running to catch a train, or returning home and climbing into a hot bath. All these activities shift temperature from its preferred setting, and their effects are neutralised to keep internal temperature constant.

It is as if a guiding hand magically regulates the control of each process in the inner workings of the body. When applied to the hundreds, if not thousands, of control mechanisms, the effect can be called *The Wisdom of the Body* [1], the title given by Walter B. Cannon (1871–1945) to his account of the concept of homeostasis. His discovery of the mechanisms by which sugar and salt, temperature and oxygen are kept constant in the living body led him to the general principle that he called 'homeostasis'. This remains the single most important concept in our understanding of the physiology of animals and humans.

Yet it was the Frenchman Claude Bernard (1813–1878) who had earlier recognised the 'constancy of the internal environment' (see [2]), and his interpretation was more eloquent. He wrote that this equilibrium 'results from a continuous and delicate compensation established as if by the most sensitive of balances', with the consequence that 'the perpetual changes in the cosmic environment do not touch it; it is not chained by them, it is free and independent'. It is this meaning of homeostasis that we shall take from Bernard's writing, with an outcome that, significantly, has less to do with the constancy of control than the capacity to resist external variation. Throughout evolution, the gradual liberation of living organisms from their susceptibility to the continuous fluctuation of the major environmental variables has been made possible by their increasingly complex functionality. By such adaptations, their internal workings became protected more completely from everything that varies outside the body. Control mechanisms work continuously to achieve internal constancy by countering external change, just like the 'windhover' in a breeze.

Of course, both definitions mean the same thing, and the illusion of constancy obscures what must be done to achieve it.

PROCESSES AND THINGS

Some aspects of natural features in the world around us are more accessible than others. An oak tree is a large and immovable 'thing', with elements of which we are well aware: a trunk, branches and leaves. But 'processes' are less accessible, such as the rate at which water is transpired from roots to leaves, before being evaporated into the air. At any particular moment in time, an oak tree is perceived as a 'thing'; it grows slowly, and measurements are required over a long time period to determine its growth curve. Growth is a 'process', as it can only be measured with respect to the passage of time. So 'things' are tangible, but 'processes' are invisible and happen over time, and so are much less accessible. To access a 'process' requires that we make observations of a 'thing' at intervals over time, and make a graph of its progress with respect to time. We then draw a line between the points, and assume that interpolation can tell us what happened in the intervals between measurements. This involves an assumption, but we can make more frequent observations to give a more accurate depiction of the rate of growth, and so visualise the 'process' more clearly. After all, it is what organisms *do*, rather than what they *are*, that is more important, so the focus of our study should be change due to 'processes', rather than a static 'thing' approach. Conrad Waddington (1905–1975), the inspirational biologist, made a clear distinction between a 'thing' approach and a 'process' approach to understanding life, saying that the 'things' we see are simply 'stills' in a movie, which we can access in various ways [3]. To understand growth, we must turn changes in size or number into a movie.

The second problem in studying growth is that the agents of the control of growth are not well known. That is not strictly true because endocrinology is all about the agents of control, and we know that various hormones are involved in the control of growth. But the feedback mechanisms responsible for control consist of several essential components that include a sensor of the controlled process, an adjustable goal setting and a comparator where a sensed rate and the pre-set goal come together to determine whether there is an error. In those control mechanisms that are known and understood, the seat of control may take a lifetime's work to find and involve just a few cells. The organs of control are not like other organs; they are small, obscure and are not recognisable by what they do. This makes them extremely difficult to investigate. It took 70 years to discover the mammalian thermostat, and much painstaking research to understand its operation, once its seat was known.

Such difficulties legitimise another approach, which is to apply the principles of control theory and assume the minimal form that a control mechanism could take, given its output. It is then possible to create a hypothetical control mechanism that is able to account for the observed output. Then, by experimentally perturbing the growth process, various forms of behaviour become apparent. These can then be used as a basis for elaborating the form of the hypothetical control mechanism required to produce each new aspect of behaviour. So this has been the approach taken here to tell the story of growth as an invisible process controlled by inaccessible mechanisms.

CONTROL MECHANISMS AND THEIR ORIGINS

Maia is the story of an idea, and its development over 30 years, which provides a different approach to understanding the control of biological growth. Key to this approach is an experimental method of accessing growth control output which would be normal in control engineering, but is much less common in biology. To observe the output of a homeodynamic system when it is at rest is uninformative, because the controlled process or state is stable and constant. Perturbation is necessary in order to deviate the system from equilibrium and then observe the response of the control mechanism as it restores the equilibrium. Understanding then lies in interpreting the characteristic oscillatory output.

As already mentioned, a control mechanism has three basic components. The first is a goal, or preferred setting, that represents some ideal rate or state that is optimal for the organism. The second is a comparator by which the actual state of affairs in the present is compared with the goal setting. The difference between them constitutes an error. The third is the controller that provides the means of minimising the error. It is unclear how such a mechanism might have evolved, because the components are interdependent and it appears that none would have survival value on its own.

Homeostatic mechanisms have grown in number and complexity throughout evolution, giving mastery of the environment to those animals, such as the mammals, in which internal control is highly developed. It has enabled animals to withstand change, keeping processes constant; as steady as the eyes of the 'windhover', focused on its prey, before it plunges to the ground. Homeodynamic mechanisms control the inner workings of all organisms, from protozoans to mankind. The workings of a single cell are reputed to be regulated by a thousand control mechanisms, regulating the operation of the genome, metabolic pathways, osmoregulation and so on. But it is the increase in the number and sophistication of homeodynamic systems that enables the most complex organisms to become 'free and independent' of the changes around them. From the Khoikhoi in the tropics to the Inuit of the Arctic, the human brain is regulated at an equable 37°C. This trend of increasing freedom from environmental change is called 'anagenesis' and has been apparent for at least half a billion years: a driving force in the progress of evolution towards greater complexity and adaptive functionality.

Each cybernetic mechanism is defined by a goal setting, a preset state or rate or a set point. Its purpose for life is to maintain some state or process close to a setting. Each goal is an item of memory that is not necessarily stored in the brain, and is not simply remembered, but is referred to frequently to determine errors and maintain control. Control mechanisms are sometimes referred to as 'error minimisation systems', always working to reduce the difference between actuality and the goal setting. What is more, these goals are self-adjustable to meet increases in work load, adapting to change throughout the lifetime of the organism.

Together, the organism's homeodynamic goals constitute a diffuse body of extra-genetic information. Collectively these preferenda are life's purposes, providing the template by which an organism maps to its ever-changing habitats. It is this body of cybernetic information, and the mechanisms that continually refer to it, that fit an organism to its particular ecological niche and make its life possible. If one needs to reconsider the idea, one does not have to look further for a 'life force', for when prescribed preferenda, along with their coupled feedback mechanisms, no longer refer to their pre-set goals, there can be no life. What is curious, but completely understandable, is that evolution is blind and mechanistic, and so cannot create anything capable of meeting a future purpose; has it created myriads of homeodynamic machines that each incorporate direction, drive and future purpose? These are teleological mechanisms.

The science of control is cybernetics and all homeodynamic systems operate according to its principles. A.A. Lyapunov (1911–1973), the Russian mathematician, recognised that 'control, in its broadest sense, is the most universal property of life, independent of form' and that it must be possible 'to describe living systems from a cybernetic point of view'. He saw cybernetics as a concept central to biology, by providing the underlying principle of self-regulation.

Norbert Wiener (1894–1964) was an eccentric, stout and short-sighted polymath; a child prodigy driven by his father. He became a leading mathematician of his time, who wanted to make a telling contribution to the outcome of World War II. This he did by his work on the automatic directing of anti-aircraft fire, which led him to develop his ideas on the theory of control. In 1948 he brought them together in the book *Cybernetics*, for which he is best known. It also marked the birth of a discipline and the name he gave it, in what was the first theoretical treatment of the subject. Wiener's book had the subtitle *Control and Communication in the Animal and the Machine*, which linked the theory of control in technology with that in biology. The link was primarily due to Wiener's collaboration over many years with his friend Arturo Rosenblueth (1900–1970). Significantly, he had previously been the colleague and collaborator of Walter Cannon, and the connection ensured that there would remain close links between cybernetics and homeostasis.

THE EMERGENCE OF BIOLOGICAL CYBERNETICS

The Macy Conferences established cybernetics as a multidisciplinary concept. They were held in New York from 1946 until 1953 and were designed to explore the implications of the recent discovery of cybernetics. There were three British-based scientists who contributed to these meetings, and who brought news of the fast-growing field of cybernetics back across the Atlantic. They were William Grey Walter, W. Ross Ashby and J.Z. Young; each of them spread the word and incorporated the central ideas into their work and through their books. Grey Walter's paper at the 1953 Macy Conference was on *Studies on the activity of the brain*, Young's paper at the conference in 1952 was on *Discrimination and learning in Octopus*, while Ashby's paper at the same conference was on *Homeostasis*. Grey Walter was the first to show that simple control devices could learn and produce lifelike behaviour, while Ashby developed an electronic analogue which he called a Homeostat. For Young, homeostatic principles became important in his work on the brain of the octopus. He made homeostasis the overarching concept in his great volume *The Life of Mammals* [4], which has been a textbook for generations of zoologists. In the first chapter he traced its significance for the reader, as a thread that runs throughout the volume. Over 40 years later, Elaine Marieb adopted a similar focus in her medical textbook *Human Anatomy and Physiology* [5], with sections on homeostasis in each chapter. In this way, generations of students have been introduced to homeostasis as a vital concept with which to understand life.

Nevertheless, the term 'cybernetics' has become less used, because the word has been corrupted to create jargon such as 'cyberspace', 'cyborg', 'cybermen' and 'cybernaut'. Cybernetics is now more often thought of as a branch of control theory, but Wiener chose his term so carefully that it would be a pity to lose one so full of meaning. 'Cybernetics' is derived from the Greek word *kybernetes*, meaning steersman, recognising that the earliest feedback mechanisms were the 'steering engines', or servo-mechanisms, designed to steer ships. He also wanted to acknowledge the first theoretical work on feedback mechanisms, published in 1868 by James Clerk Maxwell (1831–1879) in a paper entitled *On governors* [6]; 'governor' was a Latin corruption of *kybernetes*. Maxwell's paper was on the centrifugal governor, credited to James Watt, which controlled the steam engines of the Industrial Revolution, preventing 'hunting' and 'runaway'. So one cannot avoid the conclusion that there are good grounds for reverting to Wiener's original term.

Cybernetics is an important part of 'systems theory', which includes the study of any system that can be thought of as a group of related elements organised for a particular purpose. The abstraction is valuable in recognising the properties of systems like networks, hierarchies and feedback loops. The concepts of systems theory transcends many disciplines, as they are portable and can be applied wherever control is involved. Norbert Wiener developed the theory of cybernetics to direct anti-aircraft guns, but the same theoretical understanding was soon applied to the control of distribution of electricity from power stations, and provides insights into Parkinson's disease. It is the extraordinary versatility of the central ideas of cybernetics that makes them so powerful. Here examples from technology and everyday life will be used to make the cybernetic behaviour in animal and human biology more accessible.

FOCUS ON GROWTH AND ITS CONTROL

Despite the generality of cybernetics in biology, here the focus is on growth and its control in biology. The tendency of living things to increase in size is one of the defining features of life. J.Z. Young (1907–1997) wrote, 'The study of growth consists largely in the study of the control and limitation of growth.' Failure to impose control is potentially disastrous, whether the unfettered growth of cells that causes cancer, to growth in the number of people who inhabit the Earth. Each shows the consequences of regulatory systems out of control. Nevertheless, cybernetics and biological growth took a longer time to come together than might have been expected. In the 1940s, when the young Peter Medawar (1915–1987) was boldly assembling the 'laws of biological growth' [7], he showed why the measure of growth called the 'specific rate' is physiologically the most appropriate measure of growth. He hinted that, as it is constant in populations of single cells, it reflects the rate of the underlying biosynthetic processes, when teased from their cumulative product. Yet in the 1970s, cancer researchers were still trying to understand growth in a cumulative sense rather than as a rate-controlled process. Nor did they see the value of perturbation in order to understand control and its failure. Ross Ashby's Law of Requisite Variety stated that, for any homeodynamic mechanism, disturbance is neutralised by a response that gives a stabilised outcome. To neutralise disturbance, and restore the equilibrium, a counter-response is required to achieve a balance of opposing forces. Perturbation is the novel element to the approach taken here: an approach that is natural to a cyberneticist, but has not been so for students of biological growth.

The idea developed here became the Maia hypothesis, named after the Roman goddess of growth and increase. In temperate northern climes, the spring outburst of growth occurs predominantly in the month of May, to which Maia gave her name. The approach considers growth as a controlled process: a hypothesis that demonstrates and aims to explain growth control from a cybernetic perspective. Apart from perturbation, a subtle difference in approach is to think of growth not so much as the maintenance of constancy, but rather as the resistance of internal processes to external change.

The first few chapters deal with the problem of biological growth in general terms, together with the origin and application of cybernetics to biological processes. 'Growth' means increase in cell size together with the multiplication of cells by which animals and plants grow in size, but also the growth of populations of free-living cells and organisms. Enlargement in size and increase in population number are both referred to as growth, and involve processes that are analogous and homologous. Rather than distinguishing between them, the aim is to consider growth in its broadest sense, and to draw together the physiology and ecology such that the understanding of one can inform the other.

The products of growth and replication grow arithmetically in the same way (Chapter 2); growth is a multiplicative process of the kind seen in nuclear fission or in the increase in money through compound interest. Growth accelerates at an increasing rate because the products of growth also grow. The products of replication also replicate, whether through the increase in number of cells within a tissue, or the individuals within a population. This is exponential growth, which makes populations double in size at regular intervals. However, unlimited exponential growth becomes unsustainable and can have dire implications for survival. For this reason, mechanisms have evolved which limit growth and multiplication, thus avoiding problems of excess and instability. Everyday examples abound, from the growth of the bacteria that sour milk, to the spread of duckweed over a pond in summer. Such growth can be compared to a forest fire, which spreads at an accelerating rate, doubling in size every 30 seconds or so at first, although it must eventually burn out for lack of fuel. We know that biological growth must ultimately be constrained, as exponential growth cannot be sustained indefinitely. But is limitation due to external factors, or to some internal mechanism, or both?

We must begin to look at the part played by self-regulating systems, first by considering the simplest kind of control systems (Chapter 3). The familiar systems for heating a home are used to illustrate the analogous thermoregulatory system of warm-blooded animals (Chapter 4). A wealth of examples of homeostasis are drawn upon in athletes and astronauts, racehorses and bumble bees, to illustrate the subtleties of adaptation.

THE MAIA HYPOTHESIS

Then we see Maia as a single loop control mechanism. Using simple model organisms, and an experimental design incorporating perturbation, it is possible to observe the oscillatory output of the control mechanism (Chapter 5). So the approach is primarily experimental, and the principal novelty is the development of a method that makes it possible to isolate the output of growth control mechanisms. The work was carried out on cultures of simple organisms in the laboratory. The first is the marine hydroid *Laomedea flexuosa* (frontispiece) and the second a marine yeast *Rhodotorula rubra*. The essential part of the experimental design is the requirement to perturb the growth of the organisms, and then isolate the consequences of having done so. Using low levels of toxic inhibitors proved the simplest way to deviate growth from its preferred or goal setting, so that graduated loadings could be imposed upon the control mechanism to reveal variations in the control responses to inhibitory load. Initially the Maia hypothesis is a method by which growth data can be expressed as the oscillatory output that is typical of feedback mechanisms (Chapter 5). This reveals the richness of output behaviour by which organisms may overcorrect at low levels of loading, neutralise inhibition at higher levels or become overloaded when the capacity to counteract is exceeded.

The concept of homeostasis was central from the outset. Biological cybernetics focuses upon homeodynamic systems which are responsible for control in biological systems. Such systems are dedicated to serving the inbuilt purpose of maintaining the process at some preferred rate or state. For the life of the organism, they maintain a homeodynamic equilibrium but, for the biologist, this condition is uninformative as to the workings of the mechanism responsible. The equilibrium must be disturbed and deviated from its goal in order to see the response necessary to restore it to its goal setting. This reveals the characteristic oscillatory output of the feedback mechanisms responsible for homeodynamic control of growth. The growth rate output of perturbed organisms, and the output of simulation models incorporating feedback, were so similar and characteristic of cybernetic mechanisms that there was little doubt that the Maia hypothesis was correct. At low levels of perturbation, equilibrium was quickly restored; at higher levels, recovery was slower; and at even higher levels, the capacity of the control mechanism was exceeded and finally overwhelmed at a level that coincided with lethal levels of the toxic inhibitor.

Most of what is known about cybernetics comes from man-made feedback mechanisms, whose designers start with a blank sheet of paper when creating such systems. The study of biological control mechanisms requires what is called 'reverse engineering'; the system exists within the organism and the researcher must find out its properties by deducing them from the output of the system. As investigation proceeds, the model comes to represent more accurately the system within the organism. A systems approach is a necessary escape from the limitations of reductionism, as the properties of the whole system are not to be found in its constituent components, but require the output of the intact system. Knowing the key properties of the whole system, the investigator can design a feedback mechanism that accounts for the control behaviour that is observed. Simple models account for what is first observed, but as other and more subtle behaviour is observed, the control models need to be more complex.

A wide-ranging search is made of likely subcellular control systems that might account for the characteristic oscillatory behaviour of feedback mechanisms (Chapter 6). Control systems have a finite capacity to resist perturbation, which counteracts the effect of any agent that creates an error between some preferred condition (input) and the actual condition (output). Growth control mechanisms probably originated as minimal systems with such basic features. The ability to increase in numbers by exponential growth is harnessed to the complementary, though antagonistic, ability to inhibit growth, reminiscent of the oriental concept of yin and yang. In its simplest form, the two elements become two halves of a single control mechanism. The Hungarian-born American Albert Szent-Györgyi (1893–1986) received the Nobel Prize in 1937 for his work on vitamin C. He fought in Hungary for the resistance during World War II and was offered the Presidency, but as soon as the communists seized power, he and his family fled to America. He explained that 'To regulate something always requires two opposing forces. You cannot regulate with a single factor.' The imperative to grow is due to a positive feedback as the products of growth also grow, causing the rate of growth to accelerate in a way that must inevitably be contained. An inhibitor operating within a negative feedback loop provides the complementary antagonistic factor. The development of the idea of a dual control mechanism is explored in detail in a search for a simple molecular system that might represent the simplest kind of control mechanism that emerged from the modelling.

In the early stages, control was represented by a single loop feedback mechanism responsible only for controlling growth rate, although it was realised that, to curb exponential increase, the mechanism must also incorporate size or density limitation. Of greater importance to organism fitness is an enhanced capacity to withstand external inhibitors, and the load they impose (Chapter 7). Ashby's Law of Requisite Variety tells us that organisms might be expected to have a capacity adequate to neutralise the effects of the external variables to be found in the habitat in which they live. The outcome of any perturbation is the sum of the inhibition, and the antagonistic counter-response. Capacity can be expressed as the range of concentrations of an inhibitor that can be neutralised, from the lowest level that prompts an oscillatory response, to the point at which the system becomes overloaded. Whatever the cause of the load, homeodynamic systems provide a measurable capacity to counteract and neutralise the effects. What is more, in the longer term an enhanced capacity to counteract inhibition slowly adapts to events, gradually adjusting its capacity to neutralise similar challenges more effectively in the event of greater toxic challenges in the future. There exists an understanding in us all that most activities, such as running or cycling, increase our capacity for physical work, and a capacity that is not used decreases. We tend to 'use it or lose it'. Thus capacity varies in relation to recent loads, requiring some simple form of rolling 'memory' that is neither genetic or cerebral, but is embedded within the control mechanism. It is a cybernetic kind of memory involving the adjustment of preferenda in relation to variation in load over time.

The simplest kind of external population control is by the limitation of the raw materials necessary for further growth (Chapter 8). We see this in single-celled autotrophs, which multiply until they have used up the available nutrients, and then die back as a result, leaving dormant spores to regenerate their populations when sufficient nutrients and the right conditions return. They cannot help but lead a life of boom and bust. All other creatures are organotrophs and depend, in one way or another, on living prey. Clearly, if they lived their lives as autotrophs do, they would quickly extinguish the food species on which they depend for survival. So they have evolved the means of self-limitation at a level that allows prey species to survive, and their own kind to thrive indefinitely.

The evolution of thinking about population growth is seen through the history of the logistic equation from its publication in 1838 by Pierre Verhulst (1804–1849) to the discovery by Robert May of chaos in the logistic map (Chapter 8). The realisation that the logistic equation was also a formal description of a control mechanism was made by G. Evelyn Hutchinson (1903–1991) in 1948. Knowledge that this is so appears to have been forgotten, but the properties of the logistic as a model for population control, with its double feedback loops, are explored. In more recent times, Robert May and others discovered chaos in the logistic map. The onset of chaos by what is essentially a doubling in the level of goal setting coincides first with limit cycling and then increasingly chaotic oscillations.

Biological systems are arranged in a nested hierarchy (Chapter 9), which is a functional as well as a structural organisation. Different aspects of growth are controlled by feedback mechanisms that regulate specific growth rates, cell density, the size of organs in relation to functional load and overall size. The hierarchy of control mechanisms that exists, even in the simplest organisms, became more complex during evolution with the addition of new levels of organisation. From the perspective of growth control, each level capable of replication has its own growth control mechanism(s), creating a hierarchy of coupled control mechanisms in which none acts autonomously, as this would risk the order and stability of the organism. The combined effect of a multi-layered organisation provides an enhanced stability and increased resistance to perturbation. During evolution, the resilience of organisms has increased with each new layer of biological complexity, as the capacity of the responses of an organism are summed to provide a stratified stability. So it is that more complex organisms, with more levels of organisation, may be expected to have more powerful responses to meet environmental challenge and change. At the same time they have become increasingly independent of the natural environment: unending stresses, extreme conditions and continual fluctuations. Evolution is progressive, and improvements in the independence of internal processes from environmental variation have led to a trend of increasing complexity.

GROWTH PHENOMENA FOR WHICH MAIA OFFERS AN EXPLANATION

Later chapters describe those phenomena that Maia helps to explain. In particular the curious effect called 'hormesis': the paradoxical stimulatory effect of low concentrations of toxic substances (Chapters 10). This is an effect probably known since early humans first fermented beverages that created alcohol. It was first known formally to toxicologists over a century ago, but has never been satisfactorily explained. Many thousands of examples of hormesis have now been documented by toxicologists. It had been sidelined, due partly to its historical association with homeopathy, but recent times have seen a resurgence of interest. A homeodynamic explanation for hormesis, as given here, is becoming accepted as plausible, and suggests that low levels of any inhibitor may be expected to stimulate biological processes over a range of concentrations, as we know from our everyday experience of appropriate levels of intake of alcohol, caffeine and nicotine.

The relationship between hormesis and homeopathy is an interesting one. At one time hormesis was accepted as the scientific principle underpinning homeopathy. But homeopathy depends on the assumption that the potency of homeopathic drugs increases through a series of dilutions so great that none of the original drug remains. Hormesis occurs over a limited range of concentrations immediately below those that inhibit processes. It is at this point that hormesis and homeopathy part company, and while there is much scientific evidence for hormesis, there is none for homeopathy that withstands scientific scrutiny. Hormesis should be part of the common knowledge that low concentrations of agents that are toxic at higher concentrations have the ability to stimulate over a limited range of lower concentrations. Here effects are found to be a byproduct of adaptive homeodynamic adjustment. Thus it is commonly assumed that toxic agents in everyday use stimulate us in a way with which we are familiar. In using them we know, but tend not to rationalise, that higher concentrations have a toxic effect.

Adaptation to low levels of toxic agents comes about by resetting the pre-set rate of growth control mechanisms to a higher level, conferring a greater tolerance to future toxic exposure. One consequence is that such organisms grow faster and larger than those that had not adapted. In this way, low levels of toxic agents, whose effects are typically inhibitory, have the surprising effect of stimulating growth (Chapter 11). When adaptation is not followed by further exposure to a toxic agent, growth is stimulated. The phenomenon has been widely exploited for many years in promoting the growth of agricultural livestock and crop plants by agents as diverse as antibiotics, metals and radiation.

'Catch-up growth' is another growth phenomenon that finds a ready explanation in a cybernetic interpretation of growth. After equilibration to some toxic load, the removal of load causes 'relaxation stimulation'. This is an inevitable response to the sudden removal of load to which a control system had equilibrated. This typical feature of control mechanism output may well explain the origins of 'catch-up' growth in mammals, whose growth accelerates rapidly after being temporarily held back by sickness or deficiency (Chapter 11).

WHAT HAPPENS WHEN MAIA FAILS?

In the final chapters, the consequences of Maian failure are explored. Two examples of the failure of control are considered, seen from the perspective of cybernetics. When one considers all the trillions of cells in animals and humans, with the potential to replicate exponentially, it is remarkable that more of us do not get cancer (Chapter 12). The potential exists because many kinds of cells, when removed from an organism and cultured in the laboratory, multiply exponentially. What is curious is that a more explicitly cybernetic approach to the study of cancer and its causes has not been adopted. Cancer would occur more frequently were it not for the ability of DNA to repair itself when damaged, implying that susceptibility may be due to the fitness and capacity of the repair mechanism and not simply to the original cause of DNA damage. Again we see a homeodynamic system which operates as a repair mechanism, with the goal of complete genetic fidelity.

Overpopulation by mankind has long been perceived as a pressing problem for humanity, but it is rarely asked how it is that other highly evolved species regulate their density and numbers, while our own species does not (Chapter 13). Given the control of replication and reproduction at all levels of biological organisation, how is it that modern man has not retained the control mechanisms of his biological ancestors? Innate controls that existed in primitive humans, even to the last descendents of Stone Age people, have apparently been lost in our evolutionary past. It now seems that the growth of the global human population is out of control and does not have any innate natural limit to its growth. A carrying capacity for the human population determined by ecological footprint analysis has already been exceeded, and there seems little hope of the population levelling out until 2100, by which time another 2 billion inhabitants will have been added to the present population. Overpopulation not only aggravates other man-made global problems, but we return to the problem with which we started (Chapters 2 and 3). Exponential growth is a positive feedback process resulting in the exponential growth rate of the global population, currently increasing at 77 million each year. We lack the intrinsic population control of animals, and although fertility is now beginning to fall, we will continue to increase exponentially for the foreseeable future.

In the penultimate chapter we consider the implications of Maia for the future of mankind. It appears that during evolution animals solved the problem of overexploiting their resources, and so avoided the so-called 'tragedy of the commons' (Chapter 14). People, on the other hand, continue to exploit the commons in an unsustainable way. It is more than a decade since a group of environmental economists provided a realistic value of ecological goods and services. They arrived at a valuation of services that mankind could not do without, and could not afford to pay for, yet continued to overexploit. We simply undervalue the true worth of the global environmental commons, not just of the resources they provide in terms of crops, timber and fish from the sea. We also exceed the capacity of the land, sea and air to assimilate the waste byproducts of human production and consumption. The group concluded that, as the global economy could not do without the ecological services, 'their total value to the economy is infinite'.

Worse still, our species has created many systems that grow at accelerating rates which cannot be sustained. These include air travel, meat production and motor vehicle manufacture, which are good indicators of the global economy and continue to grow at an exponential rate. Like living systems, the economic metabolism respires. Oxygen is used to combust fuels to release their energy, with carbon dioxide as a byproduct. So much carbon dioxide has been released from burning fossil fuels to provide energy to power the economy that climate change now threatens to impose its own limit on economic growth. A comparison of biological and human strategies for survival indicates that we are ill-prepared for a future on what we are beginning to accept as a finite Earth.

2

Growth unlimited: blooms, swarms and plagues

In taking a view of animated nature, we cannot fail to be struck with a prodigious power of increase in plants and animals.

Population, when unchecked, increases in a geometrical ratio.

Thomas Malthus (1766–1834)

If in successive time intervals of the same length a number doubles itself by the same factor, we have an example of exponential growth. Self-reproduction or autocatalysis underlies this law. The quantity at hand – whether we are dealing with neutrons in a block of uranium, bacteria in a culture, people, capital, information or knowledge – will catalyse, program and regulate its own reproduction.

Manfred Eigen and Ruthild Winkler

THE GROWTH OF MICROALGAL POPULATIONS

In 1868 the Victorian scientist Thomas H. Huxley (1825–1895) gave a lecture to the working men of Norwich *On a piece of chalk* [1]. He first described the size of the chalk deposits, stretching from Flamborough Head to Lulworth Cove, which is 280 miles as the crow flies. But these are merely the western extremity of the Cretaceous chalk deposits laid down some 140 million years ago, and extending 3,000 miles across Europe. Having convinced his audience of the scale of the geological chalk strata beneath their feet, he then teased them as to whether they might know their origins.

Huxley described how he had looked at thin slices of chalk under the microscope and found innumerable fossilised remains of marine organisms. He likened the calcareous shells of Foraminifera of the genus *Globigerina* (single-celled organisms with globular chambers and shells) to 'a badly grown raspberry'. The intricate siliceous tests of these diatoms were abundant. Huxley also found in the chalk the tiny granules he named 'coccoliths', before the discovery of the single-celled organisms that created them. The chalk deposits were the remains of countless microscopic fossil organisms. But how could such vast deposits have been formed by such tiny creatures?

In 1858 *HMS Cyclops* surveyed the bottom of the Atlantic from Ireland to Newfoundland along the proposed track of a new submarine cable. The aim was to find a soft level bottom for the cable. A few years earlier a young American, Lieutenant Brookes, had thought of a simple way to bring small amounts of sediment up from the abyssal depths. He attached a quill to the sounding lead, so that the quill penetrated the sediment just before the fall of the lead was halted by the sea bottom. Where the sediments were fine and soft, the hollow quill filled and could be returned to the surface along with a few grams of the muddy sediment.

These precious samples were among the first to be brought up from the ocean deeps, and provided an early glimpse of life in the abyss. The samples were passed to Huxley for microscopic analysis. He was soon able to show that for 1,700 miles across the North Atlantic, the abyssal plain at 3,000 metres deep was covered by fine calcareous sediments packed with the shells of globigerinae, such that the mud – known as 'globigerine ooze' – was found to cover nearly 50% of the deep-sea floor. It consisted almost exclusively of calcium carbonate derived from the remains of planktonic life forms that had settled after the death of the plankton living in the surface waters far above.

Huxley also found coccoliths in the quill-caught samples from the abyss, and sometimes 'coccospheres' (loose spherical aggregates of coccoliths), which convinced him that they were the remains of 'independent organisms'. If there was a *frisson* of excitement, the record does not show it, as it was only a few years after his good friend Edward Forbes (1815–1854) had put forward his belief of a lifeless azoic zone below a depth of 300–700 metres. Not without good reason, Forbes had believed all deeper waters to be lifeless. In this way, Huxley brought together the evidence for his audience of Norfolk worthies, showing that the processes that had created the massive chalk deposits, laid down over the 30 million years of the late Cretaceous era, were still going on in the ocean deeps.

In the perverse way that scientific evidence sometimes comes together, the life forms that Huxley first suggested were responsible for creating coccoliths were not discovered until some years later. Coccolithophores, the living cells that make up the innumerable

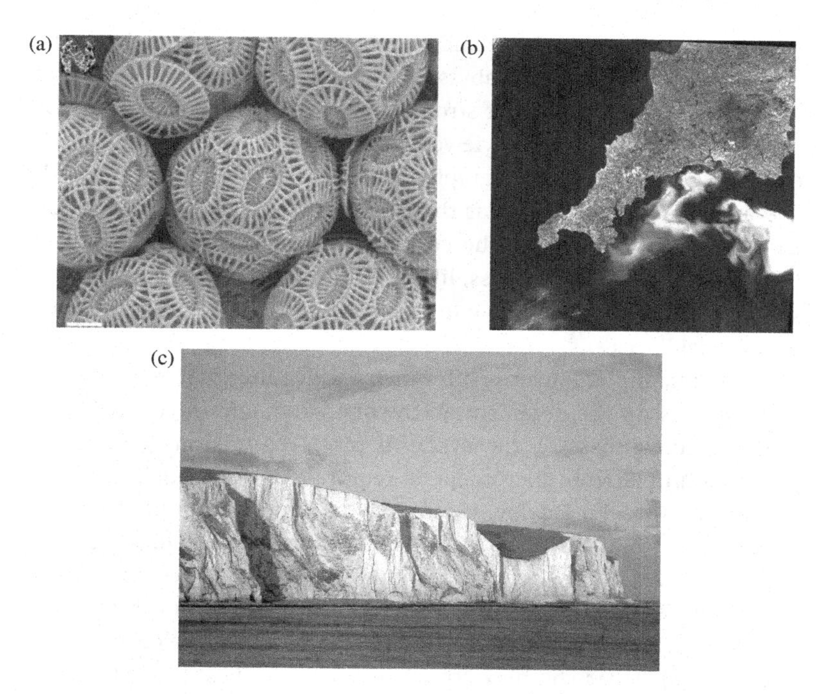

Figure 2.1. A scanning electron micrograph of *Emiliania huxleyi* cells covered in the coccoliths that reflect light (a), making their blooms visible from space (b). They then fall to the seabed, and when compacted on the ocean floor over many millions of years with the shells of microscopic Foraminifera and other organisms, they create the chalk deposits found across Europe, and the White Cliffs on the south coast of England (c).
(a) Courtesy of Jeremy R. Young at the Palaeontology Department of the Natural History Museum; (b) satellite image courtesy of Steve Groom of the Remote Sensing Group, Plymouth Marine Laboratory; (c) courtesy of 'Maki', www.flickr.com/photos/makiwi/290719612/.

coccoliths of the chalk deposits and the ocean deeps, were discovered in plankton samples taken from surface waters by *HMS Challenger*. But this ship did not return to Spithead from her 70,000-mile expedition until 1876. Coccolithophores are so called because 10–30 calcareous coccoliths cover each cell; the most common living species of their kind is *Emiliania huxleyi* (Figure 2.1a), named after Huxley to mark his contribution to our understanding of the species.

Huxley's observations on the abyssal sediments and Forbes' hypothesis must have given rise to the question in Huxley's mind as to just how far down these organisms might have lived. We know now that they are photosynthetic and live in the photic zone, and then as they die fall slowly into the abyss.

Coccoliths reflect light strongly, and their milky-green blooms near the sea surface can be viewed from space and monitored by satellites (Figure 2.1b). Oceanographers, guided by satellite images, follow and sample the blooms as they are carried by the North Atlantic Drift. South of Greenland, the current spins off parts of the bloom, hundreds of kilometres across, like a spiral galaxy wheeling slowly in space. Each year blooms grow to cover 1.4 million square kilometres of the world's oceans.

Few examples in nature offer such a wide range of scales between the smallness of the organism on the one hand, and the enormity of their collective mass on the other. At just 5–20 microns across, the power of an electron microscope is required to make out their individual structure. To Huxley, viewing them under the limited magnification of a light microscope, illuminated by gaslight in his study, they appeared simply as tiny, glassy ovoid plates. When magnified 10,000 times under an electron microscope, each coccolith appears as a spoked wheel whose rim is made of overlapping calcite crystals (Figure 2.1a). Their complex structure begs the question as to what selective pressures had led to the evolution of such intricacy at a magnification only made possible by the electron microscope.

The plankton in the sunlit surface waters contains immense numbers of single-celled organisms – including globigerinae, radiolarians and coccolithophores – that bloom and die each year. As they die, their remains – innumerable shells, tests and liths – fall like snow, taking weeks to reach the ocean floor thousands of metres below. Over the immensity of time, each millennium is marked by the addition of just a few centimetres of the accumulated remains to the muddy sediments. They form the calcareous and siliceous sediments that cover much of the bottom of the world's oceans. But, as Huxley first described to the working men of Norwich in 1868, these fine calcareous sediments, compacted by the weight of sediment above, then uplifted from the Cretaceous seas by some later geological period of mountain-building long ago, formed massive structures up to a thousand feet thick, like that of the White Cliffs, which mark the English Channel coastline (Figure 2.1c).

The multiplication of microalgae – such as *Emiliania* – is explosive. In the summer months, with sunshine and warmth, blooms of microscopic algae grow to cover large areas of the ocean. As densities can reach 5–6 million per litre, it is clear that blooms covering over 100,000 square kilometres of ocean contain immense numbers of cells: too many to provide a figure that would have meaning. What is more, the growth of such huge populations of cells occurs in just a week or so, arising from a dormant background population of just a few cells per litre. It is a paradox of nature that these animals and plants, which may seem insignificant due to their microscopic size, are giants in ecology for the part they play in the global flux of energy and materials.

A man rose at the conclusion of Huxley's address to say that 'they had never heard anything like that in Norwich before'. That the Cretaceous chalk deposits of the Channel are the cumulative remains of innumerable small organisms deposited over aeons of time remains just as thought-provoking today as when the story was first given life by Thomas Huxley to a Victorian audience in 1868. How is it that such vast numbers of organisms can spring up every year to create dense blooms of algal cells, before dying back to repeat the process again next year, and the next, repeating the process tens of thousands of times?

EXPONENTIAL INCREASE

If at equal intervals of time a number doubles itself or multiplies by the same factor, growth is exponential (Figure 2.2). This applies to all such processes as the accelerating growth of a forest fire, the increase of information available to us on the internet, the growth in the use of

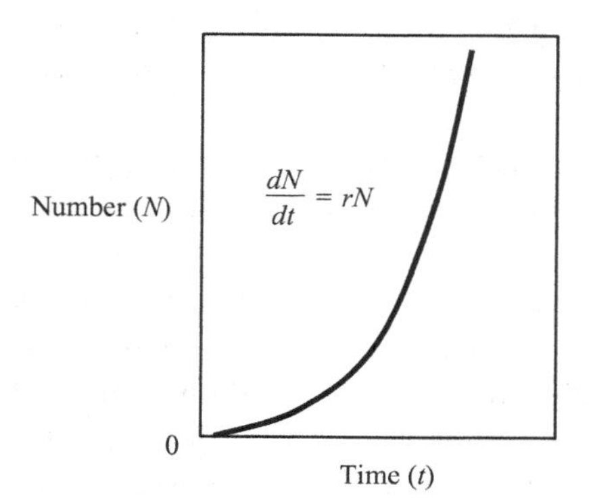

Figure 2.2. Population growth curve expressed as numbers (N) against time (t). The simplest model is provided as a differential equation, giving the rate of increase of the population as the population size multiplied by the per capita growth rate.

computers, the growth of the human population, and nuclear fission. In living processes it is typified by the souring of milk by bacteria, by the multiplication of duckweed that covers a pond in summer and by replication of the influenza virus in the Great Pandemic that killed at least five times more people than World War I, which preceded it. Here we need to examine the underlying property of such systems that grow at ever-increasing rates. Specifically, we must ask, how do such vast numbers of organisms like *Emiliania huxleyi* come about?

It is due to simple arithmetic. A good illustration of the large numbers that are rapidly created by repeated doublings is the legend of King Shirham of India. He wished to reward his Grand Vizier Sissa Ben Dahir for devising the game of chess. What would he like in recompense? The clever vizier told the King that he had modest needs which might easily be met. He placed his chessboard in front of the King and asked that a grain of wheat be placed on the first square, then two grains on the second and four grains on the third and so on; with twice the number of grains on each square as the one before, until the 64th and last square on the board.

'Your wish shall be granted,' said the King, and immediately sent for a sack of wheat, surprised that the vizier had not asked for something more as his reward. Before long the King's servants appeared, carrying the sack as commanded, and the counting began. By the twentieth square the sack was empty and the servants went to get another. With each square the number of grains required was increasing so rapidly that one granary had been emptied and then another. Eventually all the King's granaries were empty and only a fraction of the vizier's request had been met. It transpired that the wheat to meet the clever vizier's 'modest' request amounted to 4,000 billion bushels.

I first read this fable in a book by George Gamow (1904–1968), the Russian-born American physicist who delighted in making the basic concepts of science accessible with such stories [2]. Much later, the fable was retold in an essay written by Carl Sagan (1934–1996), in which he had reworked the calculations. The total weight of grains due to the Grand Vizier was actually 75 billion tonnes, which Sagan estimated was equivalent to the world's wheat production for 150 years. To really drive the point home that exponential increase produces huge numbers, Sagan calculated that had the vizier used a board of ten by ten squares instead of eight by eight, the amount of wheat owed to him would have weighed as much as the Earth.

Table 2.1. *Exponential growth of* Emiliania huxleyi, *based on a doubling rate (t) of 2.8 times a day.*

Days	Divisions	Cells
1	0	1
5	14	8,192
10	28	134,000,000

In the simplest case, single-celled species reproduce by binary fission, each dividing at intervals to create two new cells. Thus, populations grow with each division by the series 1, 2, 4, 8, 16, 32, 64, 128 and so on, like the wheat grains on the vizier's chess board. This is exponential growth, where numbers double and redouble, then double again, with the result that the population size expands at an ever-increasing rate. *Emiliania huxleyi* reproduces asexually at a high rate, doubling 2.8 times each day.[1]

Coccolithophores grow more slowly than diatoms, but faster than dinoflagellates. Using this information alone, we can get an idea of how quickly phytoplankton populations can increase and then, as species reach a peak and decline, they are replaced by another group of species. Theoretically, in 10 days, 134 million cells of *Emiliania* could result from one cell (Table 2.1). Their multiplication accelerates as numbers grow, until the conditions (e.g. optimal light, nutrients, temperature) required for growth cease, or the resources needed for growth are used up. This is growth at its most spectacular and helps explain how cell numbers that start so small may grow in weeks to densities of 10 million cells per litre, to dominate the summer surface waters of great tracts of the world's ocean.

Bacteria are primitive life forms, usually multiplying even more rapidly than microalgae. They are probably the earliest life forms to colonise the planet and occur almost everywhere in their unseen myriads. In natural waters, densities of thousands per millilitre are typical; fertile soils have millions per gram, while in faeces there are billions per gram. Thus, given their ubiquity and speed of growth, they can exploit ephemeral niches rapidly, whether a mucky farmyard puddle

[1] The doubling time (t) can be derived from the Law of Exponential Growth as $t = \log_e 2/r = 0.6931/r$.

Table 2.2. *Doubling times of exponentially growing populations.*

Clostridium	10 minutes	*Hydra*	3 days
Escherichia coli	20 minutes	*Lemna*	2.5 days
Saccharomyces	37 minutes	*Aphis*	5 days
Euplotes	18.5 hours	*Flustra*	150 days

or a litre of milk. Higher temperatures are better for accelerated growth, so more rapid souring is likely if milk is warm. Deliberate inoculation with particular strains of bacteria is necessary to make milk products such as buttermilk and yoghurt.

The faecal bacterium *Escherichia coli* divides once every 20 minutes and will continue to do so for 10 hours in a laboratory culture (Table 2.2). *Clostridium perfringens* multiplies even more rapidly, with a doubling time of 10 minutes. Densities may reach 10 billion per millilitre before the exponential phase of growth slows down when nutrients are depleted and the byproducts of metabolism accumulate to toxic concentrations, bringing cell division to a halt.

Bacterial diseases, such as plague, typhoid fever, cholera, salmonellosis and dysentery take effect rapidly, following an incubation time, as bacterial multiplication passes to the steeper part of the J-shaped curve that characterises exponential growth (Figure 2.2). Either the pathogen itself, or the toxins produced, then rapidly overwhelm the body's natural defences. Bubonic plague killed one person in four throughout Europe in the fourteenth century, amounting to 25 million deaths in all. The bacillus responsible, *Yersinia pestis*, originally caused a disease of rats which was transmitted by fleas, but as the rats died of the disease, so the fleas moved on to the human population, rapidly spreading the disease.

Yeast cells are somewhat larger than bacteria and reproduce more slowly. *Saccharomyces cerevisiae* has a doubling time of 1.6 times an hour. The Egyptians brewed an acid beer called 'boozah' in 3500 BC, which was made from the fermentation of dough into which they kneaded germinating barley. For bread, the rising of dough is caused by the production of carbon dioxide by trapped yeast. By 1200 BC, the distinction between leavened and unleavened bread was well known, as was the need to use a portion of today's bread to inoculate tomorrow's. Different species of *Saccharomyces* are used to ferment wine, cider, beer and bread. Concentrations of alcohol up to 15% can be achieved in some wines, but at these levels ethanol becomes toxic to the yeast, so distillation becomes necessary for the higher concentrations of alcohol needed to make spirits.

Bacteria and yeast were domesticated and used for millennia before they were made visible to the naked eye, and it was even longer before the roles of the different kinds of micro-organisms could be linked to what they actually do in souring, fermenting and leavening. In 1680 a linen draper of Delft named Antonie van Leeuwenhoek (1632–1723), developed a simple microscope which he used to describe and illustrate micro-organisms. The lens was just a single bead of glass, yet it gave magnifications of up to 275 times: enough to see the protozoan *Amoeba* and spermatozoa for the first time, although it remains difficult to understand how he was able to see bacteria clearly enough to illustrate them. It took a century and more before yeast was recognised as the living organism responsible for fermentation, and longer still before Louis Pasteur (1822–1895) in his *Études sur le Vin* [3) was able to show that yeast was responsible for the fermentation of sugars to alcohol.

The ciliate *Euplotes* is unusually large for a protozoan and is just visible to the naked eye. Most ciliates swim by the passage of waves over their cilia, like wind passing over a field of corn, propelling them along at high speed. For *Paramecium*, which used to be called the slipper animalcule due to its shape, 'high speed' means as much as 2–3 millimetres per second. However, *Euplotes* has cilia fused in bunches which function as stick-like 'legs' enabling it to crawl, but at a much slower pace, as it grazes on bacteria and other micro-organisms. It is also easy to culture in the laboratory, doubling in numbers every 18.5 hours (Table 2.2).

It is the much larger multicellular *Hydra* to which we turn next. It can easily be seen with the naked eye, as can be judged from the fact that a ciliate closely related to *Euplotes*, and of a similar size, is a commensal living on the polyps of *Hydra*. As many as 50–100 of these ciliates may be found grazing on the outer surfaces of its body and tentacles. *Hydra* is a primitive invertebrate polyp, among the simplest of metazoans, measuring 5–10 millimetres high when fully extended, which feeds by trapping small crustaceans that swim into its stinging tentacles. It reproduces asexually by budding new polyps, which become detached from the parent when fully formed and are able to live independently. Carolus Linnaeus (1707–1778), who devised the binomial system by which all plants and animals are named, gave them the name *Hydra*, indicating their regenerative ability. One of the mythical labours of Hercules was to kill the gigantic many-headed hydra which, when decapitated, grew two more heads in its place. A description by Abraham Trembley (1710–1784) shows how appropriate the name is:

> I cut off the heads of one that had seven, and in a few days I saw in it a prodigy scarcely inferior to the fabulous Hydre of Lernaea. It acquired seven new heads; and if I had continued to cut them off as often as they grew, it is indubitable that I should have seen seven others grow. But there is something more than the legend dared to invent: the seven heads that I cut off from this Hydre, after being fed, became perfect animals; and it only remained for me to make a new monster from each of them [4].

Abraham Trembley added so much to the story of asexual reproduction that it is worth saying more about him. In the 1740s the young Swiss was tutor to the two sons of Count Bentinck in Holland. Educating the young boys gave him the opportunity to use his dip net to search for aquatic life in the ornamental ponds of the Count's estate. He was perhaps the first true experimental zoologist who, with a few jars on the window sill of his study, conducted a series of experiments with *Hydra* and made many novel and important observations. His simple microscope was a single lens, held firmly by a series of adjustable ball joints, enabling careful observation of the contents of the culture jars. Nevertheless, Trembley was able to describe the fission of a single cell. The long, narrow cells of the diatom *Synedra* attach to water weeds, which he observed over a period of 24 hours. A single cell first split from the unattached end, then detached and moved away from the parent cell. This was the first account of cell division.

Apart from his study of regeneration in *Hydra*, Trembley was able to demonstrate the incredible flexibility of its primitive organisation. When a polyp was turned inside out on a pig's bristle, it could reorganise its two layers of cells within 48 hours. By everting the polyp, the endoderm cells were on the outside and the ectoderm on the inside, but they rapidly reverted to normal and the polyp was able to function as before. He also carried out early grafting experiments by tying two polyps together with a single hair. It is not surprising that *Hydra* became a popular species for research, and is still used as an experimental model to study the processes of growth, regeneration and development. Trembley's discoveries made him famous and the world of 'polypes' became a topic of conversation in the fashionable salons of Paris.

Trembley made a careful comparison of the multiplication of the duckweed *Lemna* with that of *Hydra*. He calculated the rate of multiplication of *Hydra*, observing that a single individual gave rise to 45 new individuals in just 56 days. A typical growth rate in laboratory culture is from 10 to 120 individuals in 11 days, with a doubling time of about 3 days (Table 2.2). Its rate of reproduction is similar to the duckweed *Lemna*, to which *Hydra* is often attached in ponds. This small floating plant has just a pair of leaflets and roots. Trembley noted how the leaflets of *Lemna* form new leaves and roots, which then become separated to create a new plant. So both species reproduce 'vegetatively' by fission, with *Lemna* doubling every 2.5 days. Such rapid reproduction explains why one rarely sees a half-covered pond in nature. If their division is synchronised, a half-covered pond is just one doubling away from being completely covered. In the late summer, ponds are often completely covered by *Lemna*, sometimes with tracks made by ducks or moorhens. It is said that sufficient *Lemna* to cover 1 square inch (6.5 cm^2) of the water surface would be capable of covering a one-acre (0.4 ha) pond in 55 days.

When walking along a sandy shore after a gale, searching among the flotsam will sometimes reveal the crisp, dun-coloured fronds of a colonial bryozoan. *Flustra foliacea* lives attached to the rocky bottom offshore, and the thousands of identical zooids feed on the passing phytoplankton. Many are torn off during storms and cast ashore. A hand lens reveals that vegetative growth produces a frond of zooids, lying in tiny rectangular boxes arranged like the bricks in a wall. Each frond consists of two such layers of zooids, arranged back to back. New zooids, each with its tentacular polyp, are added to the growing edge and the colonies grow exponentially to form colonies of 150,000 zooids over 7 years [5]. Unlike the other examples given so far, the population of asexually produced zooids, although capable of independent existence, grows as a plant-like population, such that the history of their growth is spelled out in their fronds. Each year in winter they leave a growth check line on the frond, indicating where growth ceased during the winter months. This allows the size of the colony to be calculated retrospectively at annual ages, and the growth curves of each year class back-plotted over time. This provides a visible demonstration of exponential growth in relation to each passing year (Figure 2.3).

Again, it was Abraham Trembley who first discovered the animal nature of bryozoans such as *Flustra*, by correctly identifying their polyps with his simple microscope. These and many other sessile colonial animals, such as the hydroids, were still regarded in the nineteenth

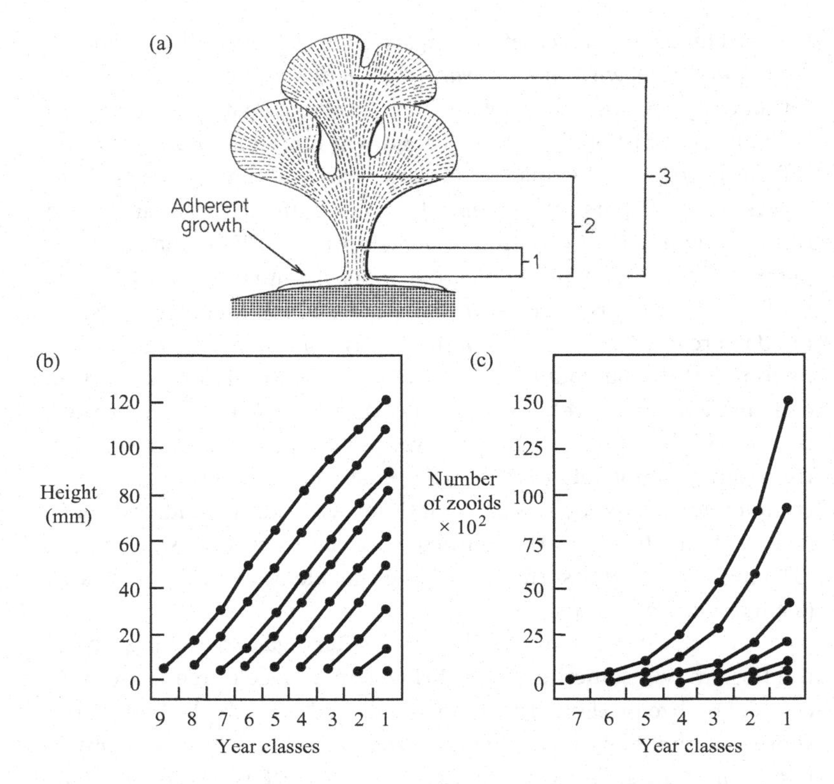

Figure 2.3. The growth of different year classes of the colonial bryozoan *Flustra*: (a) a young frond showing annual growth check lines, from which age can be determined and growth can be back-plotted; (b) back-plotted growth of the colonies in terms of height (mm) for nine year classes; (c) growth of the colonies in terms of number of zooids ($\times 10^2$) for seven year classes. Reproduced or redrawn with the kind permission of Springer Science+Business Media [5].

century as zoophytes, or plant-like animals. I remember learning that the herbarium collection at the British Museum of Natural History included many pressed bryozoans that were not passed on to the zoologists until the 1960s!

Some insects also reproduce asexually, and aphids in particular have spectacular population growth rates. They overwinter in small numbers on the guelder rose and spindle tree, and begin to reproduce in the spring. They are adapted to reproduce rapidly; the young are born alive and develop from the egg without it being fertilised, and even grow without wings to increase the growth rate. In this way the early generations can attain the steeper parts of the exponential curve more rapidly. Aphid densities can increase from a single aphid to 16,000 in 8 weeks, doubling in number every few days.

RATES OF REPRODUCTION IN SEXUAL ANIMALS

All the examples of exponential growth of populations given so far reproduce asexually, or vegetatively by budding. Such growth is more rapid than sexual reproduction, since the whole population replicates. The price of creating genetic variety for natural selection to act upon is high in terms of lost reproductive potential, as the male half of the population do not give birth. Some invertebrates have evolved to reproduce by asexual budding as well as sexually, switching between one and the other, depending on which form offers the greatest contribution to survival fitness. Thus, in colonial hydroids, asexual growth of colonies predominates when conditions are ideal, but when the environment is challenging, or even stressful, investment in the sexual phase predominates. This results in a greater variety of genotypes from which natural selection will choose those best fitted for the new and more difficult conditions.

These examples of exponential population growth, from bacteria to yeasts, and from *Hydra* to *Lemna*, come most often from those species that can be grown in laboratory culture, where multiplication of small organisms can easily be maintained and monitored; *Flustra*, which cannot easily be kept in the laboratory, but conveniently tells us its age; and aphid data from a major insect-monitoring programme.

The best known examples of exponential, explosive growth in nature occur when species are introduced into a new country and removed from their natural competitors and predators [6]. The American slipper limpet arrived by accident on the Essex coast in 1890, and has gradually spread around the coastline. It was introduced with American oysters to the east coast oyster beds, and has now reached densities that compete with oysters for planktonic food and smother them with the mass of their shells. This chain-forming limpet has the fitting name of *Crepidula fornicata* because the males, which are smaller, wander over the chain of individual shells, fertilising all the females, which are larger. The species has spread widely and multiplied rapidly, with high densities of 450 per square metre that are unknown in its native American habitats.

The European starling, with a far greater mobility, has established itself all over the USA and Canada over a period of 60 years, from about 80 birds introduced into Central Park, New York in 1891. By 1954 it could be found almost everywhere in the USA. The starling has similarly colonised Africa, Australia and New Zealand, with populations almost certainly increasing exponentially after establishing itself in each new country.

Given the fecundity of the robin, a single pair could theoretically produce ten young per year, increasing to a population of 120 million in 10 years. However, this potential can never be realised by the robin, which fiercely defends a considerable territory; the robin population therefore runs out of space much sooner than non-territorial species.

Species of mammals such as the rabbit, European hare, fox, grey squirrel, coypu and various species of deer have at times exhibited explosive population growth. In Australia, rabbit populations reproduced to plague proportions in the absence of predators like the fox or buzzard of its native countries. They became such a problem that a substantial prize was offered for a solution. The virus disease, myxomatosis, was introduced into Europe and Australia from natural populations in South American rabbits, where the virus infected and killed vast numbers.

THE PLAGUES OF EGYPT

Explosive growth in numbers of plants and animals may result from the destabilisation of ecosystems, causing a cascade of linked consequences. The biblical account of the Plagues of Egypt given in Exodus provides some good examples. Of the ten plagues, at least seven were due to population explosions. There were plagues of frogs, flies, gnats and locusts, and three diseases caused by pathogenic micro-organisms.

The so-called Plague of Blood is thought to have been caused by the death of fish, perhaps due to the micro-organism *Pfiesteria*, which causes toxicity and bleeding. The Plague of Frogs that followed may have been caused by rotting fish driving the amphibians from the water. The Plague of Flies is thought to have been *Stomoxys calcitrans*, which multiplied rapidly on the accumulated fish and frog corpses. The Plague of Gnats were likely to have bred in the flooded fields in late autumn, while the Plague of Livestock Death was a disease of domesticated animals, possibly due to the flies of the earlier plague acting as carriers of the *Anthrax* bacillus. The Plague of Boils that followed may have been due to skin anthrax in both humans and animals. But perhaps the most devastating was the Plague of Locusts, for which the biblical account in Exodus Chapter 10 provides a vivid impression:

> By morning the wind had brought the locusts; they invaded all Egypt and settled down in every area of the country in great numbers. Never before had there been such a plague of locusts, nor will there ever be again. They covered all the ground until it was black. They devoured all that was left ... everything growing in the fields and the fruit on the trees. Nothing green remained on tree or plant in all the land of Egypt.

Locusta migratoria swarms when crowding makes conditions for the species in its original habitat intolerable. Vast numbers may gather and then migrate as a swarm to cover great distances in the search for new habitats. In 1869 such a swarm of locusts reached England, but perhaps the greatest swarm ever described is that of a swarm observed on 25 November 1889 from the ship *MV Golconda* in the Red Sea. On board was the naturalist G.T. Carruthers, who wrote an account of what he saw in a letter to the journal *Nature*. The swarm was vast and high-flying, and given that it took the ship steaming at 12 miles an hour at least 4 hours to pass the swarm, it was estimated to cover an area of 2,300 square miles. Although Carruthers' estimates of numbers assumed a far higher density for the swarm than is now considered possible, the size of the swarm was not questioned.

These observations had a profound effect on a young Russian mineralogist, Vladimir Vernadsky (1863–1945), at the University of Moscow [7]. He calculated from Carruthers' figures the mass of minerals and chemical elements contained in the swarm and realised that biological processes can play a major part in the transport of geologically significant quantities of materials – perhaps even on a comparable scale to physical processes such as erosion by water and transport by rivers. And so he came to coin the term 'biosphere' to capture the concept of living processes, at the same level of abstraction as the terms 'atmosphere' and 'geosphere'. This enabled him and later researchers to consider the role of living systems in driving geophysical processes. It is a concept that T.H. Huxley would have approved of, as is clear from his study of the formation of geological deposits of chalk from marine micro-organisms.

THE EXPONENTIAL EQUATION AND ITS APPLICATIONS

The growth process is responsible for the increase in biomass, which we may quantify in terms of numbers of cells, or numbers of asexually produced organisms. Colonial organisms provide an intermediate step, as their multicellular members reproduce asexually but they remain organically attached to one another as a colony, like *Flustra* and hydroid colonies. The growth of a population, or the growth of a colony, involves essentially the same process. The population dynamics are, in principle, indistinguishable, whether considering the growth of a population of budding hydra or the growth of a hydroid colony. What is of interest here is that the growth of a metazoan and of a population of individuals is similar, and colonial species can be used as model organisms for either or both. As already indicated, the form of a *Flustra* colony is of particular interest because it provides a map of exponential growth, marked as it is by growth check lines which delimit the parts of the colony that grow during each year (Figure 2.3a).

My purpose is to use the ambiguity of coloniality as a model organism in relation both to the growth of individual organisms and the growth of populations. The former has traditionally been thought of as physiology and the latter as population ecology. Whether individuals or populations, they consist of groupings of modules of life, whose cohesion or dispersion creates different forms of life which may aggregate as organisms or which are dispersed as populations of free-living organisms.

The key is that the products of multiplication are also able to replicate or reproduce; all populations of plants and animals, from *Emiliania* to humans, have the capacity to multiply at accelerating rates. The simplest form of growth rate (R) may be expressed as $R = N/t$ but, as we shall see in Chapter 5, such growth is not independent of population size; the rate increases with population size. For reasons that will become clear, it is preferable to separate the two factors that contribute to overall growth rate. The most appropriate measure is the relative growth rate, given as $r = 1/N.dN/dt$ or $d\log_e N/dt$. This is a per capita growth rate which is independent of population size and is, in principle, constant for each species. In the Law of Exponential Growth ($dN/dt = rN$), r denotes the relative growth rate. This has also been given the cumbersome name 'the intrinsic rate of natural increase', or the reproduction rate of an organism, which we will encounter frequently in the following chapters. The use of the relative growth rate has the effect of normalising growth rates with respect to the cumulative raw data (Figure 2.2). So for exponential growth, the relative growth rate becomes constant, and the growth curve is viewed without its predominant feature of unceasing acceleration.

The relative growth rate, r, is a measure of the rate of processes within the organism that are responsible for reproduction. It is fastest when individuals of a species can replicate under conditions that are optimal, so is sometimes given by a maximum, r_{max}. This index of reproductive rate was first identified by R.N. Chapman, who worked on populations of flour beetles at Rothamsted Experimental Station in England in the 1920s. Although it was intended to indicate the 'biotic potential' as a measure of the maximum reproductive power, *r* represents a maximum rate of reproduction of an individual when resources are unlimited and when there is no predation. This makes it a rather artificial value in that organisms in their natural environment rarely reproduce at their maximum rate for long. It provides a measure of the growth potential of a population vested in each individual, and is typically measured experimentally, as only in the laboratory can optimal conditions be guaranteed.

The important feature of *r* is that it is a measure of growth rate that is a constant, a measure that is intrinsic to the species, and independent of the number of individuals in the population. This leads to the term *rN*, which describes the growth rate of a population, showing that while *r* may be a constant, it is the increase in population size (*N*) that is the multiplier and is responsible for accelerating population growth. When *r* is multiplied by the actual size of the population (*N*), it gives the growth of the population. This creates a positive feedback and is responsible for creating the exponential form of the population growth curve (Figure 2.2). This is due to the recruitment of new individuals to the population, which themselves soon become reproductive, such that population growth rate accelerates, creating the exponential curve. Exponential growth is adaptive for species in situations where an ephemeral or seasonal niche opens up new opportunities. But, ultimately, exponential increase becomes maladaptive, in ways that will be discussed later.

While this argument has been set in the context of population ecology, the same problem arises in the physiology of growth. For similar reasons, physiologists have a 'specific growth rate', which is also given by *r*. It represents the growth rate per unit of biomass and can be calculated in the same way using $r = 1/W \cdot dW/dt$ or $d\log_e W/dt$ where *W* represents biomass. This measure of growth has its own origin and history, and was set out by Peter Medawar in his analysis of the Laws of Biological Growth. His contribution to the study of growth is summarised in Chapter 5. Here too, *r* is recognised as physiologically the most appropriate measure of growth, indicating the rate of biosynthetic processes within the cell, which is a constant and is independent of the size of the organism.

To return to the relative growth rate, and the ecological interpretation of r, the Law of Exponential Growth is given by $dN/dt = rN$, which has the solution $N_t = N_0e^{rt}$, where N_0 is the number of individuals at time zero and N_t is the number at time t, and e is the base of natural logarithms. In the solution to the original equation, we look at the final expression e^{rt}, where the first exponent is r and the second is t, which is the elapsed time. To return to e, this is the base of natural logarithms, which are written as $\log_e$ or ln. Like π, e is an irrational number, with a value of 2.71828 … Its particular property as a base for natural logarithms is that it is the only base with a rate of growth equal to its size. Since the products of reproduction also reproduce, or the products of growth also grow, natural logarithms provide a measure of growth that is independent of the size of the population or organism, as if size were not a factor. This is important, as it allows the rate to be calculated easily, irrespective of population or organism size. The doubling time (t), as has been mentioned, can also be derived from the exponential equation as $t = 0.6931/r$.

An economic analogy will help in making some of these ideas more accessible. The relative growth rate r is analogous to the rate of interest, and population size N is analogous to the sum of money invested. Biological growth behaves like compound interest in that, for natural populations, the newborn in time join the breeding population, just as the interest is returned to join the principal sum in compound interest.

There is an important application of r that is one of the more important generalisations in ecology. It was first used by the American ecologist F.E. Smith in 1954 [8], and has been elaborated several times by ecologists since then. It is used here in a conceptual form (see Figure 2.4). A graph of r against generation time, for organisms of all sizes from micro-organisms to humans, shows a declining linear trend from the smallest to the largest. It is the smallest animals, with the shortest generation time, that have the highest relative growth rates. By plotting the relative growth rate, or r, against the generation period for species of a wide range of sizes and complexity (Figure 2.4), it can be seen how the relative rate declines and generation time increases in approximate order of size. The graph of relative growth rate against generation time produces a remarkable result: the fastest-growing animals, with the shortest generation time, tend to be smaller and simpler organisms, while the slower-growing animals with longer generation times tend to be larger and more complex organisms. This suggests that an evolutionary trend in evolution is the tendency to

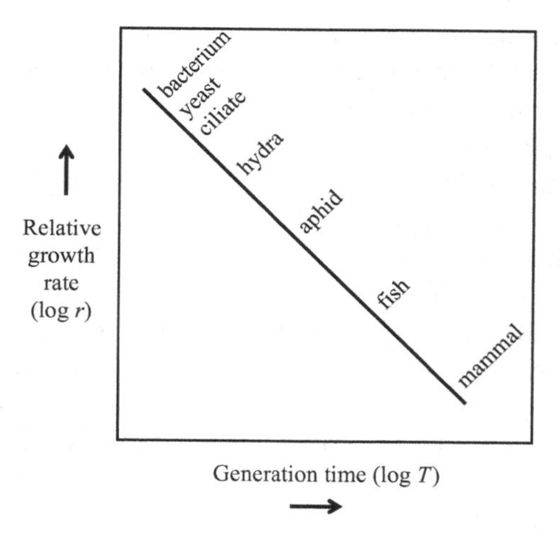

Figure 2.4. Graph showing the relationship between relative growth rate (r) and generation time (T) for a range of biological types (see also Figure 8.5).

progress from small to large, from simple to complex, and from short-lived to long-lived. A more detailed version of this plot will be used in Chapter 8.

THOMAS MALTHUS AND HIS IMPACT UPON EVOLUTIONARY THINKING

The economist Thomas Malthus (1766–1834) graduated from the University of Cambridge in 1788. He had monitored the size of the population of the UK during the Industrial Revolution, but feared that progress towards a better society was not possible due to the rapid increase in population. As we have seen, it was Malthus who first appreciated the explosive nature of biological reproduction, as part of his wider theory. His thesis was that, while human population growth would be expected to increase by geometric (= exponential) progression (1, 2, 4, 8, 16, 32, ...), the means of supporting such population growth with food would only be expected to increase linearly by arithmetic progression (1, 2, 3, 4, 5, 6, ...). Thus, he forecast that the increasing disparity between population growth and the means of supporting it would create hardship and poverty. Malthus expected that population growth would always outrun food supply and that, in the end, numbers would be kept down by famine, disease or war. He maintained that 'Population growth will always tend to outgrow food resources unless strict limitations are placed on human population'. His *Essay on the Principle of Population* [9] soon became widely read and led to impassioned debate. Forebodings of a crisis became a national issue that influenced political thinking.

This chapter has provided examples of exponential reproduction in organisms kept for limited periods in experimental systems, with additional examples from nature. The actuality is that populations in nature generally maintain relatively stable, though fluctuating, populations. Various factors, such as mortality due to predation and competition, suppress population growth in such a way that it actually remains relatively stable. The other arm of Malthus's theory places unjustifiable emphasis on the arithmetical increase in agricultural productivity as opposed to human geometric population growth, which has proved to be wrong.

Despite the flaws in Malthus's theory, reading his essay was a catalyst to both Charles Darwin (1809–1882) and Alfred Russel Wallace (1823–1913), as they independently developed their theories of evolution by natural selection. Darwin wrote in his *Autobiography* (edited by his son, Francis [10]) that in 1838, 18 months after starting his first notebook on the variations of animals and plants under domestication and nature, he read Malthus on population 'for amusement'. He recognised that species have the capacity to reproduce geometrically, even though the actuality is that populations of animals remain relatively stable. The loss of unfavourable variations in the struggle for existence would tend to conserve fitter individuals and favour the creation of new species. 'Here then', Darwin wrote, 'I had at last got a theory by which to work'. His theory was first written as a brief abstract in 1842, and then at greater length two years later. In 1856 his friend, the geologist Charles Lyell (1797–1875), advised Darwin to write up his ideas more fully. But in 1858, as had been feared, his 'plans were overthrown' on receipt of an essay by the young Alfred Russel Wallace, then on an expedition in the Moluccas.

Wallace had his revelation when suffering from a severe attack of malaria, lying wrapped in blankets during a cold fit, even though the ambient temperature was 31°C. He too had been impressed by Malthus and his essay, which he had read some years before. While thinking of the various checks and constraints that prevent the potential for reproduction being realised, 'there suddenly flashed upon me the idea of the survival of the fittest – that the individuals removed by these checks must be on the whole inferior to those that survived'. In the hours before his fever subsided, Wallace thought out his theory of natural selection. During the following evenings he drafted his essay *On the tendency of varieties to depart indefinitely from the original type*. This was sent to Darwin, with a request to forward it to Lyell for his opinion. Upon the arrival of Wallace's letter, Darwin immediately wrote to Lyell that his worst fears had been realised.

The outcome brings credit to both Wallace and Darwin, who went to particular pains to act honourably. Darwin wrote to Lyell saying 'I would far rather burn the whole book (*On the Origin of Species*), than that he or any other man should think that I behaved in a paltry spirit'. The result was that Wallace's essay and Darwin's 1844 sketch were both presented to the Linnean Society on 1 July 1858. Lyell and the botanist John Henslow (1796–1861) spoke in support and later wrote to Wallace defending Darwin's priority, which Wallace accepted completely. Darwin worked without let-up on the 'abstract' of his theory, which was published as *On the Origin of Species by Means of Natural Selection, or the Preservation of Favoured Races in the Struggle for Life* on 24 November 1859. Its publication was an immediate success and, with the support of a powerful review in *The Times* by T.H. Huxley, the first print run of 1,250 copies was sold out on the day of publication, and a second edition of 3,000 copies soon afterwards. Wallace later wrote, 'I feel much satisfaction in having thus aided in bringing about the publication of this celebrated book, and, with the ample recognition by Darwin himself, of my independent discovery of "natural selection"'.

What is so important here is that both Darwin and Wallace had taken the same point about exponential increase from Malthus. They realised that it was the difference between the potential and exponential increase, and the actual rates of reproduction, that implied large mortalities. Not only was the reproductive potential made possible by large numbers of gametes, seeds and eggs, but the survival of the young during their development into adults was far less than would be expected. It followed that there must be considerable mortality in the struggle for survival. For Darwin, and for Wallace, here was the raw material by which natural processes might select those individuals with more favourable genes, in preference to those that were less well adapted, in a way that would lead to the evolution of new species. Without significant mortality, the numbers would inevitably become impossibly large and unsustainable.

Darwin drove home his point about reproductive excess by calculating the exponential increase of the elephant as the slowest-breeding animal (Figure 2.5). He reckoned that, at a minimum, it might begin breeding at 30 years old and produce about six young before ceasing at

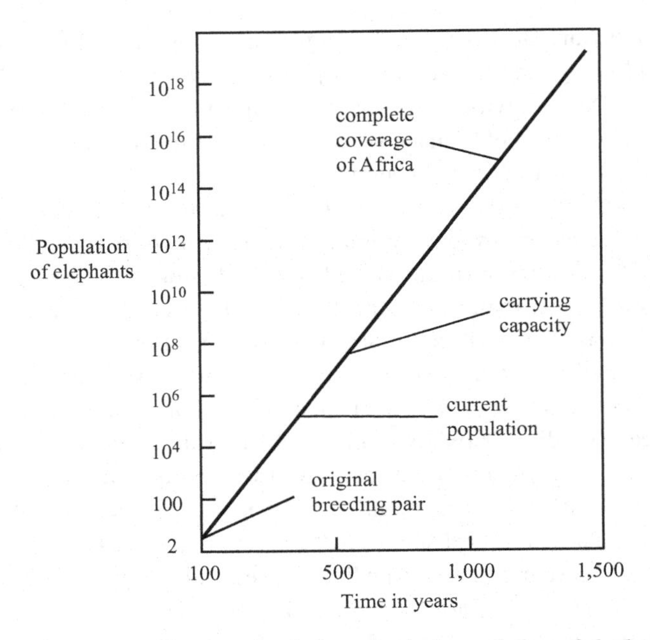

Figure 2.5. The growth of a hypothetical population of elephants from a single breeding pair, illustrating Darwin's example from *On the Origin of Species*. It shows that, in principle, even a slow-growing animal is capable of producing populations of huge numbers if it multiplies without constraint.

the age of 90. He calculated that after a period of 740–750 years there would be 19 million elephants alive, descended from the first pair. It is plain to see that actuality is different, and mortality due to predation, competition and disease reduces population increase to sustainable numbers. While this example could hardly be more different than those already given, it demonstrates that all plants and animals have the ability to multiply exponentially, at rates that would inevitably outstrip the capacity of the environment to sustain such numbers. It was the difference between the reproductive capacity of organisms and their survival that indicated the gulf between the numbers that could survive and those that do, revealing the high proportion that are lost. The inevitable outcome is that the few that remain have been selected from the many by virtue of their effectiveness as competitors, surviving all the hazards that might contribute to mortality.

It has now been shown that the ability to grow exponentially is of great value to organisms in colonising new habitats, like the blooming of algae in spring, or the increase in aphid numbers which feed and multiply in farms and gardens with the first growth of plants each year. Then there is the rapid colonisation of new habitats and countries by invasive species. In competition for space, it is the fastest to grow that occupy empty space first, and thereby exclude others from doing so. This is most obvious in competition for space, whether the competitors are lichens on a gravestone, or barnacles on a rocky shore. Clearly prior occupation of space is an advantage to these species. It is as true of the first microalgae that bloom in spring, which utilise the nutrients. The first not only wins, but it also denies other competitors the prize. The annual competition for space on the spaces on rocky shores is mostly won by colonial species, simply because the exponential growth of bryozoans, hydroids and tunicates enables them to cover the available space more rapidly than solitary species can.

SELF-LIMITING POPULATIONS?

The overwhelming observation that we are left with is that exponential growth is a double-edged sword. It is adaptive in its early stages, but becomes increasingly maladaptive as growth rates accelerate. This is because the products of replication also replicate, and join the original population; the inevitable consequence is that the growth rates accelerate and the overall increase is exponential.

Although growth rate, r, tends to remain constant, the positive feedback effect of replication and reproduction (Figure 2.6) creates a population, or tissue size, that can only grow faster. This becomes maladaptive, and must be constrained, for otherwise the future of any species is at risk due to reproductive excess and the growing competition

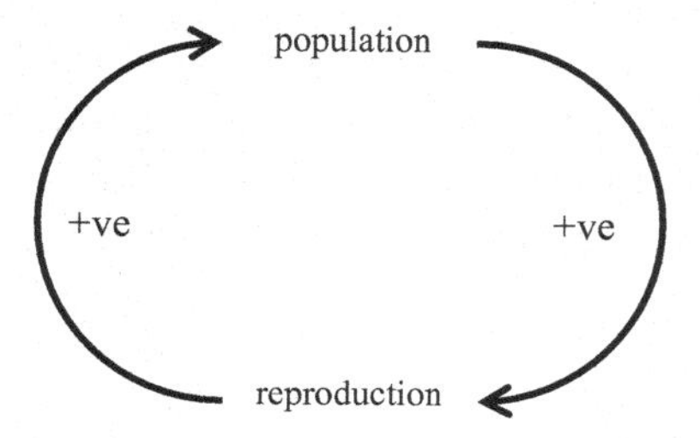

Figure 2.6. The positive feedback of biological growth. Exponential increase is created by the addition of new individuals to the growing population which, in becoming reproductive themselves, create an exponential increase.

for resources. It is inescapable that the positive feedback of exponential growth required the evolution of self-limitation, for exponential growth is potentially lethal, and in evolution there is nothing more important than survival. Survivorship ranks higher in the lives of organisms, and their evolution, than maximising their population size at potential risk to themselves, or maximising the exploitation of their habitat or prey populations.

Cyberneticists argue that positive feedback may be useful, but it is inherently unstable, capable of causing loss of control and runaway. A higher level of control must therefore be imposed upon any positive feedback mechanism: self-stabilising properties of a negative feedback loop constrain the explosive tendencies of positive feedback. This is the starting point of our journey to explore the role of cybernetics in the control of biological growth. That is the assumption that the evolution of self-limitation has been an absolute necessity for life forms with exponential growth.

SUMMARY

There is a natural tendency of populations to grow with a constant doubling time and exponential growth. The unlimited population growth of all species tends to reflect a constant relative growth rate, reflecting an underlying constant rate of reproduction in the individual organism. This is the 'intrinsic rate of natural increase'. New individuals join the population, and in time become reproductive themselves. This positive feedback of the young back into the population acts as multiplier, such that the larger the population becomes the faster it grows. In the long term, exponential growth is unsustainable, even for animals with the slowest rate of reproduction. It is argued that, however adaptive the ability to grow exponentially might be, the superabundance of any species becomes a threat to itself. As the basis of all biological growth is exponential, there is a requirement that it is self-limited. It is better to avoid extinction by limiting reproduction than to overpopulate, overexploit and risk extinction.

3

Self-regulating systems: from machines to humans

When, perhaps half a century ago, the fecundity of this concept [feedback] was seen, it was sheer euphoria to philosophize, epistemologize, and theorize about its consequences, its ramifications into various fields, and its unifying power.

Heinz von Foerster

EARLY MAN-MADE CONTROL MECHANISMS

I was drawn towards the sight of a whitewashed stone windmill. It was high on the top of a conical hill in central Spain, its sails turning steadily in the breeze. The miller was pleased to show me around. The rotating sails, wooden gearing and turning millstones rumbled and creaked beneath the conical roof: a sound that cannot have changed for centuries. I knew that such windmills incorporated a primitive control mechanism to regulate the flow of corn, and I wanted to see it in operation. Suspended from the bottom of the hopper, a shallow sloping trough fed corn to a hole in the centre of the 'running' millstone. If the corn flows too fast the corn is incompletely ground; too slow and the flow of corn fails to keep the millstones apart, resulting in grit in the flour and millstones that soon need 'dressing'. All this is prevented by a simple device which regulates the rate at which the corn flows. Attached to the hopper is a wooden 'fiddle' which rests on the rough-hewn surface of the 'running' millstone. As it turns, the fiddle dances over the surface, transmitting vibrations to the sloping trough. The faster the stone revolves, the more actively the fiddle dances and shakes the trough, and the more corn is shuffled into the hole in the millstone. In this way the feed of corn is regulated in relation to the speed of rotation of the sails and stones, saving the miller from having to continuously oversee the process.

(a)

(b)

Figure 3.1. *In situ* centrifugal governor in a windmill (a) suggesting the kind of mechanism, dating from the eighteenth century, from which James Watt derived his governor for use on steam engines (b).
(a) © Peter Cock, re-used with kind permission; courtesy of Warwick University, www2.warwick.ac.uk/fac/sci/moac/currentstudents/peter_cock/photos/berkswell_windmill/; (b) by courtesy of 'Modern Relics', Flickr at http://flickr.com/photos/modernrelics/750074948/.

Such crude control mechanisms have been incorporated in windmills since the Middle Ages or even earlier, the knowledge passed on orally from one miller to another. When more powerful windmills were developed in the seventeenth century, there was a danger that their power could destroy the machine that produced it, unless some means could be devised to limit the power produced. One device was a 'whirling regulator', or flyball governor, to control the sail area of the windmill, an idea originally patented by T. Mead in 1787. A different application used essentially the same device to resist the tendency of the gap between the stationary bedstone and the upper running stone to widen with the speed of rotation (Figure 3.1a).

James Watt (1736–1819) was repairing a model Newcomen engine when he realised there was a serious defect. Steam was wasted with a loss of latent heat, so he incorporated a separate condenser. He also added another crucial refinement. Watt took out a patent for his centrifugal governor in 1788 (Figure 3.1b), the year after Mead took out his patent, which incorporated a similar governor as a component of his mechanism for furling and unfurling the sails on windmills.

Operating the earliest steam engines always required a hand on the steam regulator; as speed was determined by the amount of steam entering the engine, there was always the danger that they would run too fast - and they often did. Watt's governor tamed the steam engine, and took some of the unpredictability and danger out of working nearby.

As the engine went faster, the ball weights flew outwards under the influence of centrifugal force. This pulled a sleeve upward on the shaft linked to the steam inlet valve, reducing the amount of steam admitted to the cylinder, thus slowing the speed of the engine. With the governor spinning more slowly, the bob weights fell back under their own weight and opened the inlet valve, so that the engine begun to increase in speed. A governed engine could be set to run by itself at the desired rate and then left, as it stabilised the engine speed automatically. Before this simple invention, steam engines could be dangerous, as they sometimes ran out of control, and literally shook themselves to pieces. It is not surprising that the flyball governor became an icon of the controlled power of the steam engines that propelled the Industrial Revolution. However, even steam engines controlled by a governor could be unruly, and engineers approached the eminent Scottish physicist James Clerk Maxwell (1831–1879) to help them understand the theory of governors.

A crucial step in the emergence of the principles of feedback mechanisms was the first description of their operation in mathematical terms. It is seldom recognised that Maxwell wrote the founding paper on cybernetics. The paper was published in 1868 and dealt with the theory of Watt's governor. Maxwell attempted to provide engineers with a theoretical basis for the centrifugal governor to help answer their questions about its sometimes unforeseen dynamics. Problems of unexpected behaviour needed to be solved: from 'ringing' to 'runaway', from sustained oscillations, hunting between a high and a low level, to engines that simply increased to their maximum speed. Such problems were ultimately solved by damping, or adjusting the weight of the balls on the governor. Maxwell's equations could be used by engineers to account for the behaviour of steam engines, avoiding the use of trial and error. This was the first theoretical analysis of a control system, to which the father of cybernetics, Norbert Wiener, later paid homage.

An important property of the governor becomes apparent when the engine comes under load. Imagine a steam traction engine setting out on a road journey at a constant speed; on reaching a hill the work load increases significantly, speed reduces and the governor opens the inlet valve allowing more steam into the cylinder. The engine automatically works harder to maintain its speed when under load. On the other side of the hill, the steam engine accelerates downhill under its own weight. Engine speed increases, the governor weights fly out, shutting the inlet valve and reducing steam to the engine, so that the engine-braking helps to reduce road speed. The point is that the maintenance of a constant speed, as is sometimes assumed, is not the entire purpose of the governor. More important is to increase the output to counter the effect of changing load. In the case of a steam engine, or even a modern tractor, any increase in workload that might slow the engine is automatically compensated through the adjustment of the governor, which initiates an increase in work output. The speed of the engine may actually change little, but as the engine works harder, a wider throttle opening is required to increase the torque. Such adjustments due to the operation of the governor can be heard when a tractor reaches heavier soil when ploughing, or when a pneumatic drill is operated and its compressor comes under load.

The automatic adjustment of the engine power output in relation to workload is the most important property of the governor. It was a crucial invention because it became the means of automatically controlling the work output of steam engines that, in their tens of thousands, hauled loads by road, ploughed and cultivated the land and drove threshing machines. Stationary steam engines powered cotton mills and sawmills, and replaced windmills. They drove pumps for water and sewage, powered the manufacture of goods and were used in engineering works to power lathes. Steam engines were the prime movers of the nineteenth century and all required control to regulate and maintain their speed under load to a predetermined setting, which was made possible by Watt's centrifugal governor.

Regulation of a process by using its own output was a novel idea, and in time the governor became an icon for automation. The idea had been adapted by Watt from the device used to regulate the gap between millstones, and was refined by Maxwell's theoretical work, although it was still 80 years before the birth of cybernetics.

THE GOVERNOR AS METAPHOR

It is interesting to ask at what point did the principles of automatic control emerge as an abstraction from mechanical devices such as the governor. When Alfred Wallace wrote to Charles Darwin from his sickbed in the jungles of the Moluccas, explaining his idea about evolution that virtually paraphrased Darwin's own unpublished theory, he used the metaphor of a governor. He wrote in a letter, which arrived at Down House on 18 June 1858:

> The action of this principle is exactly like that of a centrifugal governor of a steam engine, which checks and corrects any irregularities almost before they become evident; and in like manner no unbalanced deficiency in the animal kingdom can ever reach any conspicuous magnitude because it would make itself felt at the very first step, by rendering existence difficult and extinction sure to follow.

The principle he alludes to would not win acceptance for another 80 years, yet Wallace understood it well enough to use the concept of self-correction in presenting his argument to Darwin. Those who later examined Wallace's letter could not see how natural selection could be related to the operation of a centrifugal governor. But Wallace was not alone in thinking of the governor in this way; a few years later Herbert Spencer (1820–1903) used the metaphor of a steam engine in the chapter on equilibration in his *First Principles* [1]. Spencer was a Victorian thinker and writer, who saw philosophy as the science of sciences. His self-appointed role was to seek out those generalisations that unify disciplines. His weakness was that he had an unquestioning confidence in his own views. In the words of his protagonist, Thomas Carlyle (1795–1881), he carried 'more sail than ballast'. His good friend Thomas Huxley was a necessary foil who burst many of his higher-flying theoretical balloons. Spencer began to write a series of volumes on synthetic philosophy, starting with *First Principles* in 1862, exploring the evolutionary landscape beyond Darwinism. Huxley reviewed the book and thought it was good. This surprised Spencer, who was used to scant praise from Huxley. Robert K. Merton writes about Spencer, 'of whom it can be said that never before had anyone written so much with so little knowledge of what others before him had written'. But Peter Medawar put his finger on what Spencer was doing when he wrote of the content of *First Principles* as 'primordial truths' arrived at by deduction from 'the elementary datum of consciousness'. Pre-existing knowledge is sometimes a handicap to original thought.

It is Chapter XVI on *Equilibration* that is of greatest interest here. Spencer explores an explanation for equilibration where forces producing change in one direction are equalled by those antagonising it. His discussion of instances and applications ranges widely, from spinning tops and gyroscopes to the solar system, and from physiology to ecology and society in general. With respect to such systems, he wrote in the final revision of *First Principles* (1900):

> Providing the disturbance is not such as to overturn the balance of functions, and destroy life, the ordinary balance is by and by re-established ... Not even in those extreme cases where some excess has wrought a derangement that is never wholly rectified, is there an exception to the general law; for in such cases the cycle of functions is, after a time, equilibrated about a new mean state, which thenceforth becomes the normal state of the individual. Thus among the rhythmical changes constituting organic life, any disturbing force that works an excess of change in some direction, is gradually diminished and finally neutralised by antagonistic forces; which thereupon work a compensating change in the opposite direction, and so, after more or less of oscillation, restore the medium condition.

Spencer's fascinating thought experiment describes much that we could accept today about the dynamic behaviour of homeodynamic mechanisms. Spencer defined life, with such a system in mind, as 'the continuous adjustment of internal relations to external relations'. A search to reveal what influence the work of Claude Bernard had on the thinking of Herbert Spencer revealed that the reverse was true. Bernard's use of the terms 'equilibrium' and 'equilibration' with respect to homeodynamic systems is actually drawn from Spencer [2].

Spencer did refer to the work of Maxwell, and may have read his paper *On governors*. In the chapter on *Equilibration* he classifies equilibria of different kinds, referring to a 'dependent moving equilibrium' illustrated by the steam engine and commonly met with in evolution. Wherever the idea came from, it is an inspired description, a century before its time.

THE ORIGINS OF THE THERMOSTAT

Consideration of mechanical controllers has taken us ahead of other kinds of mechanisms that were invented much earlier, in particular those that controlled temperature. A Dutchman Cornelius Drebbel (1572–1633), who was patronised by King James I, invented a diving boat that was demonstrated travelling up the Thames at about 4 metres deep. He is also credited with inventing a self-regulating oven, used not for cooking but as an incubator to hatch and rear chicks. It incorporated the first thermostat, which for many epitomises what a feedback mechanism is. Mimicry of homeothermy of a hen on eggs, provided the archetypal model of a control mechanism.

The Frenchman Denis Papin (1647–c.1712) in 1680 achieved automatic pressure regulation with his steam safety valve, the forerunner of the modern pressure cooker. This was not a direct control of temperature, but by controlling the pressure, water could be induced to rise to a higher temperature before boiling, and so cook food more quickly. He was made a Fellow of the Royal Society and cooked a meal for the Fellows, and later for Charles II.

The rate of combustion of an open fire is determined by the amount of fuel and the 'draw' of the chimney, but for control over temperature a closed system is required that incorporates a temperature sensor and uses some measure of it to control the fire. The distinction is made clear by the first thermostatically controlled kitchen range. A Swedish scientist, Gustaf Dalén (1869–1937), was an inventor of automatic systems for lighthouses, coastal beacons and lighted buoys. He devised acetylene lights that could operate untended, turning on automatically at dusk and off at dawn, flashing at intervals to aid navigation. For this invention in 1912 Dalén was awarded the Nobel Prize for physics. Dalén was injured that same year by an explosion in a quarry, while testing safety devices on acetylene cylinders. He lost his eyesight, but otherwise made a full recovery. It was during his convalescence at home that Dalén became aware of his wife's frustration in coping with an inefficient kitchen range with a temperature that varied and was difficult to regulate. Dalén decided to build a thermostatically controlled stove, which would make cooking more predictable. He designed the cooker during his convalescence, with his wife's help. The result was an efficient cooker that used only 3.5 kg of coal a day, maintaining a range of stable oven and hob temperatures for every kind of cooking.

The Aga cooker was the first domestic cooker employing a 'closed loop system' to control temperature. Control was by a mercury-filled thermostat that activated a flap controlling the air supply to a small, intense and efficient fire. If the temperature dropped, the thermostat admitted more air, and when it rose too high the flap closed the airway. Without feedback regulation of temperature, the Aga could never have lived up to its reputation. The smooth lines and covered hobs made the cooker safe for a blind man to use, while a constant temperature was achieved and Mrs Dalén no longer burned her cakes. The 'Aga' stove was named after the company of which Dalén was Managing Director – the Aktiebolaget Gasaccumulator. The key features of the design were patented in 1922 and the stove went into production and remains so, with only cosmetic changes to Dalén's original concept.

It is certain that the inventors of the miller's fiddle, the centrifugal governor and the thermostat had no idea what each had found in common. They had not considered each system at the level of abstraction that viewed them as systems whose operational logic ran in a causal cycle to achieve a regulatory purpose. The common feature is that some aspect of their output is fed back to adjust their input. Another kind of example is economic and has nothing to do with any physical mechanism, but with the flow of information within it.

THE 'INVISIBLE HAND' OF MARKET CONTROL

Adam Smith (1723–1790) was a curious man and comically absent-minded. His speech was stumbling until he warmed up and then he couldn't be stopped; his gait was described as 'vermicular'. He is known primarily for his work *An Inquiry into the Nature and Causes of the Wealth of Nations* [3], generally known by the last three words of the title. He posed the question, 'How does a system of perfect liberty, operating under the drives and constraints of human nature... give rise to an orderly society?' Smith's particular achievement was that he recognised the importance of the 'invisible hand' of market forces. He saw that the price mechanism in effect brought about a tacit agreement between millions of producers and consumers over the prices at which goods exchanged hands. The efforts of those selling products to maximise their own profits brings about the quantity of goods available to match the quantity that people wanted to buy. Smith realised that the market is the key; yet it is controlled by no-one, is automatic, flexible and self-regulating. At one time, economists even considered it was the 'Eighth Wonder of the World'. The idea that such a system could manage a task that 'no wisdom or human knowledge' could undertake gave it magical properties that Smith obviously held in awe.

Let us look in more detail at the market system. The laws of supply and demand operate by a simple feedback mechanism (Figure 3.2). A vendor might take to market a product and, given the production costs and a profit margin, have some idea of the anticipated price at which to sell it. The market itself then takes over, achieving an integration of the wishes of vendors of the product in relation to the overall demand for it by the buyers. The market finds its own price by way of the interplay of opposing forces. Prices may be controlled by supply and demand, such that if there are many vendors with large amounts of some product, or there is little demand for the product, prices fall.

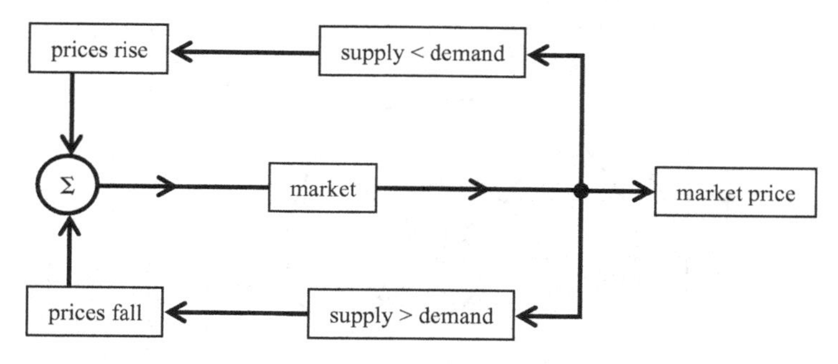

Figure 3.2. The regulation of market prices depicted as a simple feedback mechanism of opposing forces of supply and demand.

Conversely if there is little of the product and great demand, prices rise. In this way supply and demand interact, and the relationship between them regulates prices independently of any notional prices of the players.

The nineteenth-century tea trade illustrates the extreme fluctuations of markets, and clearly shows the rapid adjustments in prices as supply and demand change. Before the cultivation of tea spread around the world, it was grown primarily in China. During the era of the tea clippers, they raced from China, around the Cape of Good Hope, to London to bring to market the first tea of the season. Tea from the first ship commanded high prices, because there was great demand, and scarcity in supply. The fashionable demand for the first tea of the new season was soon satisfied. With the arrival of other clippers, supply soon exceeded demand, the market became saturated, and prices plummeted. The cost of losing the race to bring tea to the major markets was high, and led to the fastest and finest sailing clippers ever made. Above all else, they were built for speed, and driven to win.

If we look at the situation where supply and demand are relatively stable, we can better understand how markets typically control prices. The point at which the supply curve intersects the demand curve represents the level of goods where consumers require exactly the quantity that producers want to sell (Figure 3.3). This is the market equilibrium, to which the operation of the market will tend to draw prices. The intercept is the 'attractor' and the further the price and/or quantity are away from this point, the greater the attraction for them to move towards it. In this way, prices change, moving towards the attractor, spiralling in on the market equilibrium. As Adam Smith

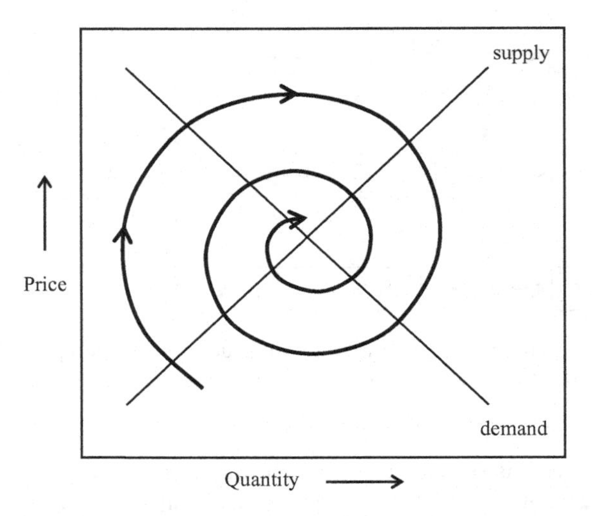

Figure 3.3. Diagram illustrating the self-stabilising behaviour of market prices in relation to supply and demand.

discovered, even though sellers and buyers are all acting in their own self-interest, the market operates in such a way that production and prices are drawn to the intercept of supply and demand.

GUIDANCE BY MOVING THE GOAL

Servo-mechanisms probably represent the most common use of feedback control, such as the servo-assisted steering and brakes in motor vehicles. The system differs in that the goal setting is adjusted to meet the operator's requirements. The mechanism then works to maintain the required adjustment at the setting, whatever the deviation or load placed upon it. The performance of a mechanism is controlled by some error-detecting feedback; an error is derived from the comparison of some preferred position in relation to the actual position of a tiller, control column or steering wheel. A signal is then amplified such that a small control mechanism is able to control a large device, such as a ship or a jumbo jet.

Joseph Farcot (1823–1906) invented the servo-motor in 1868, in the age of the large tea clippers, when a number of men were often required to control the helm in heavy weather, in order to maintain speed and hold course. The first servo-mechanisms were 'steering engines' for ships, making it possible for a ship of any size to be steered by one man. Today ships are steered by descendents of Farcot's steering engine, while modern aircraft require a multiplicity of servo-mechanisms to position their control surfaces. Their servo-mechanisms operate continuously and automatically to maintain a course or altitude and aerodynamic stability, responding to any disturbance of sea state or air turbulence. The development of servo-mechanisms made significant advances in naval gunnery, with the advent of large guns mounted in turrets during World War I, for without automated control the hit rate was just a few per cent. Their large weight and the pitching and rolling of the warship made the automation of gun aiming and the control of firing essential. Similar devices maintain the aim of a gun on a tank as it travels over undulating ground, its effectiveness indicated by the uncanny stability of the gun barrel, reminiscent of the stillness of a kestrel's head in a hover.

During the 1950s, control system development progressed rapidly, as the multitude of applications grew. Many military applications of control theory were recognised, and large budgets funded academic research into electronic control mechanisms. Servo-mechanisms are used to control satellite tracking, and telescope aiming, not to mention their innumerable uses in the aiming of guns and guiding missiles and bombs. Satellite tracking dishes weighing many tons can now be aimed at satellites with the accuracy of hundredths of a centimetre. The ratio between the power of the control signal and the size of the device can be of the order of billions to one.

BIOLOGICAL CONTROL MECHANISMS AND HOMEOSTASIS

We now focus on the earliest discoveries of control in biology, generally referred to today as homeodynamic mechanisms. Here the process of investigation is reversed, in that an unknown homeodynamic mechanism is implicated in the constancy of internal conditions under variable external conditions. The engineer knows exactly what he is trying to invent, but the biologist may never uncover all the subtleties of the system he has found.

Claude Bernard (1813–1878) was the father of experimental physiology. He first put forward the idea of control in physiology, and is best known for recognising the importance of the internal constancy of physiological processes, saying that 'All the vital mechanisms have only one object: that of preserving constant conditions in the *milieu intérieur*'. Bernard observed that we flush on hot days, as blood vessels just under the skin dilate to allow excess heat to dissipate, and we blanch with cold due to the reverse response as the body works to conserve heat. In mammals, this and many other such responses, controlled by the nervous system, maintain a constant internal temperature over a range of ambient temperatures.

However, Bernard's interpretation was not merely his repeated observation of constancy. He clearly recognised that the internal fluctuations of 'oscillating life' occur in relation to those of the external environment. Bernard stated that 'the perpetual changes in the cosmic environment do not touch it [the organism]; it is not chained by them, it is free and independent'. He went further with the famous expression *La fixité du milieu intérieur c'est la condition de la vie libre*. Underlying mechanisms are implicated 'such that external variations are at every instant compensated and brought into balance ... so that its equilibrium results from a continuous and delicate compensation established as if by the most sensitive of balances'.

A century later, in his 'articles of belief', Walter Cannon (1871–1945) gave this principle the name 'homeostasis' [4]. He explained his reasons as follows:

> The coordinated physiological processes which maintain most of the steady states in the organism are so complex and so peculiar to living things ... that I have suggested a special designation for these states, *homeostasis*.

He accepted constancy as evidence of physiological responses acting to maintain it, as Bernard had also observed, implying some automatic regulation by responses that resist change in the controlled state. Cannon noted that, as the intensity of disturbance increases, so the reactions become more pronounced, and subside quickly when the disturbance ends. He also recognised that homeostasis may involve a number of cooperating responses, which may take effect at the same time or sequentially. Cannon could not have known that the common denominator of homeodynamic systems was the feedback mechanism. It would be another 25 years or so before his collaborator Arturo Rosenblueth collaborated with Norbert Wiener, and helped Wiener recognise that his concept of cybernetics was genuinely transdisciplinary and included homeostatic systems. The concept has remained central to our understanding of physiological control of internal processes.

Cannon's final 'article of belief' was especially telling. He observed that when a factor perturbs a homeostatic state in one direction, it is to be expected that it will elicit a response in the opposite direction. Thus we may expect to observe a stimulatory response to an inhibitory perturbation, and vice versa.

MOLECULAR CYBERNETICS

The early insights came from the founding years of molecular biology, and the discovery that inside cells many biochemical pathways act as feedback mechanisms in the control of the cell's metabolic pathways. One could not find a greater contrast than between the flying balls of the centrifugal governor and the interplay of molecular agents controlling biosynthesis. Such is the portability of cybernetic principles. The French biochemist Jacques Monod (1910–1976) made the point that the various biochemical pathways which make up cellular metabolism would be chaotic, were they not coordinated by some coherent system of controls [5]. At this level, chemical processes are controlled by enzymes; they are the agents that act as catalysts to modulate a reaction, even controlling the progress of sequential reactions. A substrate molecule binds to a specific site on the enzyme molecule, as a key fits to a lock, to form an enzyme–substrate complex that speeds up the reaction by millions of times to create a product. Monod drew parallels between the operation of these molecular feedback mechanisms and man-made cybernetic systems. For example, the final product of a sequence of reactions may catalyse the first reaction of the sequence, such that the concentration of the product in the cell actually controls its overall production. The more of a product that is used, the more is produced. There are several similar examples whereby molecular processes are regulated by feedback activation, or activation of a parallel sequence of related reactions. Pathways converge as they provide components of the same macromolecule.

The amount of enzyme necessary to control transfers of energy is very small. Monod likened the efficiency of molecular control mechanisms to servo-mechanisms, where a small amount of energy is required to initiate something on a much larger scale. To emphasise the efficiency - and thus the power - of these control mechanisms, he compared the controlling enzymes with electronic relays, calculating that molecular systems were a thousand billion times more efficient. 'That astronomical figure affords some idea of the "cybernetic" power at the disposal of a cell equipped with hundreds or thousands of these microscopic entities.'

In recent years, D.Y. Zhang and his colleagues have shown new levels of cybernetic complexity in molecular circuitry, involving signals conveyed by nucleic acids [6]. They see nucleic acids as 'designable substrates for the regulation of biochemical reactions'. They visualise a design strategy, which allows a specific input oligonuclide, and another as output, which serves as a catalyst for other reactions. Such circuits include a feedforward cascade, and 'a positive feedback circuit with exponential growth kinetics'.

Jacques Monod was fascinated by amplification at the subcellular level, comparing a chemical factory and a cell. In manufacturing industry, minute amounts of energy are used in servo-mechanisms that control processes involving massive amounts of energy. A good example of amplification in a mammal comes from the hypothalamus. A minute amount of a hypothalamic hormone is passed to the anterior pituitary and stimulates the production of another hormone, which leads to the deposition of glycogen in the liver. The specific steps are unimportant, but intermediate glands act as a series of amplifiers, each magnifying the input further. Typically a signal originates in a cell or gland, spreading by dispersal and amplification to have a much wider effect; by means of a signal passing through several glands before reaching its target, amplification is achieved at each step, so that the final signal to the target tissues is 56,000 times stronger than the initial signal.

It is clear that we must think of biological control systems as servo-mechanisms too, amplifying signals to allow populations of cells to become synchronised, enabling the smallest glands to control the largest organs. Amplification, and the cell communication that makes it possible, creates synchrony among large populations of cells, allowing them to adopt the same metabolic beat. What this means is that, due to the flow of information by the passage of signals between cells, and their amplification, the cells become part of something much larger and more powerful.

The aim here has been to introduce the reader to the early work on homeostasis, and to provide some detail of the workings of a generalised homeodynamic mechanism. Before doing so, we need to appreciate that biological control mechanisms are not simply physiological, of the kind that Bernard and Cannon first discovered. We now know that the largest number of feedback mechanisms are molecular, so we look now at some of the early work at this level of organisation, so as to realise just how powerful biological control mechanisms can be.

THE ORIGINS OF CYBERNETICS

It required the far-sighted thinking of Norbert Wiener (1894–1964), working in the 1940s, to establish cybernetics. His father was a Russian

Jew who had a major influence on the education of his son. Wiener was an infant prodigy, and when a boy he was taught by his father for some years. After graduating in mathematics from college at the age of 14, he went to Harvard to study zoology. This was against his father's advice and did not work out, so after a year he began to study philosophy. Aged 25, Wiener accepted a chair in mathematics at the Massachusetts Institute of Technology, which he held for the rest of his career.

During World War II, he wished to make a contribution to the defeat of Germany and felt that an improvement in the accuracy of anti-aircraft fire would help tip the balance. The problem was to fire a shell into the flight path of an aircraft by first tracking it, and calculating the time it took a shell to reach the position of the aircraft. This involved the development of flight prediction models that could quickly estimate a probable track, despite pilot intervention. Wiener worked with a mathematician named Julian Bigelow. Their study led them to also consider the gunner and the pilot, as living components of the system. It seemed that voluntary control behaviour attributable to the human operators, in particular the feedback of errors, was as important as those attributable to the control mechanism itself. They wondered if it was possible that feedback loops existed in the human nervous system, and that they too might give rise to an oscillation in voluntary movements, such that an intention movement might first overcorrect and then undercorrect its target, leading to an uncontrollable oscillation.

The Mexican, Arturo Rosenblueth, held the chair in physiology at Harvard Medical School at the same time that Norbert Wiener held the chair of mathematics at the Massachusetts Institute of Technology. Rosenblueth organised monthly dinners at the Vanderbilt Hall in the early 1940s, and it was at one such dinner that they met. They had hoped for some time to collaborate, but it was the war effort that really brought them together. The link was also important in bringing together the legacy of Walter Cannon with the early work on cybernetics.

Wiener wondered whether there might be a medical condition due to excessive feedback, as this can result in 'hunting' and persistent oscillation about a desired set point, whether in the aiming of a gun or controlling the hunting of an idling engine. Rosenblueth was able to confirm that such 'purpose tremors', or locomotor ataxia, are found in some people, particularly if they have suffered injury of the cerebellum. Wiener was delighted to find that ataxia or 'purpose tremor' in humans was due to excessive feedback. This offered a step forward in neurophysiology which was not simply a property of nerves and synapses, but was a property of the nervous system as a whole. Their paper, published in 1943, has been referred to as the 'birth certificate of cybernetics'. Later Wiener collaborated with Rosenblueth and others in the use of nerve–muscle preparations to study their 'hunting' behaviour, as though it was a mechanical or electrical mechanism. The same approach and expectations led them to research the heartbeat and to develop theories to account for heart flutter and fibrillation. The same cybernetic principles could be used to understand the behaviour of both machines and men.

Wiener decided to call the entire field of control and communications theory 'cybernetics', derived from the Greek word *kybernetes* meaning 'steersman'. He also wanted to acknowledge the fact that one of the earliest feedback mechanisms was designed to steer ships. The word chosen also reflected Maxwell's work on governors. Although Wiener did not work on homeostasis as such, he recognised that the narrow ranges within which physiological processes operate is such that close control was implicated. He wrote:

> In short, our inner economy must contain an assembly of thermostats, automatic hydrogen ion concentration (acidity) controls, governors and the like, which would be adequate for a great chemical plant. These are what we know collectively as our homeostatic mechanism.

Wiener's book *Cybernetics* was published in 1948, laying the foundations to this new and important field. The subtitle is informative in defining the range of cybernetics as *Control and Communication in the Animal and the Machine*. Wiener dedicated his book to his 'companion in science', Arturo Rosenblueth. The application of the same principles to technology and biology made him certain of the power and portability of his ideas.

To get an impression of how influential cybernetics became in a very few years, it is worth saying something about the development of the movement and those who were involved. The first Cybernetics Group met in May 1942. Although the subject of the meeting was *Cerebral Inhibition*, its principal outcome was a series of further meetings. The founding group included Lawrence Frank, who had been a senior executive with the Macy Foundation; Gregory Bateson and Margaret Mead, who were anthropologists; Warren McCulloch, a neurophysiologist; Arturo Rosenblueth, a physiologist; and Lawrence Kubie, a neurologist and psychotherapist. This range of interests guaranteed the future breadth of application of cybernetics. The meetings were called the Josiah Macy Conferences and occurred annually between 1946 and 1953 on *Cybernetics: Circular, Causal and Feedback Mechanisms in Biological and Social Systems*. Heinz von Foerster, as the editor of the Conference Proceedings, was the architect of this fast-growing discipline, ensuring that 'cybernetics' was in the title of each of the conferences, allied to the idea of circular causality [7]. The series of conferences involved a number of scientists who went on the become leading thinkers of their day, including Warren McCulloch, Claude Shannon, John von Neumann, Gregory Bateson and Margaret Mead, as well as Norbert Wiener. British participants included W. Ross Ashby (cybernetics), William Grey Walter (robotics pioneer) and J.Z. Young (neurophysiology of learning). Their contributions laid the conceptual foundations of self-organising systems both in the USA and the UK.

A GENERALISED HOMEODYNAMIC SYSTEM

Heinz von Foerster (1911–2002), who was born in Austria and was a student in Vienna in the 1930s, helped to lay the conceptual foundations of cybernetics. He made important contributions to the fields of electrical engineering, neurophysiology and demography. In 1958 he gave a theoretical account of a generalised homeostatic system that can hardly be bettered, so I will paraphrase his description in the account given here [8]. All homeostatic systems are self-regulating systems, and incorporate feedback which makes self-referencing possible. All living cybernetic control mechanisms are today termed 'homeodynamic', which encompasses 'homeostasis' – meaning the control of states – and 'homeorhesis' – meaning the control of rates. As von Foerster repeatedly said in one way or another, 'the causal chain is circular' and information flows around the control loop continuously. As the basic mechanism works as one, it must be understood as an entity, because no individual component has meaning on its own, or can be studied in isolation, so a reductionist approach is little, if any, help. One must inevitably adopt a systems approach, developing a theoretical understanding of the control mechanism from its behaviour or output.

What are the shared properties of the control mechanisms that undertake regulation of such diverse systems? All self-regulating processes incorporate the principle of feedback, where information relating to the output of a system is used to control the input. It is important to reiterate that there are two kinds of feedback. Exponentially growing systems, such as the explosive process of nuclear fission, cell division and compound interest, accelerate continuously, as their products are inserted into the biological growth process, creating continuous acceleration. The only difference is the time scale of events: a split second in the case of fission and years in the case of compound interest. In effect, the output contributes to future input, resulting in continuous amplification and accelerating growth. Positive feedback characterises systems that either have no control, or have lost control, and its effects are destabilising and unsustainable. Their feedback is fortuitous, and not adaptive, and errors are amplified with each cycle of the loop, as instability rapidly grows. Consequently, in biological systems, positive feedback does not operate in isolation for long, but is linked to a negative feedback loop that imposes limitation when the accumulation of growth products becomes maladaptive. The second kind is negative feedback, and characterises any system that is inherently self-stabilising. It is called negative feedback because, with each cycle, part of the error is subtracted, such that stability is progressively restored. In positive feedback, the reverse is true, as errors are accentuated and instability becomes inevitable. Negative feedback is found in the governor, the thermostat and in homeostasis.

As a thought experiment, let us imagine that we are inside the growth control mechanism following the flow of information around the control cycle; breaking the circular algorithm down into simple steps, as information is passed around the causal cycle (Figure 3.4a). The operation of such a system is expressed in terms of the flow of information required to control any process. We begin by disturbing the process with some external stimulus, deviating it from its predetermined course. A sensor of the system's output feeds errors back to the comparator. It compares the intended goal setting with actuality, or the input with the output, and determines the difference or error. The error is minimised by a neutralising response, which returns the controlled process to its goal setting. In actuality not all the error is neutralised at once, and several cycles may be necessary to eliminate the error. For this reason, feedback mechanisms are sometimes referred to as 'error minimisation' mechanisms, but as they operate continuously for the life of the organism, errors are soon eliminated.

It takes time for the information, signals and commands to cycle round the loop, causing delays such that the process and correction of errors is neither instantaneous nor complete. The first of these delays is between the disturbance, the error sensor and the error calculator. Once the error has been determined, there is another delay in the time

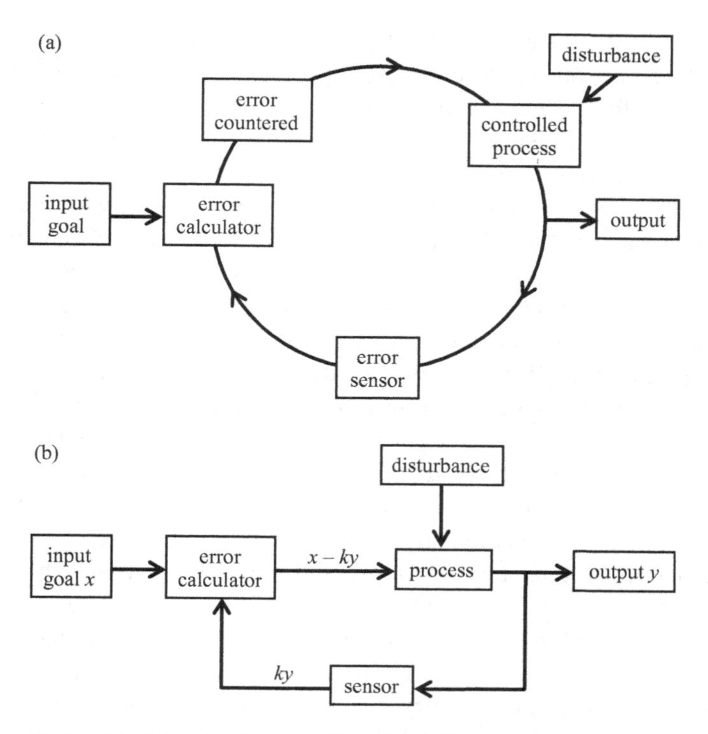

Figure 3.4. The circular causality of a feedback mechanism as seen by von Foerster (a), and the same system as depicted more conventionally as a feedback mechanism (b).
(a) Redrawn from von Foerster [8].

required to overcome the error in the controlled process due to the disturbance. Due to such delays, the response in order to neutralise an error is some time later than the detection of the error (by about one quarter of a cycle). This causes the output to oscillate, repeatedly passing through the goal setting, before the delay in the response brings the process back to its goal, and then passes by it again. The repeated overshooting of the goal setting is analogous to the behaviour of a car on an icy road. As traction is lost, the car begins to skid, the driver responds, and in their hurry to correct the skid, overcorrects. The car swings from side to side several times before the car is brought under control. There is a delay in responding to the skid, and another delay in overcorrecting, because the driver does not respond instantaneously, either to the initial skid or to its overcorrection. All such responses are retrospective and are inappropriate corrections by the time they occur, with oscillation as the inevitable outcome.

In a stabilising system, the overshooting of the goal setting decreases and the oscillations decay as stability is restored. While the lag within feedback mechanisms could be thought of as a 'design flaw' in preventing more rapid equilibration, the oscillations are helpful to the investigator interpreting the behaviour of the system.

We now look at the control loop in a more rigorous way, following von Foerster's account (Figure 3.4b). The output of a system (y) is detected by a sensor, which in effect diverts a small amount of y to produce a signal (ky). At the error calculator, the difference between the sensed information and the input or goal setting (x) is determined, giving an error (ε) from $\varepsilon = x - ky$. The difference now passes to the control process, which works to minimise the error. On sensing the error, some of the output is diverted to create a signal. To achieve control, the signal from the error calculator must be amplified, and of the opposite sign to the deviation, such that a positive response is required to neutralise a negative deviation and vice versa. It must also have a response adequate to negate the error so as to produce the output, which is determined by a 'transfer function'. As we have seen, the effect of the disturbance is eliminated in a progressive way, by the iteration of this process through several cycles and consequential oscillations. As von Foerster pointed out, the output (y) will tend to follow the input (x) to an extent that depends on how effectively the transfer function is able to reduce the error (ε). Ultimately it is the transfer function that is the engine of the mechanism, and is responsible for carrying out the work involved in neutralising errors, for which energy is required. The result is the restoration of a stable equilibrium.

What are euphemistically called 'disturbances' here include anything that disturbs the stability of the systems and deviates the process from its goal setting. This may be due to some transient disturbance, as if the system were knocked off balance and simply requires a little time to restore its equilibrium. Quite different is a perturbation that involves a sustained change, due to what is termed a 'step function' (see section on Perturbations and Oscillations). It will be noted that at rest the system exhibits a continuous series oscillations of small amplitude. The amplitude of the oscillations is slightly exaggerated here. They occur because the sensor is not of infinite sensitivity, so the controlled process oscillates between the limits of detection of deviation above and below the goal setting.

With the imposition of a stable inhibitory load as a step function, the output is counteracted and establishes an equilibrium between the amount of deviation and the response required to neutralise it. To sustain this equilibrium the control mechanism is required to maintain a constant response, sufficient to negate the constant load. In doing so, the mechanism continuously expends energy to maintain the new equilibrium.

Information cycles round the feedback loop without pause, repeatedly comparing actual output with a pre-set input, and working to reduce the difference between them. The continuous negation of errors between the actual condition and some preferred condition yields the key property of seeking to maintain some predetermined goal. In a living system, as von Foerster put it, 'the teleological trick is due to the feedback loop, by which the system can watch its own performance'. In this way, homeodynamic systems operate by a process of error minimisation, but in nature the property that matters most is the capacity to resist the effects of external change on internal processes. This then poses the question as to what the capacity of the system to sustain this load might be. The problem is considered at length in Chapter 7.

CLASSICAL CONTROL THEORY

So far we have considered systems that relate to control mechanisms responsible for homeostasis, but we now need to consider control theory, which is the province of engineers who have designed and built control mechanisms for a multitude of purposes. This shift is necessary because, in searching for the principles of control to describe what we see in biological systems, we turn to some basic aspects of control theory developed primarily by electronic and electrical engineers.

Classical theory is provided by the PID controller, meaning proportional - integral - derivative. The engineer has these three control options, which can be used separately, in combination (PI or PD) or together [9]. In each case they relate to the nature of the response to a disturbance that causes an error.

- In the case of proportional control, the response is in relation to the error. This is the main component in the control mechanism, and neutralises most of the original error. But as the error diminishes, so does the response, such that when the error is small, so is the response. The tendency is for the mechanism to achieve a steady state that does not quite reach the goal setting.

- Integral control results in a response that is based on the sum of recent errors. It tends to reduce the final error in the mechanism as it equilibrates, solving the problem left by proportional control. This too has its problems, as it tends to overshoot and oscillate about the goal setting.
- Derivative control has a response that is based on the rate at which the error is changing. It is effective in opposing rapid changes in output, such that it tends to damp down the tendency to oscillate, preventing overshoot and ringing.

It can be seen that each response tends to address the flaw in the preceding kind of control, so the natural solution is to incorporate all three within a single control mechanism (Figure 3.5). Such systems used to be designed using analogue computers in such a way that a system to meet a new control problem could be set up on the plug board that was the working area, and tuned to meet the control requirements by adjusting the gain in k_p, k_i or k_d. Now, of course, such work is carried out on a digital computer, but the problem is just the same. It is difficult to apply reason and theory in tuning a control mechanism to its purpose, and it is better to use empirical means by adjusting each of the PID components until the desired result is reached. The aim is to set the gain of each kind of feedback to ensure the required level of correction, together with speed and accuracy of equilibration. High gains tend to create a sensitive control mechanism that rapidly senses and neutralises an error, but gains that are too high will risk destabilising the mechanism. Overcorrection can cause the reverse of the desired effect, and magnify errors with respect to the set point, resulting in runaway (see below).

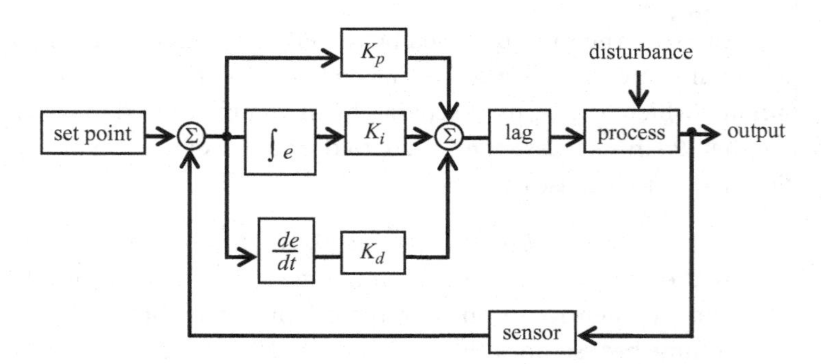

Figure 3.5. A feedback mechanism incorporating the three basic kinds of control: proportional (K_p), integral (K_i) and derivative (K_d) (PID). Redrawn from Hayes [9].

In this way, an engineer can quickly arrive at a control mechanism that is accurate in re-establishing its set point without a residual error. It needs to do so rapidly and with a minimum of overshoot and oscillation. This typically requires compromises between the different PID components. Modern methods use programmed algorithms to identify the PID settings that are optimal to meet requirements. Stochastic control involves statistical methods which assume that there is likely to be random noise and disturbance within the controller, and the mechanism is designed to minimise the effects of these random deviations.

The reader will appreciate that this is an engineer's approach to meeting a control mechanism that is required to meet a specific application. Here, the aim is to understand how nature solved these problems for herself, using the output and reason to identify the kinds of control involved. A relatively new branch of control theory is adaptive control, which allows the mechanism to learn from its own experience and so improve the effectiveness of control. As we shall see in Chapter 7, adaptive control of organisms during their lifetime has been a feature of living systems for much longer than those that are man-made.

PERTURBATIONS AND OSCILLATIONS

Even at rest, feedback mechanisms idle between the limits of detection in a series of oscillations that may not even be detected by the observer, as their amplitude is too small to measure. But even when they can be observed, the output of an idling control mechanism is unchanging and uninformative. Nevertheless, it is the oscillations following perturbation which are the most informative aspect of control mechanism behaviour. The counteractive behaviour of a control mechanism can only be observed when the process it controls is perturbed in some way. The perturbation needs to be adequate to cause a deviation sufficient to require a counter-response.

One can only expect to learn something by making the mechanism do what it evolved to do: to counteract disturbances and the deviations of the process they control. We have used toxic agents for this purpose, which makes it possible to apply the precise amount of disturbance required, and thereby the level of load, by using various concentrations of some toxic inhibitor. This creates an error which the system works to neutralise; the process of countering the inhibition can then be observed directly in the dynamic output. This method of studying growth as a controlled process is described in Chapter 5. Using toxic agents as the tool with which to impose load upon hydroids, as test organisms, it is possible to look at the effects of low concentrations that have the counter-intuitive effect of stimulating growth (Chapters 10 and 11), or high concentrations that load the control mechanism to the point at which the mechanism is overwhelmed (Chapter 7).

There are also curious and revealing facets of control system behaviour that are unexplored in biological systems; one is 'relaxation stimulation'. Equilibria are due to a balance of opposing forces, such that when the load is suddenly removed, the process is disturbed again, precipitating another series of decaying oscillations as the system re-stabilises. The removal of the load, to which the system had stabilised, creates a relaxation stimulation related to the load it had adjusted to accommodate (Chapter 11). What is instructive here is that the magnitude of the initial relaxation stimulation can be used to estimate the amount of counteraction that had been necessary to neutralise the load. With increasing levels of loading or inhibition, a greater stimulatory counter-response is required to match the load. So the disruption of any equilibrium, by adding or removing a load, causes another series of oscillations. This way of assessing the magnitude of the otherwise invisible counter-response is a great help in understanding the capacity of the control mechanism to withstand load (Chapter 7).

Another feature of the behaviour of feedback mechanisms is called 'runaway'; with increasing loss of stability, the amplitude of the oscillations increases as the system becomes more unstable. One explanation is that, with increasing load, the time lag in the system may lengthen such that the displacement of the response to the error by more than a quarter of a cycle results in the negative feedback becoming positive. Then the feedback is added to the error, rather than subtracted, resulting in runaway. This can result in increasingly wild oscillations as the system becomes more unstable. A biological example has been used from experiments with the hydroid *Laomedea flexuosa* exposed to an intermediate level of an inhibitor that causes deviation from the goal setting (Figure 3.6).

Sir Hermann Bondi (1919–2005) is best known for the Steady State theory of the universe, now superseded by the Big Bang theory. Bondi, on seeing these data, exclaimed 'Aha! The pogo stick effect!' He explained that when he was working with the European Space Research Organisation (ESRO) there had been a problem with the

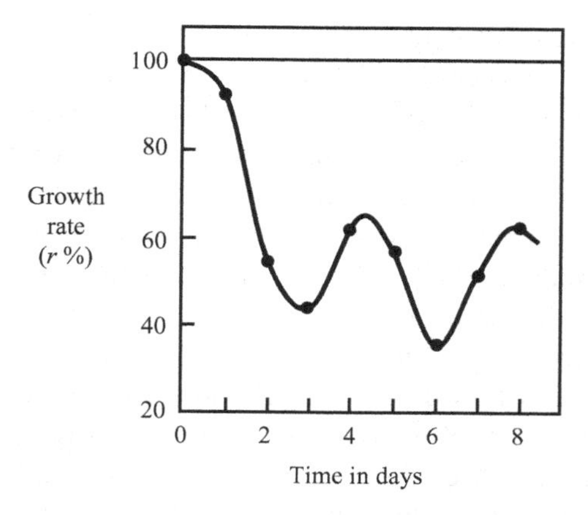

Figure 3.6. Runaway occurs when the feedback is delayed by overload to an extent that the output increases the amplitude of the oscillations, rather than reducing them. Data from Stebbing (1981), see Chapter 5, reference [4].

design of a rocket which became unstable on lift-off. The location of the accelerometer, controlling the rate of increase of speed, was in the nose of the rocket, and controlled the thrust of the engine in the tail. The accelerometer was programmed to produce a smooth and constant rate of acceleration at take-off, but the 'pogo stick effect' made the rocket leave the launch pad in a series of 'fits and starts', akin to runaway. This was due to the forces of acceleration, which caused some compression of the long, thin rocket. As the acceleration compressed its length, the accelerometer signalled to reduce thrust. Then, as the rate of acceleration reduced, the length of the rocket expanded, and the accelerometer then signalled for increased thrust. Through a series of oscillations, this resulted in instability and a jerky take-off, likened by the engineers to a bouncing pogo stick. Once the cause of the problem was understood, it was easily solved by relocating the accelerometer closer to the engine.

GOALS AND TEMPLATES

A pre-set goal sets the target by which constancy can be achieved, and at the same time defines the purpose of the control mechanism to maintain processes at constant states or rates. A goal setting is a necessary prerequisite, used as a reference point for a control mechanism, to which variation due to external factors can be related with each cycle of the feedback loop. The constancy of control alone makes one process 'predictable' to another, where two or more feedback loops need to be coordinated. It is also the way in which homeostatic systems[1] are defined as a feature of physiology maintaining a constant state or output. On close examination, this is found not to be quite true. Internal constancy indicates the effectiveness of the system in counteracting disturbance, as this is the way in which the organism becomes independent of external change.

A sustained disturbance to the controlled process is due to a step change in the level of load due to perturbation. This will cause a series of oscillations that steadily decay as equilibrium is restored, and the system adjusts to its new load (Figure 3.7). The step change may impose a positive or negative load on the controlled process, and elicits a counter-response of the opposite sign. The response negates errors, eliminating differences between the preferred state and the actual state. So in the case of mammals as homeotherms, where there is the maintenance of a constant internal temperature, the response to cold is first to conserve heat by erecting hair follicles to create a thicker insulating layer, second to reduce the peripheral blood circulation thus maintaining core temperature, and finally to generate metabolic heat by shivering.

Such goal settings define the purpose of feedback mechanisms, which are the means by which they can look at their own performance. More importantly, the goal represents an optimum state or rate which the mechanisms maintain throughout the life of the organism.

[1] It is apparent that Cannon's concept of homeostasis is inadequate to encompass the range of self-regulating processes. 'Homeostasis' may be appropriate as a term to describe states kept constant in the physiology of higher animals, such as the control of temperature, of blood glucose levels, acidity or sodium/potassium ratios and so on; but there are other cybernetic systems that go beyond Cannon's concept. Steven Rose criticises 'homeostasis' because the idea of the maintenance of constancy is not rich enough to encompass the range of control mechanism behaviour. Besides which, 'stasis' does not convey a dynamic output that fluctuates continually as it adjusts to environmental change. Even Cannon considered that the term 'homeostasis' might have its detractors, as 'stasis', he thought, might imply stagnation. The term 'homeostasis' is obviously inappropriate for processes controlled by their rates. Conrad Waddington coined the term 'homeorhesis' to mean the control of rates. The number of systems now known to embody feedback extends far beyond the limited application that Cannon intended for 'homeostasis', and to avoid this we follow Rose and adopt the term 'homeodynamic' as a collective descriptor for the behaviour of biological control systems of all kinds.

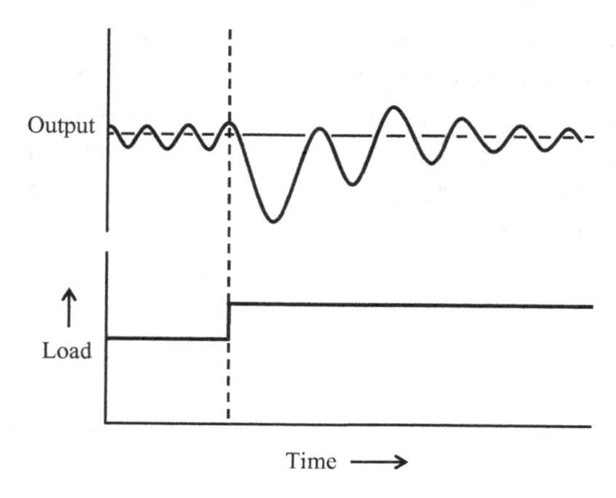

Figure 3.7. Diagram illustrating the typical oscillatory output of a feedback mechanism responding to sustained disturbance, as a step function. Note that at rest, before load is imposed, control output has a low-amplitude limit cycle, determined by the sensitivity of the sensor. See Figure 12.1 for experimental data exhibiting similar behaviour.

That said, there is some margin of variable control, such that goals may be adjusted to increase their capacity to neutralise perturbations that they encounter persistently. Feedback mechanisms can be trained, and may increase their capacity with use and lose it without use (Chapter 7). Goals are also important with respect to the evolution of physiological complexity, as the goal settings make the behaviour of the feedback mechanisms predictable or 'known' to each other. In this way the numerous feedback mechanisms can be coordinated one with another, for the whole to become an integrated system. Collectively, the sum of the goal settings represents a template by which an organism fits its environment.

SUMMARY

Like other self-regulating processes, Maia depends on cybernetic mechanisms which incorporate the principle of feedback. We consider the basic properties of such systems, which are common to the physiology of organisms. Looking at everyday examples of cybernetic systems provides familiarity with the concept before turning to living systems. Homeodynamic systems are now known to be shared by all forms of life, giving them the capacity to control their internal workings in a constantly varying environment. The operation of feedback mechanisms is examined, so that we understand the need for perturbation to create an oscillatory output that can then be observed and examined. This includes the more unusual forms of behaviour such as relaxation stimulation and runaway. The logic and working of a simple feedback mechanism provides the basis for all the more complex mechanisms to be considered later.

4

The wealth of homeodynamic responses

> The broadest and most complete definition of Life will be – The continuous adjustment of internal relations to external relations.
>
> Herbert Spencer

> The maintenance of an invariant configuration in spite of variable disturbances defines the problem of homeostasis.
>
> Francis Heylighen

ADAPTATION AND CHANGE

Adaptation to variable and harsh environments is a central theme of evolution. From the North and South poles through the temperate regions to the equator, there is variability to learn to live with. Much of that variability follows a pattern and, in a sense, it is predictable. Daily, lunar and annual seasonal patterns of variability dictate the behaviour of nearly all forms of life. What is important here is that those events on Earth that relate to the cycling of the sun and the moon have remained unchanged for so long that animals have evolved adaptive rhythmical patterns of behaviour that are encoded in their genomes. The behaviour is instinctive and simply part of what they do with regularity.

But what about adaptation to erratic events that follow no pattern, fluctuating from minute to minute, and from hour to hour? In a single place, one may expect continuous variation of temperature and humidity, but if one were moving around continually in flight like a bumble bee, or running around on the ground like a mouse, the changes sensed would be quite different from those at a single location. So variation experienced by the organism is not simply about change with time, but also change from place to place. So as animals evolved to become larger and able to cover greater distances, rapid adaptation to change became essential. There is also the opportunity for adaptation to be part of moving around, by looking for less taxing places to live.

As organisms became larger, with longer generation times, the rate at which natural selection can occur has become slower (Figure 2.4). The rate of evolutionary adaptation to change in different species declines with the frequency of genetic recombination, and sexual reproduction. Larger species would have become increasingly vulnerable to change without the evolution of autonomous adaptive mechanisms. Feedback mechanisms arose that could respond automatically to change without modifying the genome; natural selection 'invented' self-regulation by cybernetic mechanisms aeons before Adam Smith's 'invisible hand', Watt's centrifugal governor or the thermostat. Our understanding of homeostasis in living systems arose from the experiments of Claude Bernard, and later Walter Cannon, each working with mammals yet trying to understand the human species.

The evolutionary innovation that deals with change is the control mechanism, which makes possible immediate and continuous adaptation to change. Self-regulation of any kind requires feedback, as we have already seen. A mechanism that watches its own operation, and uses the information by way of feedback loops to control itself, was an important evolutionary step, as it influenced all that followed. These are the autonomous systems described in the previous chapter that adapt and adjust continually and quickly to perturbation. Feedback mechanisms provided adaptive responses in a time frame that could not be matched by natural selection, which in larger, more complex animals does not provide adaptation in the short term.

OSMOREGULATION IN LARGE AND SMALL ORGANISMS

It used to be assumed that homeodynamic systems are to be found predominantly in highly evolved organisms. We now appreciate that homeostatic systems evolved much earlier, and are numerous even at the cellular level; Jacques Monod came to think of the working of cells as a 'cybernetic network'. Osmoregulation demonstrates the sophistication of homeodynamic control in a range of organisms.

Substances such as salt or sugar dissolve in water to create a solution. If two solutions are separated by membranes that are somewhat permeable, like those of animal or plant cells, there is a tendency for the solvent to pass through the membranes to equalise the concentrations of the substances in the two solutions. Aquatic animals have to keep the concentration of their internal fluids constant by counteracting the tendency for water to pass in or out of them by osmosis. In freshwater animals, water tends to enter the animal by osmosis, which is countered by pumping water out. Marine animals have the opposite problem and excrete salts. Whether a marine or freshwater organism, this is 'osmoregulation'. The marine iguana of the Galapagos Islands grazes on seaweed and sneezes a salty spray to rid itself of the excess, and even Lewis Carroll's mock turtle cried salty tears. So osmoregulation is a key process for all animals that live in water.

It was John Alwyne ('Jack') Kitching (1908–1996) who first worked out how this is achieved in microscopic protozoans [1]. Estuarine salinity varies with each tide, and with the mixing of fresh and salt water between the saline waters of the sea and the fresh water of rivers. Being microscopic increases an organism's vulnerability to osmotic pressure because, for a cell, there is so much more surface area of membrane in relation to its volume. A pea, being much larger than a cell, has a thousand times less surface area in relation to its volume, and a football has over 20,000 times less, which is why the cell is so much more vulnerable to anything that can pass through its bounding membrane.

To such changes, small naked organisms must constantly adapt. Thus, in freshwater protozoans, water enters the cell and must be expelled, or the cell would explode. The protozoan has a specialised chamber surrounded by a contractile membrane, which acts as a pump. It maintains the osmotic pressure of the cell against the influx of water by contracting every few seconds or minutes, to pump water out of the cell. To counter the constant influx that tends to dilute the concentration of those solutes necessary for life, the regular pumping of the chamber not only expels water, but regulates the concentration of solutes at an optimal level. In saline water, where the osmotic gradient is less steep, the chamber pumps more slowly, quickening as the water freshens. What is remarkable is the wide range of salinities that can be tolerated by closely related species, as protozoan species of the same genus can be found in completely fresh or fully saline waters. The same species of *Amoeba* can maintain the internal concentration of solutes and survive a range of salinities from brackish (20% seawater) to hypersaline (150% seawater). The system for maintaining water balance has all the essential properties of homeodynamic systems found in large and complex organisms.

The protozoan mechanism performs, in a simple way, the role of the mammalian kidney, whose precursor is found in fish. The fluids of a marine fish are more dilute than the surrounding water, so it is in danger of losing water to the sea, particularly across the gills. They compensate for this by drinking seawater. This creates the problem of getting rid of the excess salts, which must be excreted at higher concentrations than they were taken in. It is the gills that actively secrete salts back to seawater, while the kidneys excrete other ions as concentrated urine. By these means, more dilute tissue fluids are maintained. Freshwater fish have more concentrated tissue fluids than the surrounding water, so they have the reverse problem. A steady influx of water through the surface area of the gills is excreted as large quantities of dilute urine. The loss of up to a third of body weight per day as urine causes a loss of solutes which is compensated by active uptake across the gills. So we see that osmoregulation occurs in protozoa, and has been steadily refined in invertebrates and then fish, which possess the precursor of the mammalian kidney.

CONSERVATION OF HEAT

In many ways the control of temperature is a more accessible example of homeodynamic control. Life is possible at a temperature that would freeze water; but the formation of ice needles is prevented by the salts, and anti-freeze proteins are found in the blood and tissues of animals that live in polar waters at −2°C [2]. At higher temperatures around 45°C, proteins become denatured and inactivated, in the same way that the albumen in eggs turns white in boiling water. Apart from the so-called 'extremophiles', all life exists within this range of temperatures, but there are advantages in being at the upper limit, because the efficiency of metabolic processes increases with temperature. According to the 'Q_{10} rule', the rate of metabolism increases 2–3 times for each 10°C rise in temperature within this range, due to increases in the efficiency of enzymes and metabolic processes as temperature rises. In nature, to be fast often means to be first in the race for limiting resources; so with the rate of metabolism a priority, it is better to live at a higher temperature, at which the rates of processes can be maximised. Birds need a high rate of metabolism to fly, and a higher temperature increases energy output. We therefore find that birds tend to have a core temperature of 40–41°C, while humans and other mammals have a core temperature of about 37°C.

Heat conservation evolved before temperature control [3]. Heat is energy and requires resources to create it, so there are many adaptations to conserve heat. The more primitive animals do not regulate their temperature, are cold-blooded and are called poikilotherms. Their evolution preceded animals that regulate their temperature, are warm-blooded and are called homeotherms. Cold-blooded animals treat metabolic heat as a resource to be conserved, as energy efficiency adds to fitness. So not all poikilotherms are wholly cold-blooded, but have evolved various ways of conserving the metabolic heat they create as a byproduct of muscular activity. For example, the blue-finned tuna is the fastest-swimming fish, with short bursts of up to 90 kilometres per hour. Like engines, the activity of these powerful fish generates large amounts of heat, and the muscles used in swimming reach higher temperatures than the rest of the body. The advantage of heat conservation led to the evolution of heat exchangers that recover heat from the blood as it leaves the muscles, immediately transferring it to the blood being pumped to the muscles. Each heat exchanger is a mass of blood vessels lying side by side, with the blood in the arteries and veins flowing in opposite directions. The venous blood gives up its heat to the arterial blood, so that heat can be recycled, maintaining the temperature of the muscle mass at 14°C higher than the surrounding water. This tissue is called the *rete mirabile*, meaning 'wonderful network', and was first described by the French anatomist Georges Cuvier in 1831, but it is unlikely that he guessed its purpose. Higher temperatures increase the rate of muscle contraction, which contributes to swimming speed. Other heat exchangers maintain the digestive organs and liver at a higher temperature than the surrounding tissues which, by contributing to the efficiency of digestion and metabolism, helps to fuel a massive power output. The tuna has a system that may not actually control temperature, but as performance is improved at higher temperatures, heat is distributed to those tissues and organs that will benefit most.

Numerous other adaptations have evolved to conserve heat. Flying insects are sluggish at low temperatures, especially the large moths, butterflies and bumble bees. Before take-off, heat is generated by the paired wing muscles. Instead of using them alternately to raise and lower the wings, both sets of muscles are contracted simultaneously and repeatedly, so that they work against one another. A slight shivering indicates that the insect is doing these isometric exercises, to generate heat and warm up the wing muscles before flight. The bumble bee also has a heat exchanger [4]; heat in the blood leaving the wing muscles is transferred to the blood, to be immediately returned to the flight muscles. So with the help of such subtle adaptations, analogous to those in tuna, bumble bees are often the first insects to take flight in the morning and the first to appear in spring.

Fish are poikilotherms but, as well as heat conservation, they also have a kind of behavioural temperature regulation. The temperature of the sea in which fish are found provides an indication of their near-optimal internal temperature. Adjustment is achieved by moving north or south to keep themselves in water of near-optimal temperatures. Most of the heat due to global warming is finding its way into the oceans so, as the Atlantic warms, the range of subtropical fish species is extending poleward. The precision of control by movement is reflected in the close correlation between small increases in ocean temperature and the rate at which warm-water species are spreading north. Although the North Atlantic has warmed by only 0.5°C, there has been a steady influx of subtropical species new to the UK of about 1.5 species per year over the last 25 years [5]. This reflects the temperature adjustment of subtropical species to their optimal temperature range by moving slowly northwards. Such fish can be used as 'mobile thermometers' and are accurate indicators of the relative temperature of the oceans.

Reptiles are also poikilotherms, and it is thought that the extinct sail-back lizards were adapted to capture solar heat to warm up their bodies more quickly at sunrise, and maybe to gain some heat before sunset. The neural spines on their vertebrae grew to 12 feet (3.7 m) long, supporting a large area of thin tissue well supplied with blood vessels. In the early morning, so the thinking goes, these lizards exposed their sail-backs to the rising sun, and captured enough heat to warm their blood and be active before their competitors. We see the vestiges of this behaviour in the sunbathing of modern lizards.

The social insects such as bees, ants, termites and wasps are also poikilotherms, but there are many examples of temperature regulation of their hives or colonies [6]. While metabolic heat keeps the honey bee slightly above ambient temperature, the conservation of this heat is used to control the temperature of the hive. Each bee can generate 0.1 calorie per minute at 10°C, so a colony of 20–50 thousand bees is capable of producing thousands of calories of heat per minute. Temperatures inside the hive are controlled to within half a degree of 35°C, even when ambient air temperatures are as low as −28°C. At the other extreme, a hive in midsummer is likely to overheat when the colony is strong and the workers are busy. The workers then cool the hive by standing in their hundreds on the flight board at the entrance to the hive, fanning their wings to create a draught to ventilate the brood combs and keep the hive cool. When this is not enough, they bring water into the hive, and droplets are hung over the brood cells. Some workers even regurgitate water and spread it out as a thin film, while others evaporate the water by fanning their wings. This behaviour achieves control over both temperature and humidity. Such temperature control in colonies of social insects is extraordinary, as it not only involves the coordinated behaviour of thousands of insects, but it anticipated the evolution of homeothermy in mammals by hundreds of millions of years.

Conservation of heat has also been a factor in the evolution of birds. The insulating properties of feathers are well known, particularly those of the eider duck, which plucks its down feathers to line its nest, and gave its name to the 'eiderdown' as a bed-cover. When birds are inactive at low temperatures, their feathers are erected, and the insulating layer they create is thickened. In winter, the number of feathers insulating small birds such as finches might be double that in summer. Little blood circulates to the wings and tail, as the extremities are feathered, but the legs and feet are naked so must have a blood supply of their own. Losses are minimised by a heat exchanger that removes heat from the arterial supply to the legs so effectively that in freezing water a gull loses only 1.5% of its metabolic heat production through its legs and webbed feet. So heat is not only conserved at sites where it is most needed, but heat loss is actively prevented from sites where loss would otherwise be greatest.

Natural selection for improvements in metabolic efficiency by thermal regulation has considerable survival value. When diurnal or seasonal cold limits activity, the maintenance of higher temperatures effectively lengthens the day, or makes possible an earlier start to the season, giving a competitive advantage to warmer-blooded species. Similarly a greater latitudinal range can be colonised. With the evolution of mammals, homeothermy reaches its highest levels of sophistication, with the brain and core maintained at tropical temperatures, whether they live at the equator or within the Arctic Circle.

DOMESTIC AND BIOLOGICAL THERMOSTATS

It is interesting to compare temperature regulation in humans with a domestic heating system. The analogy helps draw out the similarities

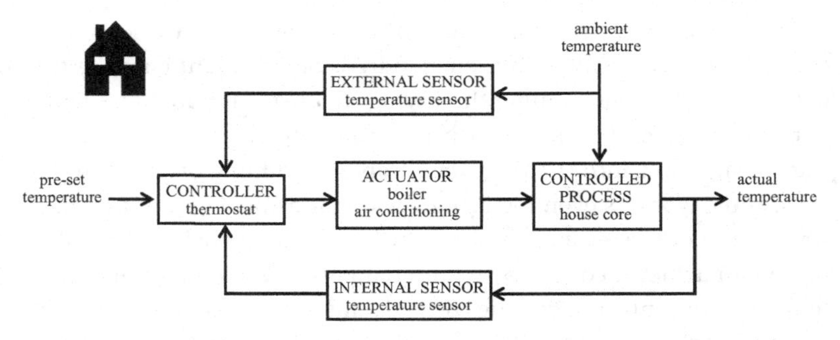

Figure 4.1. Temperature control system for a house. Information flow diagram indicates the function of a domestic temperature-regulating system for comparison with the analogous system in humans (Figure 4.2). Modified from Milsum (1966).

when they are each considered as cybernetic systems. Such systems seem to have conspired to hide their inner workings from the investigator, as often only their outputs are obvious and, if everything is in equilibrium, even that is constant. Control mechanisms are especially difficult to investigate in living systems because their essential components may be just a few cells and the vital nervous or chemical connections between components may be inaccessible. The feedback loop is closed and the information and signals passing round the causal cycle are not reflected in the output of the control mechanism. As we shall see, the challenge for physiologists trying to understand feedback mechanisms has been to 'open up the loop'.

It will serve our purposes to consider control systems as logic maps which show that the chain of causality is circular and iterative, first looking at the familiar control systems in our homes that keep the temperature at comfortable levels (Figure 4.1). A preferred temperature of say 20°C is pre-set on the controller mounted on the wall, which combines two functions. It has a sensor that provides a temperature measurement which is compared with the pre-set temperature; the difference constitutes the error. The source of heat to raise the temperature is then signalled to increase its output to reduce the error. The boiler starts up and hot water is circulated around the radiators, or fans direct heated air to the rooms. As the temperature rises and returns to the pre-set temperature, the boiler cuts out. With time, the house loses heat to the exterior and temperature falls, creating another error which is eliminated as before, and the cycle is repeated. In hotter climates, we could equally well be referring to

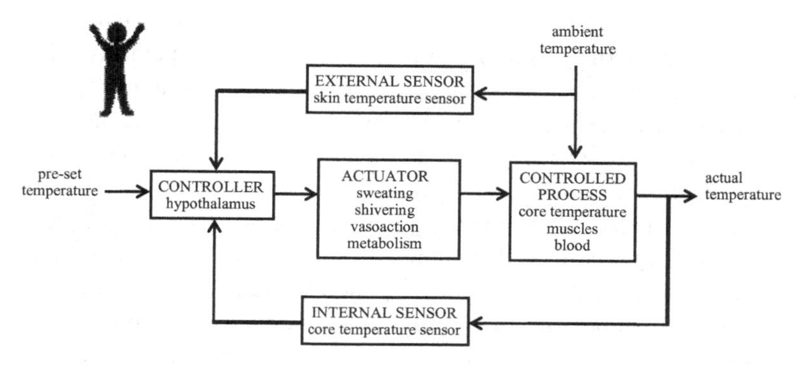

Figure 4.2. A simple human temperature control system. Information flow diagram of a human temperature control system for comparison with the analogous system for a house (Figure 4.1). Modified from Milsum (1966)

temperatures *above* a pre-set temperature that need to be reduced by cooling with an air-conditioning unit. Since the thermoregulatory system in humans and other homeotherms can control temperatures above or below those that which is pre-set, we have assumed that our house control system can warm the house in winter and cool it in summer. The analogous biological system will need to have the same capability.

More effective control can be achieved by having an ambient temperature sensor to detect any fall in external air temperature *before* it affects internal room temperature, and in advance of detection by the sensor within the controller. This may seem superfluous, but it illustrates the principle of feedforward control (Figure 4.1). A system that responds to external temperature, using feedforward control to anticipate a drop in interior temperature, has the advantage of achieving more accurate control, especially for locations that experience large and rapid falls in ambient temperature.

In humans, the central controller is the hypothalamus, a small but discrete part of the brain attached beneath its floor. Part of the hypothalamus monitors blood temperature and initiates sweating and shivering, and must also hold information on the pre-set temperature of around 37°C (Figure 4.2). This varies between individuals, and sometimes the time of day or day of the month, within a range of 36.5–37.2°C. The principal sensor is in the hypothalamus, which monitors its own blood supply continuously, providing the measurements of core body temperature. As in a house, the function of the central controller is to compare the pre-set temperature with the main sensory input from the blood, which provides the core temperature. The difference between them is the error. Once an error is detected, the hypothalamus then sends out nerve impulses to initiate physiological and behavioural responses, such as shivering or sweating, to neutralise the error.

While there are many temperature receptors, they are much less important than the detection of temperature by the hypothalamus. For many years there was considerable uncertainty, since the detection of temperature by sensors in the skin is self-evident, while the action of the hypothalamus is not. The notion of a single, central temperature sensor in the brain was far from obvious, even after its discovery in 1884 by two German medical students. E. Aronsohn and J. Sachs identified the role of the hypothalamus as a temperature regulator, by damaging it in living animals. It took another 70 years before the hypothalamus was finally established as the mammalian thermostat [7]. This required the detailed mapping of the temperature functions of the hypothalamus, together with the development of electronic instrumentation, and the use of thermocouples allied to minimally invasive surgical techniques to conduct the key experiments. The development of control system theory, and the concept of a thermostat, may also have been important, not to mention the involvement of brave human experimental subjects.

One puzzling feature of this system is that the initiation of sweating in response to sudden heating (for example, when entering a sauna), or shivering in response to sudden cold, precedes any rise or fall in core temperature. It is therefore thought that skin sensors of temperature may also have a direct input to the hypothalamus. Temperatures detected by skin sensors are fed to the hypothalamus by a feedforward loop, which anticipates the normal feedback loop (Figure 4.2). This is analogous to the feedforward loop in the domestic system (Figure 4.1), initiating a more rapid response to sudden heat or cold than would be possible from core temperature alone. Feedforward control helps to address the fundamental weakness of feedback mechanisms in responding retrospectively to changes in load. The response to any sudden change is delayed by time lags in the system, so the detection of ambient temperature by sensors in the skin recognises change long before a rise or fall in core temperature is detected by the brain.

What is interesting about the comparison of domestic and human temperature control is that the components of the two systems have nothing in common. Yet, in a thermostat, the system responsible is essentially the same. Not only are the components of the system analogous, but almost identical logic maps of information flow can be drawn for the two systems (Figures 4.1 and 4.2).

TIME DELAYS IN TEMPERATURE CONTROL

The retrospective action and delays in the response time of feedback mechanisms have been seen as 'design flaws', but what is so interesting is the way in which evolution has produced various adaptations to overcome delays, to the point that mammalian homeodynamic systems respond to change almost instantly. We need therefore to identify the kinds of delays involved. First, the controlled process has to have deviated sufficiently from its preferred path for an error to become large enough to be detected. Lack of sensitivity therefore causes delay, until a change becomes detectable. Second, the means of counteracting the deviation must be activated, and overcome the inertia in the process before any error can be neutralised. Third, as we have already seen, feedback mechanisms do not eliminate errors at a stroke, but reduce them progressively through a series of decaying oscillations.

Just a few cells in the hypothalamus respond to a rise in blood temperature, which initiates nerve impulses that set in motion mechanisms to cool the body [8]. As core temperature rises, so peripheral blood vessels are dilated and more blood flows to vessels just beneath the skin, where heat is lost to the air. This mechanism of cooling core temperature is adequate in air temperatures up to 32°C, but at higher temperatures, or when exercising, additional heat is lost by sweating. Millions of sweat glands cover the body and lose water at a rate of up to 3 litres per hour, shedding excess heat from the skin by evaporation at a rate of 2,400 calories per hour (573.4 joules). T.H. Benzinger and his colleagues found that a rise in temperature of as little as 0.1°C is sufficient to increase heat loss by sweating of one calorie per second, and to increase blood flow under the skin in order to accelerate heat loss by dilation of the peripheral blood vessels.

Modern domestic heating systems have thermostats on radiators, allowing the independent control of temperature in each room. This facility is nested within the overall system (Figure 4.1) such that separate controllers constitute subsystems. There is an analogous subsystem in the human body. The DNA in spermatozoa cannot be produced at core body temperature, but requires slightly lower temperatures.

So the scrotum and testes have a superficial location outside the body cavity, which provides a temperature about 3°C below core temperature. Like the separate thermostats on each radiator, the scrotum provides independent temperature control for the testes. When it is warm, the scrotum is loose and the testes hang lower, but when it is cold the testes are pulled up closer to the warmth of the body wall. In this way spermatozoa are not only kept cooler than the body temperature, but they are also regulated at that lower temperature, just as a radiator might be used to keep a garage just above freezing in winter. Control systems may have one or more levels, with subsystems at each.

During infections by viruses and bacteria, pyrogens (meaning literally 'fire makers') are released by white blood cells into the blood. They have the effect of turning up the thermostat by a few degrees. A moderate fever is an adaptive response, as an increase in temperature may be less than optimal for the pathogen, so that higher temperatures disadvantage the pathogen and allow the immune system to suppress the infection. It is not surprising, therefore, that an elevated temperature in response to infection is an indication of a healthy response.

The Italian physician Sanctorius (1561–1636) first adapted his colleague Galileo's thermometer to measure the temperature of a person by placing a glass bulb of air in the mouth. At that time there was no agreed scale by which accurate temperatures might be measured, so relative change was noted using an arbitrary scale. What is important is that it allowed Sanctorius to detect fevers and observe change in temperature so that he could pronounce whether a patient was ailing or recovering. As a professor of medicine at Padua, Sanctorius made some early studies on human metabolism that involved weighing a person from day to day. Using this method he was able to demonstrate that there is weight loss due to sweating. He also invented a device to measure the pulse by relating heart rates to the frequency of the swing of a pendulum. The timing of heart rate becomes important in our next example of homeodynamic adaptation.

ADAPTIVE ADJUSTMENTS TO THERMOREGULATION

We are aware from personal experience that homeodynamic control improves when the system is regularly challenged. This is what might be called secondary adaptation of the primary control mechanism. How far can the capacity of these systems be increased by use or training? The heart is a pump not dedicated to any one controlled process, but is a central component of many. The cardiovascular system not only carries oxygen from the lungs to tissues, and waste products to the kidneys, but it also carries carbon dioxide to the lungs. The rate at which the heart beats is determined primarily by the need to transport re-oxygenated blood around the body. It is the pump that determines the rate at which oxygen can be consumed, so energetic activity is ultimately limited by the flow rate of blood between the lungs and tissues. Less obvious, but just as important, is the role of the heart in pumping heat in the blood to sites where it can be dissipated or conserved as necessary.

The amount of blood that can be pumped depends on the rate at which the heart beats, together with the volume of blood that is pushed forward by each stroke, determined by the volume of the chambers of the heart. If the heart beats at 75 beats per minute and the stroke volume is 70 ml, the flow rate would be 5.25 litres per minute. As the blood volume of the average person is about 5 litres, a volume equivalent to the all the blood in the body may pass through the heart every minute. In a person of ordinary fitness, the maximum output is five times resting value, or 20–25 litres per minute. In a trained athlete in competition, a sevenfold increase in flow rate may be expected to reach 35 litres per minute.

Information on blood pressure and oxygen tension is fed to the cardiovascular control centre in the medulla oblongata of the brain, which controls heart rate by way of the nervous system and by modulating stimuli from the heart's own pacemaker. The primary demand on the heart is for re-oxygenated blood in response to the demands of the skeletal musculature, and to replenish blood glucose levels. Cramp is caused by a build-up of lactic acid and other chemicals in the muscles brought on by sustained and vigorous exercise. Lactic acid is produced when tissues break down glucose to produce energy when there is not enough oxygen in the blood. Cramps in the calf muscle also occur during sleep, when the muscular pumps in the deep veins are inactive, so lactic acid accumulates.

An interesting aside related to control is that the 'evolution' of prosthetic pacemakers seems to have followed the same course that biological evolution has taken. The invertebrate heart, like the first pacemaker, was unable to change its own rate. Now pacemakers are able to respond to vibrations, detecting physical activity of the person, which is used as a cue for heart rate to increase. Another sensor detects an increase in the rate of breathing, which correlates well enough with heart rate to be used as a surrogate cue for pacemaker rate. Patients can now live a more active life with pacemakers, as these mimic the behaviour of a normal heart more closely.

What is of overall importance is that the heart moves heat around the body with remarkable speed and, coupled with the ability to dilate or constrict blood vessels, enables body temperature to be regulated: in cold conditions, more blood can be prevented from going near the skin to reduce heat loss, and when it is hot, more blood can be directed to the skin to accelerate heat loss by the evaporation of sweat. These and other subtle adaptations ensure accurate control of mammalian temperature.

EFFECTS OF EVOLUTION AND TRAINING ON THE HEART

A comparison of the extremes of the evolutionary adaptation of this remarkable organ is revealing. Heart size in mammals is about 0.6% of the total body weight. The rate at which it beats is inversely related to the size of the organism. So a common shrew weighing about 10 grams, with a heart weighing rather less than a tenth of a gram, has a heart rate between 88 beats per minute at rest and over 1,300 beats per minute when alarmed. In contrast, a blue whale has a heart the size and weight of a small car at about half a tonne, and beats just 5 times a minute [9]. The weight of blood in mammals is about 6.5% of body weight, amounting to 400–450 gallons (1,800–2,000 litres) in a blue whale. The immense Jurassic sauropods, such as the brontosaurus, are estimated to have had hearts weighing three times as much as a whale's. Due to submersion in seawater, the blood of the whale is weightless, but the sauropod heart had to continually pump blood to the head almost 30 feet (~9 metres) off the ground. This required a much more powerful pump. There is a remarkable statistic that the number of heartbeats in a lifetime, for a wide range of animals, is about 1.5×10^9 heartbeats. Although larger animals have slower heart rates they live longer lives, while small animals have faster heart rates but live much shorter lives.

The effect of training for athletic competition is measured in terms of improved performance, but underpinning this are increased rates of physiological output. Since many sports are professional, much research has been devoted to adaptations of bodily functions that lead to improved athletic performance, best understood by those working in the field of sports medicine. Most sports depend ultimately on the output of the cardiovascular system, which delivers oxygen to muscles. Training rats for 6 weeks increased their ability to consume oxygen by 60%, due to improvements in the functioning of the cardiovascular system. Over 13 weeks of training, the weight of the heart ventricles increased by 37%. Endurance training also increases the diameter of the large arteries from the heart to the limbs. Similarly, in a group of sedentary young men and women trained over one year for a marathon, the left ventricular mass of the heart also increased by 37%. The left and right ventricles of the heart make up most of its size and mass and together fulfil the pumping role of the heart. The heart is actually two pumps side by side: the right ventricle pumps blood to the lungs, while the left ventricle pumps blood returning from the heart via the aorta to all parts of the body. The left ventricle is the more powerful pump, with walls three times as thick as those of the right ventricle, and generating far more pressure. It is not surprising, therefore, that the ventricles – and the left ventricle in particular – are most responsive to training, causing thickening of the muscular walls and an increase in volume.

One curious feature of the training of endurance athletes is that their heart rate at rest falls from 70 to 40 beats per minute and, in exceptional cases, to even slower rates. Resting heart rates of 30 beats per minute are known among some footballers, and the great cyclist and Tour de France winner Miguel Indurain had a disproportionately enlarged heart which had a resting heart rate of only 28 beats per minute. The point here is that with heart enlargement and the delivery of greater volumes of blood per stroke, fewer beats are required to deliver the body's resting oxygen requirements.

Horses evolved for life on the open plains, where their speed and stamina helped them keep their distance from predators, as zebras do today. But it is now a major international business to breed and train these equine athletes. Training racehorses for speed and endurance has led to the breeding and development of horses with highly efficient cardiovascular systems, enabling them to consume more oxygen per kilogram of weight than any other large mammal. The pumping capacity of the heart is crucial, beating at rates of 20–240 beats per minute. The heart output also depends on its volume, which relates to its overall mass. A typical thoroughbred has a heart weighing about 3 kilograms, but the breeding and training of horses selected for speed combined with endurance has given the best horses huge hearts. The most successful racehorses have hearts weighing about 6 kilograms, while the American record-breaking horse, Secretariat, had a heart weighing over 10 kilograms. It has long been thought that racing success was linked to large hearts, and a recent study of 400 horses engaged in National Hunt racing revealed a significant correlation between heart size and the official rating given to these horses based on their racing successes.

During evolution, the heart has been a highly malleable organ in that it serves the same role in animals as disparate as the shrew and the blue whale. In addition, the heart is able to adapt to the needs of the individual animal, using its history of loadings over time, or its response to training, to defer the point at which failure might occur at peak performance. Such systems are among nature's most effective adaptations, particularly in their remarkable ability to respond to training.

So it is apparent from homeothermy and the adaptation of the heart to training that there are at least two kinds of adaptation. The capacity to withstand physiological load is the primary response of homeodynamic systems, which is inherited. This is due to the basic properties of a system that have evolved and adapted in response to environmental pressures over thousands of generations. Its fundamental features are genetically inherited and immutable. However, most homeodynamic systems have evolved the ability to adjust their capacities as a result of recent history. Secondary adaptation is provided by the capacity of homeodynamic systems to adjust and adapt to physiological load, experience and training. In this way, their capacity has also evolved to be self-adjusting to withstand greater extremes of functional load. The secondary adaptation is that acquired by the individual organism's phenotype and obviously lasts no longer than its lifetime. The 'memory' of recent history of exposure to perturbation and load provides the opportunity for further adaptation that supplements the primary response, the better to meet the needs of the individual. In effect, homeodynamic systems can be trained, so that with use the capacity of these systems increases. Explorers and mountaineers become more resistant to cold as a result of sustained exposure to low temperatures [2], and the performance of athletes improves through training. The onset of overload and the possibility of high-output failure are deferred by secondary adaptation.

In effect, acquired information about disturbances and loads in the recent past is used to adjust the homeodynamic mechanisms to respond more effectively to later challenges. The system memorises information about its recent past to respond more effectively to future challenges. This learning works in both directions; with use the system increases its capacity to adapt, but in the absence of use it loses any improvement. This is the nature of homeodynamic adaptation.

CALCIUM CONTROL AND FUNCTIONAL DEMAND

Another example involves the metabolism of calcium, which differs in that it has no pre-set concentration in the blood. It has two roles; the first is to keep blood calcium at a stable concentration. Calcium is a requirement in small quantities for muscle contraction, blood clotting and the conduction of nerve impulses. The second role is more obvious; it is the regulation and use of calcium in the bony skeleton. It is a dual system which operates more slowly than any of the systems considered so far, responding to change over months and years, rather than minutes or hours.

Calcium concentrations are regulated by a control mechanism with two separate feedback loops (Figure 4.3). Virtually all of the calcium in the human body is found in bone, which is added to or drawn upon in providing a reservoir for physiological requirements, without having much influence on the mass of calcium in the skeleton. When calcium concentrations in the blood fall, a hormone is released, which promotes the release of cells called osteoclasts. These create an acid environment and dissolve the mineral matrix to liberate calcium from the bones. As concentrations of calcium rise, another hormone stimulates calcium deposition as new bone by cells called osteoblasts, thus removing calcium from the blood. Simultaneously this system regulates calcium concentrations in the blood while also responding to the need for calcification of the bones, such as is required for skeletal growth, repair or redistribution of calcium, and adaptation to physical loads upon the skeleton.

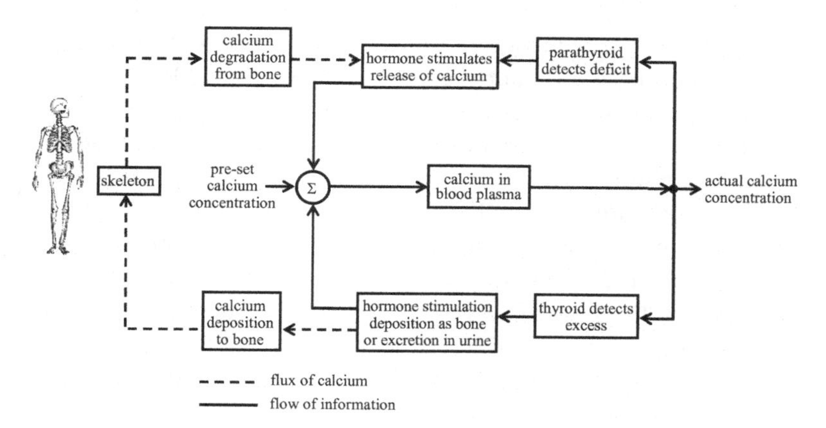

Figure 4.3. Information flow diagram of the human calcium control mechanism. Redrawn from various sources.

The continuous remodelling of bone in response to demand was first recognised by the German physiologist Kaspar Wolff (1733–1794). He discovered that bone grows in response to the physical forces put upon it. The skeleton is constantly being resorbed and rebuilt, such that cell by cell the entire structure is replaced every 7–10 years. The strength of bones is adapted to meet the stresses and strains to which they are exposed. Foetal bones are rudimentary and featureless, until loads are imposed upon them. Those who have a life of manual labour become heavily boned, while the bones of the bedridden become atrophied. Squash players who impose heavy loads on their legs, due to frequent acceleration and deceleration, have been found to increase bone density and leg bone thickness by 13–17%. As might be expected, weightlifters also experience such increases. A study of high-performance weightlifters found an increase in bone mass density (BMD) of 18% in the lumbar spine and 32% in the neck of the femur. The femurs are among the bones most stressed in the human body, because in running and jumping each must bear the weight of the whole body, in addition to any weight that is being lifted or carried.

Conversely, the absence of loads on weight-bearing bones results in the loss of bone density and bone mass. The bones of astronauts become demineralised in space, and 1–2% of the bone mass is lost for each month of weightlessness [10]. Although astronauts may spend 3–4 hours a day exercising, the nature of the exercises does not yet substitute adequately for living in Earth's gravity. Normal loss rates in men and women over 55 years old are 0.5–1.0% per annum, but in space the losses are ten times faster. On a 1-year Mir mission, astronauts lost 10% of the bone mass of their lumbar vertebrae. Losses are accompanied by a reduction in skeletal muscle, and even the heart loses mass in space without the gravitational weight of blood to work against. The reason is the same as the relatively small size of the blue whale's heart compared with that of a brontosaurus, as blood is weightless in water.

One extraordinary finding comes from a study of 21 weightlifters, in which a control group of four cyclists was provided. As was expected, the study demonstrated an increase in the bone density of weightlifters. The surprise was that during the 3-week Tour de France the control group of cyclists lost 25% of their spinal bone mass. One interpretation is that, when riding, the body weight of cyclists is shared between their handlebars, saddle and the pedals, so that their whole weight is not borne by the spine or legs. What may be more important is that during the race, much of the time not in the saddle was spent resting in bed. So, whether a cyclist or an astronaut, it seems that without the load of gravity passing through the body, even sustained vigorous exercise is not enough to prevent bone demineralisation.

Research into bone demineralisation in astronauts is a help in understanding osteoporosis, which is a frequent indirect cause of death in elderly women. Before the menopause, oestrogen inhibits the activity of osteoclasts, which are the cells responsible for releasing calcium from bone. After the menopause, with lower levels of oestrogen, there is a net loss of bone, with an increased risk of spinal fractures and loss of height in the elderly. The problem is that, whether in astronauts or the elderly, demineralisation is difficult to reverse. Five years after the Spacelab mission, remineralisation of the astronauts' bones was still incomplete.

From a cybernetic perspective, what is interesting about calcium metabolism is that the control mechanism which maintains blood calcium levels does so in response to a genetically determined, pre-set concentration (Figure 4.3), mobilising or depositing calcium in the bony skeleton. The regulation of bone mass or density, according to load, is different. The mineralisation or demineralisation of bone is determined by physical demands alone, responding positively or negatively over a wide range of conditions to the presence or absence of loading upon bones.

There are 206 bones in the human body, amounting to 13–18% of body weight. The absence of a goal setting for bone deposition becomes clear when it is reasoned that bone deposition occurs in response to physical load. According to Wolff's Law, bone tissue replaces itself in the direction of physical demand. Thus, for a manual worker involved in heavy lifting and carrying, more calcium will be deposited to strengthen the bones. Conversely, sedentary work will result in the reverse effect. For humans and animals it confers fitness to have bones of minimal strength for the lifestyle, since any excess is dead weight that must be carried without serving any purpose. It therefore contributes to overall fitness to resorb the excess calcium. Some fixed calcium goal setting would not allow such adaptation to functional demand.

The independent control of tissues and organs in the adult is not in response to requirements of size, so much as in their capacity to fulfil their function. Growth control mechanisms at this level adjust to the functional load upon them, as organs and tissues only add to fitness by way of their functionality. Compensatory and regenerative growth is important, and occurs where there is control in relation to functional demand. For example, a high-protein diet fed to rats caused a 50% enlargement of the kidneys within a week, but on return to a normal diet, the kidneys responded by shrinking to their previous size. Removal of one kidney causes a compensatory growth response of the remaining kidney, but if removal is accompanied by a protein-free diet, compensatory growth is suppressed. In humans, following recovery after the removal of a kidney, control of the diet can prevent overloading of the remaining kidney, which can then be induced to grow by increasing protein intake, and in time a normal life can be resumed. A similar response is seen in the salt glands of birds, which evolved in marine birds to eliminate excess salt. What is of interest here is that salt glands still exist in a rudimentary form in most birds. Biologists have found that the vestigial salt glands in ducklings were quickly able to develop and began to function 4 days after beginning to feed a salt-rich diet.

Endocrine glands illustrate, better than any other, the link between function and growth, because the size of the gland is proportional to its secretory capacity. The stimulus of stress induces the release of adrenalin and related hormones from the adrenal gland. The stress response syndrome includes increased heart rate and elevated blood pressure, release of glucose into the blood by the liver, dilation of the bronchioles of the lungs to improve respiratory efficiency, and increase in metabolic rate. As might be expected, sustained stress also results in the enlargement of the adrenal gland. In all these examples, from muscle mass to endocrine glands, growth is controlled in response to the functional demands put upon the organs rather than an absolute size that is genetically predetermined. Like other adaptive systems, the endocrine gland grows larger with use and smaller without.

INDUCIBLE SYSTEM TO DEGRADE POISONS

While we are prepared to accept that such examples of adapting homeodynamic systems are a part of normal physiology, it may be more difficult to accept that the degradation and excretion of foreign substances is controlled by adaptive systems of a similar kind. The difference from the previous systems is that there is no preferred level of foreign substances, because they are not necessary for life, so the goal setting for concentrations in the tissues must be zero. The inevitable intake of small amounts of foreign substances is enough to keep these systems fit and functional, if they are not to atrophy. Some chemical pathways have the function of responding adaptively to the presence of organic foreign substances by degrading them and excreting the breakdown products. There is an old saying that 'we must eat a peck of dirt before we die'[1] – a proverb used to discourage being too assiduous in avoiding dirt on food. Such a capacity can be induced by the uptake of foreign substances, and this capacity increases adaptively with use and decreases without. It is said, and may well be true, that our immune systems have become less fit over recent generations, as life is now led in much cleaner environments than in earlier times.

One detoxification system that has been well researched is the breakdown of alcohol in humans. It involves two separate pathways. The principal pathway involves several steps, with the last leaving carbon dioxide and water to be excreted along with other waste products. A second pathway is located on tiny organelles called microsomes inside liver cells, and is called the mixed-function oxidase, or MFO, system. What is interesting about these two pathways is that the capacity of the first is thought to be constant, but that of the second is adaptive, and increases its capacity in response to increased load. It is induced by the presence of alcohol, becoming more effective with 'training', and is clearly important for those who consume larger amounts of alcohol.

The MFO system evolved many millions of years ago, and it seems to have been found in every animal where biochemists have looked for it. In the shellfish that inhabit our estuaries, and even in the flatfish in the central North Sea, MFO activity is induced by the presence of toxic organic contaminants such as those that result from the combustion of fossil fuels (polycyclic aromatic hydrocarbons or PAHs). In fact, the induction of this system is used as an indicator of contamination by organic compounds by those who measure the effects of pollution of the North Sea. In each case, we see an MFO system acting like other homeodynamic systems, using its inducible capacity to metabolise and excrete toxic chemicals. The uptake of small amounts of toxic and foreign substances keeps such systems fit for their purpose.

The human MFO system metabolises alcohol and also addictive drugs. It seems improbable, but is true, that the same biochemical system in the livers of fish and humans is equally catholic in its ability to metabolise foreign and toxic agents. Such systems are able to metabolise a broad spectrum of hydrocarbons, including some they could never have previously encountered, as though they were pre-adapted to do so. The generality of such biochemical responses behaves as if

[1] A peck (an old measure of volume of dry goods) is roughly equivalent to 9 litres.

they had evolved in 'anticipation' of synthetic compounds, but it must be due to other organic compounds encountered during evolution.

Despite the capacity of such systems to degrade and detoxify poisons, and secondary adaptation that increases their capacity in response to exposure, ultimately their adaptive capacity is finite. As a toxicity threshold is approached, the greater the load any homeodynamic system is under, and the more susceptible it is to any additional load. But where different foreign compounds find their way into the bloodstream, each is 'competing' to be metabolised by the same process. Thus the presence of one organic compound makes the organism more susceptible to another. For the hard drug user, any alcohol taken makes the drug that is injected more toxic. Similarly, exposure of fish to PAHs increases their susceptibility to polychlorinated biphenyls (PCBs). Up to a point these effects are additive, but as the capacity of the system is reached, only a little more is required to exceed the threshold and make the combined effect toxic. When close to a threshold, combinations of two or more toxic agents then become much more toxic than their separate toxicities added together might suggest, creating a synergistic effect.

Each of these systems, whether involved in temperature or calcium regulation, or in the elimination of foreign and toxic compounds, reveals a common underlying system that is cybernetic and capable of primary adaptation in countering change. Most are also capable of secondary adaptation such that their capacities may increase or decrease in response to 'training' by the presence or absence of persistent load.

EVOLUTION OF RESPONSES TO ENVIRONMENTAL CHANGE

Animals have evolved many homeodynamic mechanisms to counter the effects of environmental perturbation; such adaptive systems enable them to live in a great diversity of niches. During the course of evolution, the number of responses to environmental disturbance has grown. The complexity of animals has steadily increased, as their range of responses has evolved to meet challenges of living in demanding habitats.

Unseen and unnoticed, natural selection continually makes some multiple adaptive integration one of the requirements for survivorship. No organism is perfectly adapted to its physical environment, as nearly all the relevant variables are constantly changing their rates, magnitudes and frequencies. To remain competitive and survive, an organism must keep up with the pace of environmental change, but is inevitably lagging behind the actuality. Evolutionary adaptation to change requires the organism to track the fluctuating paths of environmental variation, with the disadvantage of the inevitable delay involved in always having to follow change. Natural selection to maximise fitness to survive is an ongoing and unending challenge. The problem is the more difficult because there are many variables, so the outcome inevitably involves trade-off and compromise. There can never be a perfect match between the environmental variables that impinge upon an organism and its capacity to neutralise them. The selective pressure to escape from the external challenge to internal processes is never-ending. The most tangible consequence of this evolutionary search is the increasing complexity of organisms and their organisation, a trend that has prevailed since life on this planet began.

In parallel with adaptation to the physical and chemical environment is the progressive adaptation of organisms to their biotic environment, which constantly refines adaptations to improve fitness in competition with other organisms. The evolutionary objective is to add fitness and to become a more effective competitor, and therefore a more successful survivor. As before, the advantage lies with the initiator of innovation, and the follower is always at the disadvantage of playing catch-up. For all living things the challenge is to survive in the see-saw of the biological arms race. Whether between a parasite and its host, or a predator and its prey, or between competitors, a benevolent stalemate of equivalence is often best in creating an equilibrium that is sustainable, as the supposed contestants actually need one another. But in the struggle to optimise adaptation to meet the demands of the environment, and relationships with other organisms, there is no end. In a multifactoral world of shifting trends and unceasing change, adaptation can never be perfect, so must be a compromise.

If we take the simpler aspect of this problem, which is that of dealing with the non-living environment, we can return to Claude Bernard and his understanding of homeodynamic systems. He wrote, with respect to homeothermy:

> It is an organism that has placed itself in a hothouse. Thus the perpetual changes in the cosmic environment do not touch it; it is not chained by them, it is free and independent.

So the adaptive purpose of homeodynamic systems, and their unending adaptation to change, gives organisms the remarkable degree of independence of environmental change that has been attained since life began.

There is a scion of evolutionary thinking that related to the role of homeodynamic mechanisms but has been ignored for 50 years. Julian Huxley (1887–1975) defined evolutionary progress as increased control over, and independence of, the environment (see Chapter 15). He even linked this important concept to homeostasis, giving temperature regulation as an example. However, perfect evolutionary adaptation and complete independence of the environment could never be possible in an ever-changing planet. The answer has been for the evolution of control mechanisms of greater complexity and sophistication to operate and adapt to change within the life of the organism. The emphasis in adaptation to change has shifted from evolution by natural selection to the adjustment and adaptation of individual feedback mechanisms, which will be considered in greater depth in Chapter 7.

SUMMARY

While homeodynamic mechanisms were first understood in studies of mammals, they exist in all organisms, even among the protozoans. Comparison of a domestic heating system and biological thermoregulation reveals the robustness of cybernetic principles, as the same system is seen performing the same functions in dissimilar contexts. But adaptation is of more than one kind: there is that in homeodynamic mechanisms as they are genetically prescribed, but such systems also respond to induction and training, and are self-adjusting in response to some function of the load imposed. The adjustment in relation to functional load shows the ability of homeodynamic systems to lose functionality in the absence of load, and gain functionality in the presence of load; this is fundamental to biological control mechanisms.

5

A cybernetic approach to growth analysis

The inanimate have no goals – the living have invisible goals. Not only can they not be seen, but they are not perceived by the vast majority of observers.

Each living being, no matter how simple, has a set of innate goals embedded in it, thanks to the feedback loops that evolved over time, and that characterise its species.

Douglas Hofstadter

The tendency to increase in size is, in a sense, the most fundamental feature of living things.

J.Z. Young

THE GROWTH OF COUNT MONTBEILLARD'S SON

Parents often record the growth of their children by marking increases in height on a door or wall. So it was for Count Philibert Guéneau de Montbeillard, who during the period 1759–1777 measured the height of his son at regular intervals over the first 18 years of his life. The Count's measurements show a steady increase in height (Figure 5.1) throughout these years. The curves showing increase in size (W) over time (t) represent the traditional way that growth is depicted, and the attainment of a respectable size is what matters to a doting parent. Sadly the Count's son ended his life prematurely on the guillotine at the behest of Robespierre.

For the student of the physiology of growth, the cumulative growth curve is uninformative, revealing little about growth as a process (Figure 5.1). The curve hardly changes in slope, so shows little obvious change in rate. Much more informative is the curve showing

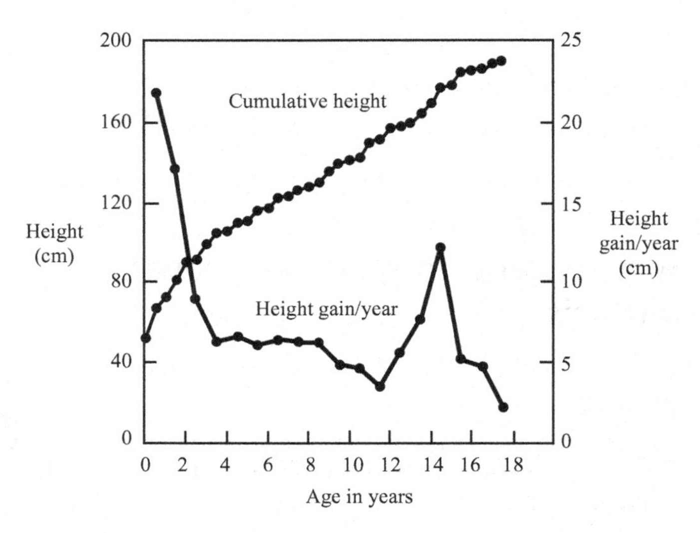

Figure 5.1. The growth of the Count de Montbeillard's son over his first 18 years of life (1759–1777). Cumulative height (cm) is given for comparison with the rate of change in height (height gain/year). Redrawn from Tanner (1978).

changes in growth rate over time (Figure 5.1). If we use increments at approximately half-yearly intervals to calculate growth rates, we see remarkable changes in rate. Growth rate (*dW*/*dt*) is the most rapid initially, then declines sharply until the age of 4 years and more slowly thereafter until the age of 11. Then, during the mid-teens, the adolescent growth spurt is seen as a sharp peak of accelerated growth spanning 4 years, when the Count's son shot up by 38 centimetres. At this stage, the most important feature of these data is that the cumulative curve, giving increase in height over time, does not indicate the changes in growth rate that we know to be present. The fluctuations in growth rate, derived from the same data, reveal important features that allow interpretation of the growth process.

Two Greek philosophers held different, but complementary, views about the world around them. Democritus (*c.* 460–*c.* 370 BC) believed that Atoms and the Void were the only things that exist. He believed that nothing comes into being or dies, as everything consists of the same eternal Atoms. The other philosopher was Heracleitus (*c.* 540–*c.* 480 BC), who took the view that the world consists of a cyclical system of continuous change, such that change in one direction is ultimately balanced by change in another. Thus he saw fire, flame and smoke rising into the ether. Ultimately it falls as rain that runs into the ocean, from which he knew that vulcanism could cause the Earth to rise from the sea and cause fire. Hence he concluded that there is unity despite change. Using the flow of a river as a metaphor, he said 'Everything is always and in all ways in constant flux'. Each in their own way was right, but Conrad Waddington (1905–1975) distilled something deeper from these contrasting interpretations, saying that Democritus adopted a 'thing' perspective of the world, while Heracleitus was concerned with the predominance of 'process'. A 'thing' approach is defined by states, while a 'process' approach is measured in rates.

This view of states and rates shows that the same two approaches can be used to consider growth. The first and traditional one, using the 'thing' approach, is to view the organism as a state, albeit one that changes over time. Traditionally, biologists have tended to express growth as a single cumulative curve, showing size over time, as the definitive expression of growth (Figure 5.1); and so it is if one is concerned primarily with the products of growth. A line through a series of points at intervals allows a mathematical expression to capture the essence of the trend. A logarithmic plot of size or weight ($\log_e W$) against time (t) often yields a straight line, which makes it possible to extend the line and so predict future growth, or to determine size for an age in the past. Here growth is depicted by well-spaced data points that may be expected to yield some lifelong trend, and a simple mathematical relationship between size and time. Interest is in the cumulative product of the growth, and the data are represented as freeze frames of a slowly changing state, so the cumulative growth curve epitomises the 'thing' approach to growth. The Count, in recording the growth of his son, saw a sequence of stills in what is essentially a movie.

The second approach is to view growth as a 'process' whose rate can only be measured indirectly by calculating the rate of growth between two measurements of size separated in time. The size of an organism is represented by measurements of its dimensions, but the velocity of growth as a process is a dimensionless number, or vector, that cannot be measured directly. Growth rates (dW/dt) are determined indirectly from measurements of size over time. So data revealing growth rates are hidden in the cumulative growth curve, whose appearance is uninformative because any size with respect to time is an integration of the growth process over the life of the organism. However, the more frequently that measurements of size are made, and the time intervals between determinations of size become shorter, so the freeze-frames begin to merge into a movie, making it possible to see growth as a changing process.

AN ANALOGY WITH TRAVEL

For greater clarity, an analogy of growth with travel is helpful. My daily journey to work by car can be represented graphically in different ways. The odometer (milometer) readings give the cumulative distance travelled (Figure 5.2a) and are as uninformative as the cumulative height data for the Count's son (Figure 5.1). Over the same journey, readings of the speedometer were also made (Figure 5.2b), which provide the best interpretation of the 45-minute journey. Low speeds through country lanes are replaced by higher speeds on the open road. Then the car was stopped after 30 minutes to pay a toll to cross a bridge, and stopped again at traffic lights on reaching the city, before finally arriving at my place of work. In both the case of growth of the Count's son and my journey to work, rates and velocities were more informative, because both growth and travel are regulated by controlling their velocities. To understand growth control,

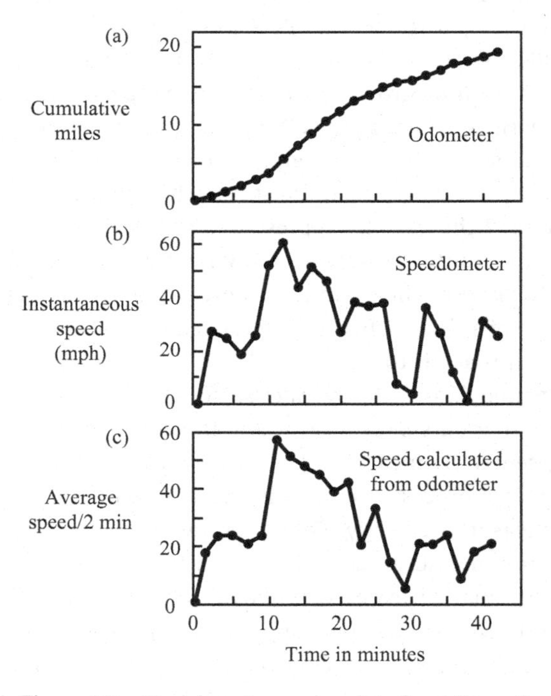

Figure 5.2. Travel analogy using data from the author's journey to work. The odometer and speedometer were read at 2-minute intervals: (a) odometer, (b) speedometer, (c) speed calculated from odometer readings.

we therefore need a measure of growth rate analogous to the car's speedometer reading.

Rate control is intrinsically more appropriate than state control, as it relates to a process in the present time. But growth measured by absolute amounts of its accumulating product is tied to the history of the process, as each measurement represents the entire cumulative growth for the life of the individual up to that point. Instantaneous rates are therefore much more sensitive to changes in the velocity of the process, by responding to rate changes in the present time, rather than some integral of the process over time. The point becomes obvious when we consider the travel analogy, in that one would never use a stopwatch and odometer readings to avoid breaking the speed limit.

Homeostasis literally means the control of states, so for the control of rates Waddington adopted the complementary term 'homeorhesis', meaning the control of rates, although together they are now referred to collectively as the product of homeodynamic processes, as the control of states and rates is closely linked. Growth control in most animals is likely to incorporate both state and rate control, since rates provide a more accurate way of controlling a process, while state control is required in the attainment of a specific size. In technology, space–time control systems that depend on both position and velocity are not uncommon, suggesting that growth control mechanisms incorporating two feedback loops for size and rate are probable. Such a system is later described for *Hydra* populations, accounting for the control of population growth rate and population density (see Chapter 8). In such systems, errors relating to the control of size and growth rate can be summed, as they must both reach zero at the same point – for example, when a sigmoid growth curve reaches its asymptote – in a way that is typical of the logistic equation. In this chapter the aim is to understand growth as a dynamically controlled process, by teasing out the process itself from the products of growth. But first we must derive rates from the cumulative growth data.

We return to the cumulative growth data which the Count gathered for his son. The problem is that at any instant in time for which we would wish to make a rate determination, there is no change in height. To measure a growth rate, we need two measurements of height separated in time, in order to calculate a rate for the intervening period. But then we only have average growth rates for the time intervals between measurements. The Count measured his son's height at intervals of about 6 months or more, which would reveal nothing about variations in growth rate in the interval between the measurements. So what of events such as a brief illness, which might affect growth between the measurements? Clearly, measurements need to be made more often in order to yield average rates informative enough to detect the effects of transient events. We must measure height more frequently, providing enough rate determinations to define the changes we are interested in.

Differential calculus treats a continuously varying quantity as if it consisted of an infinitely large number of infinitely small changes. Thus we can consider the instantaneous growth rate as a vanishingly small change in size (dW) during an equally short interval of time (dt). This is given by dW/dt, where d is an infinitely small change. However, in the examples that follow, we cannot calculate d as infinitely small changes in size and time, so a compromise is necessary. The problem is that the shorter the time interval between measurements of the size of an organism (W), the smaller the increment and the less precise the measurement becomes, until a point is reached at which any signal is lost in noise. The answer is a trade-off between minimising the interval between measurements and maximising their precision, so that there is a balance between frequency and precision that will still make it possible to define the events that interest us.

Data from experiments with a marine yeast *Rhodotorula rubra* help to illustrate the problem. We need to begin by looking at the dotted curves in three graphs (Figure 5.3), which show the growth of cell populations expressed in different ways. The first graph (Figure 5.3a) shows that when the increase in the number of cells is considered against time, growth is seen to be exponential. The second (Figure 5.3b) shows growth rate against time, and it can be seen that the rate accelerates with time because the population becomes larger. The third (Figure 5.3c) gives the specific growth rate with time which, though somewhat variable, is relatively constant over time. The reason is that at each step (from a to c), the slope of the dotted line is rendered more nearly constant.

We now return to the solid line in each of the three graphs. While the dotted curves represent data from the unperturbed control treatment, the continuous curves for the experimental treatment involved perturbation of the cell culture with low concentrations of a toxic agent. As will become clear, the use of perturbation is essential for revealing the output of feedback mechanisms. The perturbation provides a load which the control mechanism works to neutralise. If the growth rate is at the goal setting, the control mechanism is not

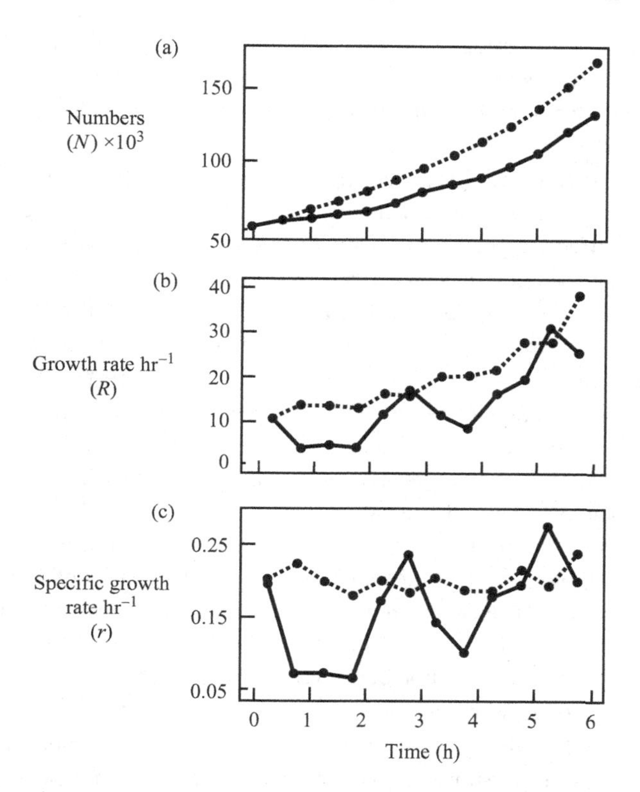

Figure 5.3. Growth of yeast cells in suspension culture, showing the effect of perturbation (*continuous lines*) in relation to the unperturbed controls (*dotted lines*). The sequence shows the steps in the treatment of the data to extract the output of the growth control mechanism. (a) Cumulative growth, (b) growth rates, (c) specific growth rates. Experimental data for yeast cultures that have been analysed in this way are given in Figure 12.1 and 12.2. Redrawn from Stebbing et al. (1984).

required to work, but as soon as there is a deviation between the actual rate and the goal setting, the control mechanism begins to neutralise the error and restore the equilibrium. In the first graph (Figure 5.3a), it can be seen that the perturbed culture fluctuates with respect to the untreated control culture. This fluctuation becomes much more obvious when the data are expressed as rates (Figure 5.3b), but both curves show a tendency to increase with time. Finally, when the data are expressed as specific rates that are independent of population size (Figure 5.3c), we see that the growth rate of the perturbed culture actually oscillates with respect to the control culture. We can now consider its specific rates as a baseline to which the more variable experimental culture can be compared. The reader will have also seen that the specific growth rate has the effect of not only expressing growth rate as a constant, but also of magnifying those fluctuations in growth rate due to perturbation.

It is clear that the three steps (Figure 5.3a–c) have had the effect of stabilising the baseline and magnifying the experimental signal, making the effect of perturbation more apparent. The magnification of such small changes in growth rate requires data of considerable precision, so a great deal of effort is required to reduce natural variability, or 'noise', to a minimum. Variation of the unperturbed population grows at each step, so the noise is magnified with the signal. Precise data are needed if the signal is not to be lost in a fog of growing variability. For this reason we have used organisms that can be cultured as clones to eliminate genetic variation, and replication is used to improve the precision of the data.

This chapter is intended to provide a general method to reveal the output of growth control mechanisms. The results of a set of experiments on the yeast *Rhodotorula rubra* perturbed by various concentrations of cadmium are given in Figure 12.1 and 12.2. Evidence in support of the wider application of this method of growth analysis will be provided by a similar analysis of growth data of the metazoan *Laomedea flexuosa*.

The travel analogy (Figure 5.2) shows that instantaneous rates are necessary to understand and control road travel, it is not possible to do this with the growth data, as at any instant there is no perceptible change in size. But if we record the odometer readings (as shown in Figure 5.2a) every two minutes, we can calculate rates as averages for each interval, and use them as if they were instantaneous rates. The plots of speed, or rates of travel, are given by the speedometer (Figure 5.2b), and can also be calculated as 'instantaneous' rates from the odometer (Figure 5.2c), revealing the same essential features. This comparison shows that it is possible to use cumulative data to determine rates that are an adequate approximation of instantaneous rates. However, even these 'instantaneous' rates may not have been based on short enough intervals to detect an emergency braking to avoid a collision. Such proxy rates may be expected to detect most of the temporal features of growth data, but they will fail to detect the most transient events; however, as we shall see, they enable us to discern the oscillatory signature of feedback mechanisms following perturbation.

The travel analogy shows that cumulative data can be used to provide a proxy for instantaneous rates. For biological data, with no access to instantaneous biochemical indices of growth rate, determinations of rate from frequently measured cumulative data can provide an adequate proxy for instantaneous rates.

SPECIFIC GROWTH RATE

We now return to say something more about growth rates, and specific growth rates in particular. Biological growth is like compound interest, in that the products of interest are added cumulatively to the principal. The products of growth also grow, and growth rates must therefore increase too. We therefore need to adopt a measure of growth rate that is independent of the contribution made by the products of growth, in the same way that the rate of interest gives the growth of money invested. This provides a measure of the rate of return in interest, indicating how our money will grow, irrespective of the sum invested. The biological equivalent of interest rate is the specific growth rate (r), and is given by the function $1/W \cdot dW/dt$. This is the increase in biomass per unit time per unit biomass; just as the interest rate is a measure of the rate of increase of the return on money per year for the number of pounds or dollars invested.

Sir Peter Medawar (1915–1987), as an undergraduate in his early twenties, became immersed in Bertrand Russell's *Principia Mathematica*, which led to a lifelong commitment to clarity of thought and logical precision. He wrote 'I wanted to be able to say to myself … that there was nothing any man had ever written or thought that I could not master if I chose to give my mind to it'. It also gave him the assurance in 1941, when he was still in his mid-20s, to write a review of the Laws of Biological Growth (see also [1]). One of the laws states that biological systems tend to grow exponentially, which we see in yeast (Figure 5.3a), but it is only by using specific growth rates that we see the underlying growth constant that remains the same irrespective of population size (Figure 5.3c). It was argued in the last chapter that the specific growth rate provides the baseline for growth, but it was Medawar [2] who recognised that only the specific growth rate is capable of indicating the underlying rate which reflects that of the biosynthetic process.[1]

It is of interest that specific growth rates (r) can also be expressed as $d \log_e W/dt$, in which the logarithm to the base e is the key part of

[1] The 'specific growth rate' is mathematically the same as the ecologist's 'intrinsic rate of natural increase' given by r. For clarity both are referred to by the term 'r'.

the function. As this is an irrational number, it cannot be expressed as a ratio of two integers, and does not have an exact value. The value of e is 2.71828 ... and is used as the base of natural logarithms. It was chosen because it is the only base that has the same rate of change as the thing that is changing. This means that for simple organisms, maintaining the accelerating pace of logarithmic growth, the specific growth rate reflects the underlying constancy of growth processes (Figure 5.3c). For these reasons, the specific growth rate is physiologically the most appropriate measure of growth. Its constancy indicates the separation of the process from its products, allowing the growth process as a rate to be studied independently of growth in terms of its products. Medawar hinted that, in its constancy, the specific growth rate provides the internal setting to which cybernetic systems keep the growth process constant and stable. With this idea, Medawar took a step towards his early ambition to study growth regulation, since the specific growth rate of a system under optimal conditions can be used as a baseline. It therefore provides the baseline against which the effects of perturbed growth can be measured (Figure 5.3c).

MINOT'S LAW

We must deal with a complication as to the 'constancy' of specific growth rates, because they are not as uniform as might be expected. While the yeast *Rhodotorula* in suspension culture maintains a stable specific growth rate, at least for the duration of our experiments, this is not generally the case. This requires a brief diversion. It is a universal property of growth that specific growth rates are the highest in the young and then slowly decline. In cumulative growth curves this tendency is barely noticeable, but when the data are expressed as specific rates, small changes in slope are magnified considerably, and a declining trend becomes apparent.

The decline in specific growth rate with age is called Minot's Law after the American anatomist Charles S. Minot (1852–1914). His law is not now well known, because he considered that the decline in growth rate with age provided a measure of senescence of an organism, as living tissue progressively loses the power to reproduce itself at the same rate as it was formed. He claimed that senescence is not only a property of the old, but happens fastest in the young. Hence Minot's enigmatic epigram 'organisms die fastest when they are young'. Medawar's work corroborated Minot's finding, and he accepted it as a law of biological growth, but questioned Minot's claim that decline in specific growth rate was due to senescence. Others since have been unwilling to accept Minot's interpretation, with the result that his observations, which are sound, have apparently been left without explanation, and are rarely referred to. Work on cultures of hydroids provided a simple explanation.

A colonial hydroid such as *Laomedea flexuosa* (see Frontispiece) can live for an indefinite period, so it is almost ageless. This is because the polyps that make up the colony live only for a week before they age and die, and then regenerate themselves anew. Senescence can only relate to the polyp and not to the colony. But when a new colony is created by taking a single upright from another colony, specific growth rate becomes rapid, before slowly declining with size as the colony becomes larger.

The decline in specific growth rate is a function of colony size. Each new colony is created by a new upright bearing polyps that grows at a fast rate, as it is able to grow in any direction, sending out stolons radially across the substratum, from which new uprights arise at regular intervals. Once the colony is established, the growth potential of an increasing proportion of the colony is locked into a network of stolons that prevents many of its members from realising their growth potential. Due to the rules that dictate the spacing of stolons and uprights, there is nowhere to grow for an increasing proportion of the colony members. The growth potential of the colony is gradually stifled by the colony's increasing size, as the colony can only grow at its periphery, where radial growth is unrestricted. Specific growth rates decline, mainly because geometry dictates that the larger the colony size, the smaller is its periphery as a proportion of colony area. Most of the colony is unable to grow, unless it is plucked from a colony to subculture a new colony. The larger the growing mass, the greater the limitation on each constituent colony member to replicate and realise its growth potential.

Medawar, in his tissue culture experiments, also demonstrated a linear decline in the specific growth rate of tissues with age, or time. The explanation for Minot's Law is not restricted to hydroid colonies, but must apply in the same way to a growing mass of cells. New cells and tissues, which are the products of growth, interfere increasingly with additional growth, whether a bacterial plaque in a petri dish or a lichen on a gravestone, so inevitably specific growth rate declines with size.

The same applies in the case of cells in culture; as the density of the cell mass increases, cells cease to divide, and a phenomenon called 'contact inhibition' takes over. Cells mutually inhibit one another's growth where they touch, so as the mass of cells grows, an increasing proportion become quiescent and cease dividing. As the volume of the mass increases exponentially, its surface area becomes a decreasing proportion of its volume. The periphery increases more slowly than its surface area in the case of a plaque growing in the form of a flattened disc. It is to be expected that only those cells at the periphery of the generative mass will replicate, so the specific growth rate of the disc will decline as it grows in area. However, when cells are used to inoculate a suspension culture that is stirred continuously, specific growth rates remain constant and a logarithmic rate can be maintained, as the space for all cells to replicate themselves is maintained by the dispersion of new cells. This was the case for the adoptation of suspension cultures in the yeast experiments described here (Figure 5.3).

Even these populations ultimately become too dense, and stop growing as nutrients become exhausted and toxic metabolites accumulate. But, until that happens, specific growth rates remain constant (Figure 5.3c), and can be used as a baseline against which to relate perturbed growth. The explanation for Minot's Law is simply that if the products of growth do not move away from one another, they soon begin to hamper each other's capacity to replicate. The increase in the size of the growing mass progressively prevents more of the population from replicating, and so specific growth rates decline. The constancy of specific rates in suspension culture shows why this interpretation is correct. Minot's Law is not due to senescence, but to the limitations that the products of growth impose upon further growth.

PERTURBATION EXPERIMENTS GIVE ACCESS TO CONTROL OUTPUT

The idea of using an inhibition for a growing tissue culture to work against was used in the early work of Peter Medawar [2]. He was advised to work for a doctorate by repeating some work on an unknown inhibitory factor in a commercial malt extract used in tissue culture. Although it might have been possible to identify the inhibitor analytically, he did not feel that his forte was biochemistry. He was more interested in the control of growth, and resolved to 'make use of the inhibitory properties of the malt factor ... in a number of largely theoretical exercises devoted to the study of growth regulation'. This led to the idea of using the resistance of a tissue to the action of the inhibitory factor, which he called 'growth energy'. In effect, Medawar anticipated the method of growth analysis described here, but circumstances were to move his own interests on to other problems before he was able to make significant progress.

After publication of this work in 1940, he turned his attention to the study of biological form, having been deeply impressed by D'Arcy Thompson's great work on *Growth and Form*. Medawar's interests moved first to biological form, and then to the more immediate demands of war casualties, using his tissue culture skills to relieve the plight of an airman who crashed near Oxford. In his autobiography, Medawar wrote of a colleague who had shocked him by saying that he should put aside his intellectual pursuits 'and take a serious interest in real life'. It was to be a piece of advice that he soon followed. It determined the course of the rest of his career, leading to the award of the Nobel Prize for medicine for his work on immunological tolerance, and an understanding of its importance to the development of transplant surgery. It will be seen that Medawar's 'growth energy' becomes the counteractive capacity of Maian growth control.

It was shown in the previous chapter that disturbance of some kind is essential in order to reveal the operation of cybernetic mechanisms, as their output at rest is constant and tells little of their behaviour. The purpose of such mechanisms is not best seen as maintaining constancy, but relates primarily to counteracting disturbance. For this reason, some kind of perturbation is essential to prompt the operation of the control mechanism. The aim therefore was an experimental design with the purpose of perturbing the controlled process, and then filtering out the consequences, in order to observe the behaviour of the system.

It must be mentioned at this stage that most of the perturbing agents used in hydroid or yeast experiments are toxic metals. For instance, Figure 5.5 shows the effects of a range of copper concentrations. However, any agent that could cause concentration-related inhibition would have served just as well. The reader is referred to similar experiments that were carried out with reduced salinity as the perturbing agent, which produced a similar set of oscillations [3]. A predictable difference was a greater tendency to adapt to deviations in growth rate caused by reduction in salinity. This confirms that the response of the growth control mechanism is not to the agent responsible, but to the deviation in the specific growth rate it causes.

A controlled experiment has the objective of pinpointing the cause of some experimental treatment by comparing an untreated 'control' with a 'treatment' that is exposed to some factor whose effect is unknown. The reasoning is that, while various unknown ambient

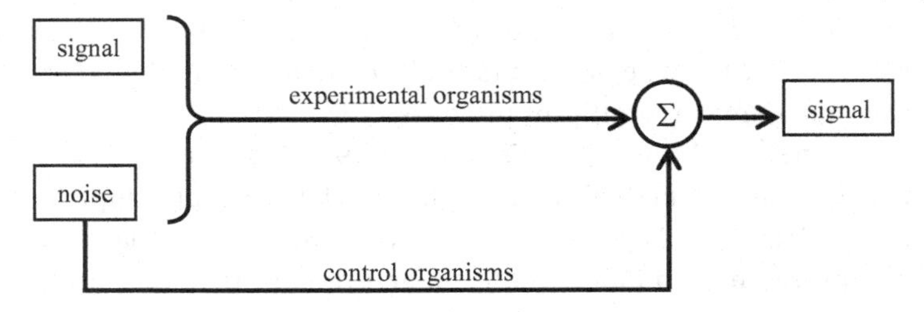

Figure 5.4. The final step in extracting the data representative of the output of the yeast growth control mechanism is to express the perturbed (experimental) data as a proportion of that of the unperturbed (control) data. This acts as a filter to remove the signal due to toxic perturbation from the noise, as specific growth rate ratios ($r\%$).

factors may influence both, the researcher is able to identify the effect of the treatment as the difference between the control cultures and the treated ones. Expressing the specific growth rates of perturbed cultures as a ratio of those of the controls provides a filter for separating the 'noise' provided by the unperturbed controls from the 'signal' provided by perturbation (Figure 5.4). It is only the response of the control mechanism that relates to neutralising the deviation of the growth process from its goal setting that is of relevance. This depends upon the assumption that the specific growth rate of the unperturbed cultures is virtually constant, as established earlier (dotted curve in Figure 5.3c). The product of this process is access to the output of the growth control mechanism that neutralises the effect of perturbations which deviate growth from its goal setting [4].

We require two parallel groups of organisms from which to derive a measure of control mechanism output; specific growth rates for perturbed organisms are expressed as a ratio of the specific growth rates for an unperturbed group. Implicit in this method is the assumption that the control, or unperturbed group, provides the goal setting for those that are perturbed. It follows that the behaviour of the perturbed organisms is due to perturbation alone. Such behaviour is typically oscillatory, due to the lag in feedback mechanisms. In time, the specific growth rate stabilises, returning to the goal rate.

GROWTH CONTROL IN HYDROIDS

The output of control mechanisms reveals idling that is uninformative until placed under load. So the control process must be disturbed in order to observe its normal responses and subsequent behaviour as it equilibrates. The choice of a toxic agent to provide perturbation is merely because it creates a simple and reproducible perturbation that can be graduated with precision in order to provide the required level of loading. The outcome is equivalent to the growth rate ratio (r%) by which we see only those effects on specific growth rates that are due to perturbation. The output data acquire an oscillatory form that can be recognised as the output of a feedback mechanism.

We now consider the growth control mechanism output of the colonial hydroid *Laomedea*, which was deliberately chosen as it is unrelated to the yeast *Rhodotorula*. The reader will notice that the yeast and hydroid data have different time frames, with a cycle period of 2.6 hours for the yeast and 6 days for the hydroid. Similarly the growth rate of the yeast is 25 times faster than that of the hydroids, so the experiments could be much shorter. Yet the results are qualitatively similar to the hydroid, and amenable to the same interpretation (Figure 5.5). Here sets of replicate hydroid colonies are exposed to various levels of load imposed by low concentrations of copper. A graduated range of concentrations is chosen, varying from levels that cause a transient response and rapid equilibration (1 μg ℓ^{-1}), to those that cause saturation and overload of the control mechanism (25 μg l^{-1}). Unlike the more primitive response of the yeast (Figure 5.3c; Figure 12.2), the response of the hydroid is immediate, and overwhelms low levels of inhibition, exceeding that necessary to neutralise the inhibition. It appears that a rapid maximal response to any level of perturbation is an adaptive refinement that quickly overcompensates at low levels, becoming more proportionate to load at higher levels. This adaptation reacts much more rapidly than the response seen in the yeast, where growth is inhibited and causes a negative error before a neutralising response can be initiated. The rapid response of the hydroid overcomes the flaw of feedback mechanisms in responding retrospectively by means of a feedforward loop (see Figures 4.1 and 4.2), which ensures a more rapid response, avoiding any initial inhibition of growth rate. Such adaptations will be considered in Chapters 7 and 11. A response that is maximal is an advantage, in that any perturbation is immediately met with a response that in most cases is overwhelming. To respond in this way neutralises a perturbation rapidly before it can take hold.

It is not difficult to visualise how a maximal initial response, unrelated to load, might be advantageous to organisms in the environment. Here, stressful variables would not increase in the same way as an experimental step function, but are likely to increase gradually

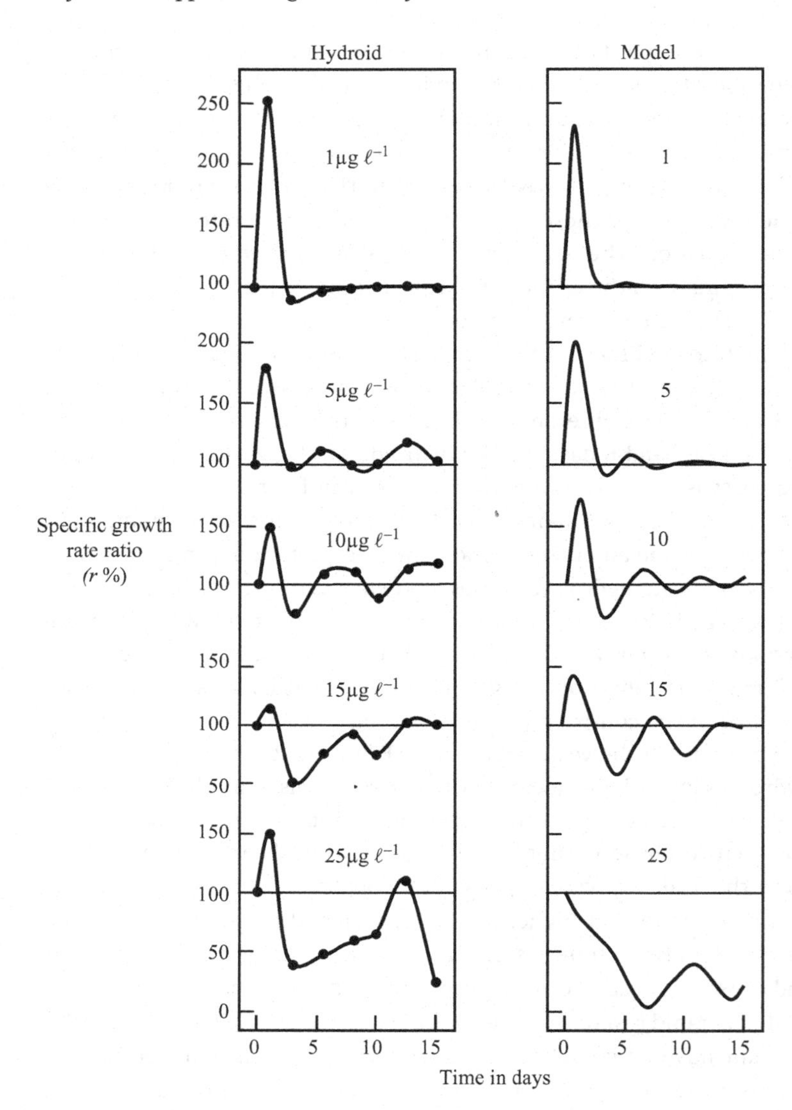

Figure 5.5. Growth control output for the colonial hydroid *Laomedea*. Here perturbation is provided by a range of low concentrations of copper. Redrawn from Stebbing and Hiby [5]. The data are compared with the similarly perturbed output of the control model (see Figure 5.6).

or in a sequence of steps. By responding maximally, such increases are in effect anticipated, but if there is no increase in load, nothing is lost by overcorrection. The refinement is an anticipatory adaptation, as if the control mechanism were aware of its own weakness in responding retrospectively to change.

After the initial maximal response, it is clear that the response of the control mechanism increases in proportion to the level of inhibition. But in the mid range of concentrations, the oscillatory behaviour indicates that although the mechanism finds it increasingly difficult to counteract the inhibition, it succeeds in recovering by the end of the experiment. At a critical level of loading, the control mechanism becomes saturated, and no higher level of inhibition results in a greater neutralising response. The control mechanism becomes overloaded and ultimately control is lost. As we shall see, in toxicological data the point of overload also determines the threshold of toxicity in concentration–effect curves (Chapter 7). For homeodynamic indices of toxicity, the concentration at which saturation comes about is a critical point in determining toxicological threshold concentrations.

It is clear that, by this stage in the work, the Maia hypothesis was becoming more than just a method to access growth control output [4]. The specific growth rate was apparently controlled in a way that resisted and neutralised the inhibitory effect of toxic agents over a range of concentrations. The oscillatory output was typical of other feedback mechanisms.

MODELLING GROWTH CONTROL MECHANISMS

The proposed growth control mechanism in hydroids that is responsible for growth rate control is unknown, as the only evidence of its existence is the output described here (Figure 5.5). Appeals to physiologist and biochemist colleagues that there must exist a known mechanism that could account for such behaviour drew a blank. One way to test the hypothesis was to see whether a mathematical control model, based on the principle of feedback, would behave in the same way as the experimental data. I was fortunate to be able to interest a mathematical colleague in the problem; Lex Hiby was keen to collaborate and turn his hand to control modelling.

For a mathematical modeller, the task of deriving a model from the output data is referred to as 'reverse engineering', as it is the reverse of the normal process of designing a control mechanism from some specification setting out its required properties. We chose an exemplary data set which exhibited all the control properties, and identified the properties that the model would need to reproduce. These included the initial maximal response to perturbation, oscillation of a constant wavelength, which increased in amplitude at higher loads. A rapid recovery and equilibration occurred at low loads and a longer, equilibration at higher loads, leading ultimately to saturation and overload.

The experimental data for *Laomedea* (Figure 5.5) were sufficient to suggest a control system structure. A linear model was postulated by choosing a closed-loop control system with the least parameter variation that could account for the main features of the result. It was to be expected that they should resemble the simple feedback mechanism of von Foerster (Figure 3.4); and similarly the control model was based on a box diagram with feedback representing the flow of information responsible for the causal cycle. Any control model is required to mimic the flow of information, and its transition from sensed actuality to the difference between actuality and some goal setting. This constitutes an error that provides the basis for a response to minimise the error and so restore equilibrium.

The overall layout of the control model is given in a diagram (Figure 5.6). Its design required the selection of appropriate 'transfer functions' indicated as the 'control elements'. It describes mathematically the processes by which inputs are modulated to become outputs; so in effect it is the control mechanism. The parameters, including the time constants and exponents of the transfer function, determine the magnitude and shape of the modulation induced by the system operators. The inner secondary loop causes the delays within the mechanism that create an oscillatory output.

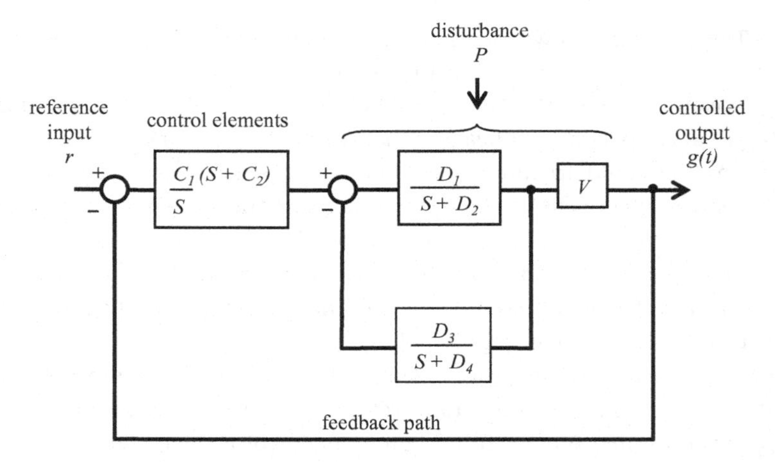

Figure 5.6. A diagram of the control model used to simulate the behaviour of hydroid data: *r* goal-specific growth rate; *g*(*t*) growth over time; *P* perturbation by inhibitor; *V* gain. Other terms are Laplace transforms which simplify differentiation and integration into multiplication and division. See Stebbing and Hiby [5], from which the diagram was redrawn.

Rate terms such as dW/dt have been used generally here, but tend to be replaced by Laplace transformation for greater rigour by modellers. This method was adopted by Lex Hiby in developing a model for the hydroid data (Figure 5.6). The subtleties of Laplace transformation are beyond the scope of this book, but the rationale for their use should be explained. It involves taking some function of time $f(t)$and converting it into a complex variable. This simplifies the procedures of differentiation and integration into multiplication and division using the Laplace operator S. A good introduction to the use of Laplace transformation is provided by Peter Calow (1976) in his book *Biological Machines: A Cybernetic Approach to Life*; although it is now out of print.

As we have come to expect, the feedback of sensed information about the control process is compared with some preferred condition at the comparator, which by summation determines the error. Control in this model is of two kinds: proportional control and integral control. Proportional control involves a corrective response that is proportional to the perturbation, whose magnitude is directly related to the size and sign of the error. Integral control means that the response is not instantaneous, but is the result of the summing of information about errors over a period of time, as has been described in Chapter 3. If the error is measured continuously, and the results added together, the error is larger and more easily detected. The restoration of stability may take longer but, in time, integral control (Figure 3.5) helps to ensure that oscillations are eliminated. The result is that when proportional and integral control are combined, it is possible to achieve stability more rapidly than with integral control alone.

The processes involved in the control mechanism (as shown in Figure 5.6) are subject to additional delays, on top of those due to the sensory feedback loop. These involve initiating a response to negate an error, overcoming the inertia of a process that must be reversed for inhibition to be neutralised. The control signal must be amplified by increasing the gain (V) sufficiently to overcome the perturbation (P), giving an output ($g(t)$) that has a reduced error with respect to the goal growth rate (r).

The control model incorporating these features was then used to simulate a range of loads, equivalent to metal concentrations in the experiment with hydroids (Figure 5.5). Overall the data supported the interpretation that the stability of a control system under load represents a balance of opposing forces, in that the imposed load is countered by a neutralising response. In other words, the effect of external inhibition is neutralised internally by a response that is stimulatory [5].

The initial wild overcorrections at low concentrations were at first interpreted as the result of instability, and only later did an adaptive interpretation arise, involving a maximal initial response triggered by feedforward control to provide an adaptive response. The hydroid data needed to be compared with the contrasting yeast data (Figure 12.2), where sustained inhibition and a negative error occurred before a neutralising response could be initiated. Given this interpretation, the initial hydroid response appeared to be a sophisticated adaptation to avoid the disadvantage of a delayed response. From the outset, the yeast cells appeared to be struggling to claw their way back from the initial inhibition. With greater concentrations, overcorrection was replaced by overload. Overload in the hydroid output suggested that the capacity of the control mechanisms to resist inhibition could not be sustained, indicating that the capacity to resist inhibition required work, and output ultimately failed due to lack of resources.

Rather than reproducing the data and simulation against a time base (as in Figure 5.5), this is a good point at which to introduce both data and simulation in phase space (Figure 5.8). Phase space diagrams are helpful, as they best show the process of stabilisation; and have also proved useful for ease of communicating later ideas about the behaviour of growth control mechanisms. The focus becomes the attractor at the centre of the plot to which the controlled process is drawn as it stabilises. Such diagrams also make it possible to directly compare control mechanisms that operate on dissimilar time bases, like the yeasts and hydroids.

Phase space diagrams are best illustrated using the example of a swinging pendulum. To begin with, our pendulum, unlike a clock, has no weights to drive it, and is set in motion by hand. When its motion is plotted, it swings through a series of oscillations of decreasing amplitude as its energy is dissipated, until it comes to a stop. In phase space the depiction of a pendulum moving from side to side is a spiral that finally becomes stationary at the attractor. We can see a process approaching equilibrium as one of continuous cycling and convergence, rather than oscillations, which suggest repeated changes in direction. So this system has a 'fixed point attractor', to which the pendulum returns after disturbance (Figure 5.7a), as its energy is dissipated. If the clock is wound up and the pendulum is set in motion

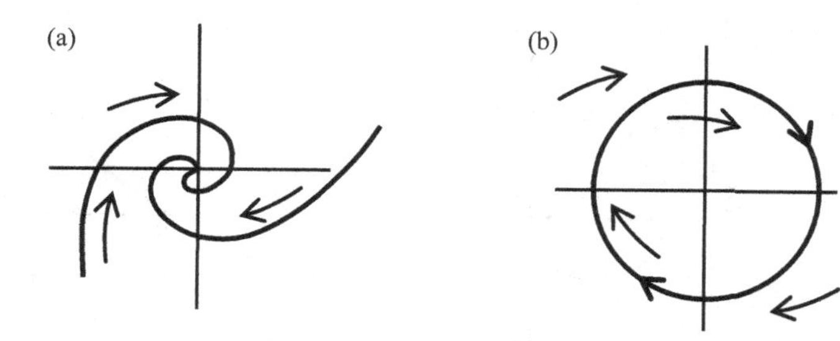

Figure 5.7. Diagrams showing the behaviour of a pendulum, which in phase space spirals into a stable equilibrium as it stops (a) or, when the clock is running, stabilises at a limit cycle (b).

again, it soon settles into a swing of constant amplitude. The amplitude is drawn to its limit cycle. In phase space this is also an attractor, because if the pendulum is started with a slightly lesser or greater amplitude, it soon adopts its own preferred amplitude. In a phase space this is called a 'limit cycle attractor' (Figure 5.7b), and represents the behaviour of systems that oscillate continuously with the same amplitude and frequency.

The description can be applied equally well to the hydroid data (Figure 5.8a), or to the simulation (Figure 5.8b), as they vary little in the form of the cycles as they restore equilibrium, or become overloaded [4]. We then see the initial overcorrection as a widely sweeping ellipse, which at low levels quickly equilibrates as the process is drawn inward to the attractor. At higher concentrations, more cycles are required before reaching the attractor. But with overload the process falls away from the attractor, never to reach an equilibrium. As the system becomes overburdened by increasing concentrations, the control mechanism can no longer draw itself to the attractor, and declines into negative growth rates. When the hydroid data as rate ratios ($r\%$) are compared with the Maian interpretation of output of a growth control mechanism, the data provide evidence that the control model produces output that replicates all the essential features of the experimental data. Given that two dissimilar and unrelated organisms, the yeast and the hydroid, yield similar outputs using the same method, suggests that other organisms may also behave in this way, and that their growth may be shown to be regulated. Later chapters will describe other growth phenomena and experimental data in support of this interpretation.

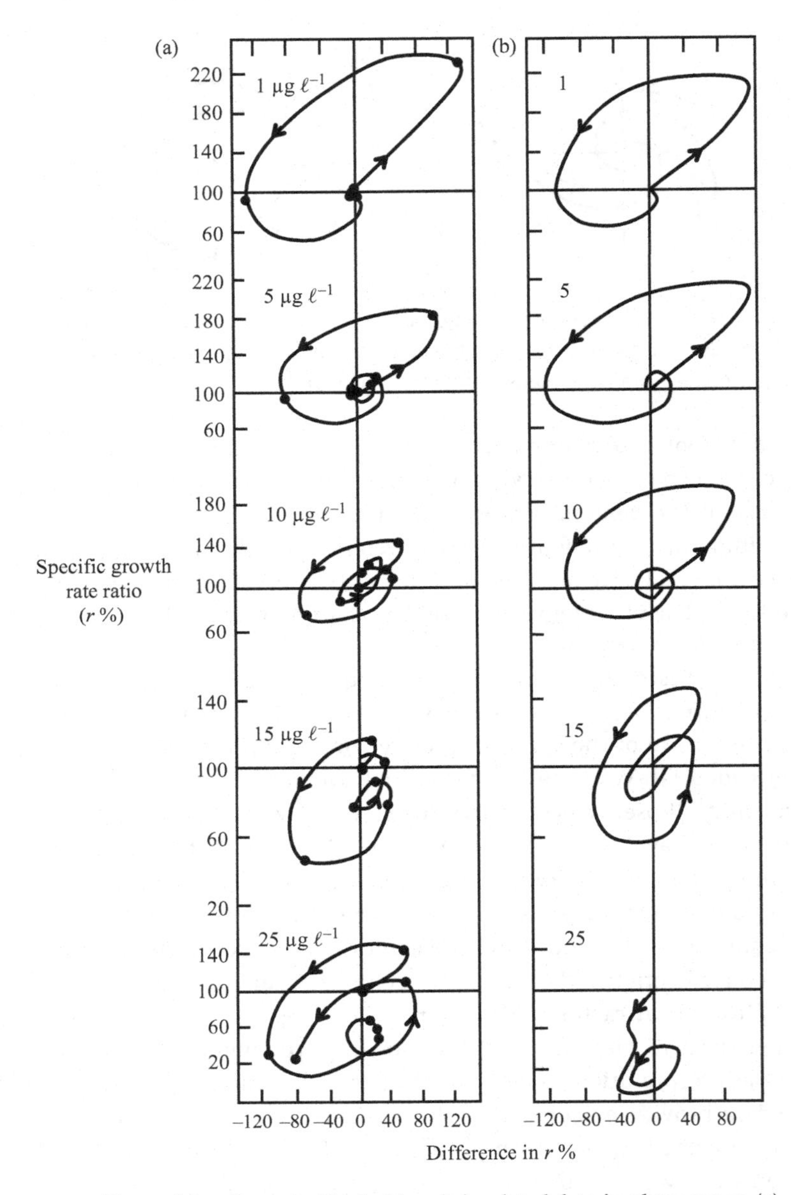

Figure 5.8. Output of hydroid and simulated data in phase space. (a) Hydroid growth control data, (b) output of the simulation model. Data are the same as in Figure 5.5. Redrawn from Stebbing [4].

SUMMARY

A method is described by which the output of a rate-sensitive growth control mechanism can be derived from cumulative growth data. The technique has been applied to two dissimilar systems, a yeast cell suspension and hydroid colonies, suggesting the method may have general application. The perturbed rates are expressed as a ratio of the unperturbed rates, and are calculated at frequent intervals in time, to reveal the dynamic output of the growth control mechanism. This interpretation is corroborated by developing a mathematical model using the principles of control theory. The experimental data and model output concur, inferring that it is valid to derive growth control mechanism output in this way. Perturbation is an essential feature of the approach in all that follows, as it is only when growth rates are deviated from their goal setting that the control mechanism has to work to restore equilibrium, revealing its characteristic oscillatory output. This approach to the study of the rate control of growth was given the name Maia, although, the hypothesis becomes more complex in the following chapters.

6

A control mechanism for Maia

> The study of growth consists largely in the study of the control and limitation of growth.
>
> J.Z. Young

> To regulate something always requires two opposing factors.
>
> Albert Szent-Györgyi

> What is true for E. coli is also true for the elephant.
>
> Jacques Monod

DYNAMIC EQUILIBRIA IN INERT SYSTEMS

The principle of homeostasis became a central and abiding theme in the physiology of animals. Once adopted, it was never abandoned, because for all life homeodynamic systems have great survival value, conferring control over external perturbation that would otherwise disrupt internal processes. The result was a balance of opposing forces of perturbation and counter-response that characterises homeodynamic systems. But how did such systems originate?

Maia emerges as a mechanism that creates stability in growth rate, but what within the organism is the particular mechanism that accounts for the observed behaviour? What are the specific elements which create the kind of output that we have come to recognise as typical of growth control mechanisms? Before looking in more detail at living control mechanisms, we should be aware that some kinds of regulatory behaviour can also be found in non-living physical and chemical systems. As they preceded whatever biological evolution added later, such systems may have provided the precursors of homeostasis.

About a hundred sentences of what the Greek philosopher Heracleitus wrote still survive; his one book was lost, yet his ideas on cosmology continue to have an influence. He derided men for failing to understand the *logos*, meaning the reason, or the order that pervades in universal processes. One manifestation of the *logos* is the underlying connection between opposites, such as health and disease, or hot and cold. Heracleitus believed there was a fundamental unity in the coexistence of opposites; he maintained that change in one direction is ultimately balanced by change in the opposite direction. Some classical concepts two millennia ago implied self-regulation. Such thoughts were not unique to Greek philosophy; similar ideas arose in the Chinese Yin–Yang school of cosmology in the third century BC. The Chinese believed that two opposing but complementary forces, or principles, occur in many aspects of life. Their interplay is such that, as one increases, the other decreases in a way that describes the processes of the universe. We take from Heracleitus the idea of the dual nature of processes that have antagonistic, yet complementary, elements that fluctuate continually in a dynamic stability.

Le Chatelier's Principle was discovered in 1888 and named after the nineteenth-century French chemist Henri Le Chatelier (1850–1936), who taught at the Sorbonne in Paris and to whom we are indebted for the invention of the thermocouple and oxyacetylene welding. His principle states that if the conditions of a reversible chemical reaction are changed, for example in temperature, pressure or volume, then the equilibrium will shift in a direction that will tend to oppose that change. In effect, the reaction adjusts itself and resists change, so minimising the effect of external disturbance. Similarly Lenz's Law, which is named after the Russian physicist Heinrich Lenz (1804–1865), states that the current induced by electromagnetic force always flows in the direction to oppose whatever is inducing the current in the first place.[1]

The American mathematician Alfred Lotka (1880–1949) was interested in Le Chatelier's principle, where equilibria were set up by pairs of opposing chemical reactions proceeding simultaneously in the same reaction vessel. While the reactions are proceeding, there may be no overall speed or direction because one is cancelling out the consequences of the other. In this way they create a dynamic 'equilibrium'. The word is derived from the Latin *aequus* (meaning equal) and *libra* (meaning scales), in which opposing forces are so well balanced

[1] Both laws have been criticised because they fall within the ambit of other more general laws: the First Law of Thermodynamics and the Law of Conservation of Energy.

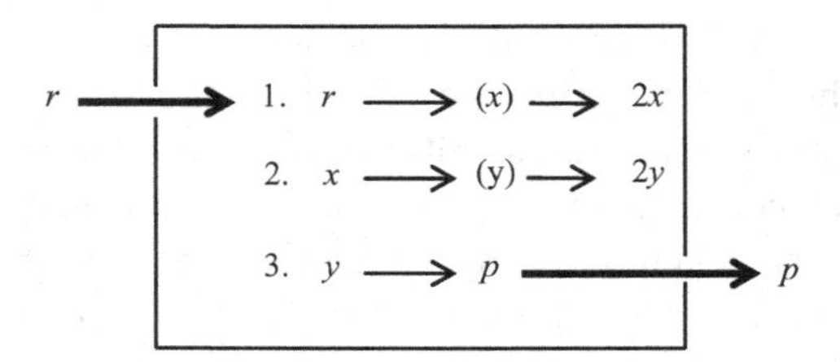

Figure 6.1. Diagram of Lotka's chemical oscillator. A diagram of a vessel into which a reactant (*r*) flows, and from which a product (*p*) emerges.

that there is no outcome. Chemical reactions tend of their own accord to proceed towards an equilibrium, typically helped by catalysts in chemical reactions and by enzymes with an equivalent role in biological reactions.

In 1910 Lotka published a theoretical paper describing opposing chemical reactions [1]. Equilibrium of the system is approached through a series of damped oscillations, settling at a fixed point attractor. Later he showed that simple opposed reactions in the same vessel could stabilise at a dynamic equilibrium of sustained oscillations or a limit cycle, even returning to the cycling equilibrium after disturbance. The following scheme illustrates a simple mechanism that creates such behaviour.

The mechanism for chemical oscillators has just three steps, so can easily be described in sufficient detail for its behaviour to be seen. Imagine a vessel into which a reactant flows, and from which a product emerges at the same steady rate. The first two steps are catalysed such that the amount of x and y is doubled, accelerating their own rates of production. In the first step, $r + x \rightarrow 2x$ is catalysed by x, and the second step, $x + y \rightarrow 2y$ is catalysed by y, which is also the final product. The overall effect is to reduce the concentration of x, which reduces the rate of step 1, which in turn reduces the rate of step 2. When less x has been used, step 1 can increase, and so in turn does step 2. In this way, x and y oscillate synchronously, but out of phase, as y tracks the fluctuations of x (Figure 6.1).

THE BELOUSOV–ZHABOTINSKY REACTION AS A CHEMICAL OSCILLATOR

One of the best known examples of a chemical oscillator is a reaction with a curious history [2,3]. Its relative simplicity belies its capacity to shed light on far more complex biological systems. The Belousov–Zhabotinsky reaction, or B–Z reaction for short, was first seen as an impossible oddity when put forward by the Russian biochemist Boris Belousov in 1958. He had apparently been trying to understand catalysis in the Krebs cycle, and was seeking a simple model, but discovered something of much greater interest. The reaction with which his name is associated is sometimes called the malonic acid reaction (see Appendix 3 in [4]). The B–Z reaction, when first described, alternated between two states at regular intervals. In fact the reaction oscillated from yellow to colourless every 30 seconds through a hundred or more cycles, until eventually one of the reagents became exhausted. Such experiments were carried out in reaction chambers; today the B–Z reaction tends to be associated with images of spirals in shallow dishes, but these were to come much later.

One scientific journal after another rejected papers describing Belousov's discovery. The very idea was not credible, as it was at odds with the laws of physics. Belousov not only had difficulty in publishing his findings in chemistry journals but was also ridiculed by his peers, because it was claimed that his malonic acid reaction contravened the Second Law of Thermodynamics. With entropy, disorder should increase, rather than create order in the form of regular oscillations over time; but his critics did not realise that the reaction does not oscillate indefinitely, as disorder eventually takes over. He did publish a brief abstract in the proceedings of an obscure meeting, but the work remained forgotten for years. His discovery rejected, and in frustration at his peers, he decided that no-one would share what he had seen so many times with his own eyes.

Nevertheless, in the 1960s, rumours about Belousov's discovery spread among Russian chemists. A young research student named Anatol Zhabotinsky was asked by his supervisor to look into what Belousov had found, and he was soon able to repeat Belousov's experiments. During the 1960s, Zhabotinsky found a dye to create a more marked colour change for the reaction. He presented the work at an international conference in Prague in 1968, just before the Russian occupation. Through this closing chink in the Iron Curtain, East briefly met West, and Art Winfree met fellow research student Anatol Zhabotinsky. Biological rhythmicity and biochemical oscillators were of growing interest, and the B–Z reaction was a simple spontaneous oscillator. Simple, that is, in relation to brain waves and other complex

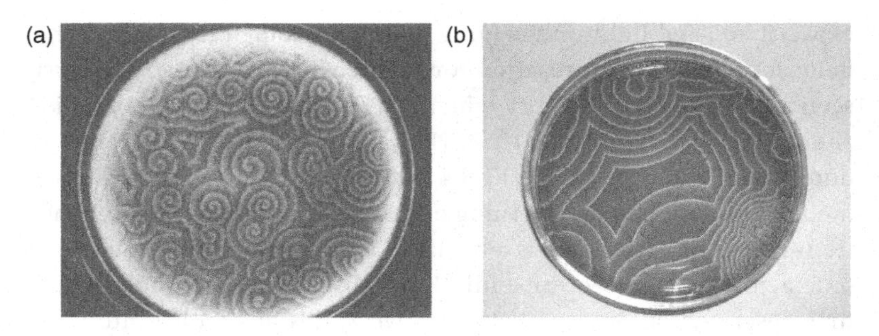

Figure 6.2. Photos of growth patterns of the slime mould *Dictyostelium* (a) and the Belousov–Zhabotinsksy reaction (b) in petri dishes.
(a) Courtesy of Drs. J.C.W. Bax, www.metafysica.nl/index.html;
(b) courtesy of Stephen W. Morris, www.flickr.com/photos/nonlin/3572095252/.

wave phenomena for which, in time, the B–Z reaction became a useful experimental model.

It was not until 1970 that Zaikin and Zhabotinsky published their findings in the journal *Nature* [5], showing the striking patterns of concentric rings. The reported observations were of circular chemical waves passing through a modified version of the original oscillating reagents. Vivid blue waves emerged from a ground of red ochre at points where dust or disturbance created ‘pacemakers’ and initiated the cyclic reaction. The immediate effect is one of closely packed concentric rings but, on closer inspection, the rings are seen to be spirals (Figure 6.2a), which rotate once every minute; each wave growing outward at the same pace, while the rotating pacemaker creates the new wave continuously from the centre. The overall effect is a beautiful pattern that has a similar but ever-changing form. This extraordinary discovery reveals in space the rhythmic behaviour of a chemical reaction alternating process in time. Iteration and oscillation in time create repetition and regularity in space, and remind us of the output of control mechanisms.

Before the Zaikin and Zhabotinsky paper was published, Winfree wrote to Zhabotinsky asking whether he had seen the spirals, as he had observed similar spirals in fungal cultures (Figure 6.2a) and wondered whether Zhabotinsky had observed spirals in the B–Z reaction. Frustrated by the delays in mail between the USA and Russia, Winfree eventually made up the recipe for the Zaikin and Zhabotinsky experiments in shallow dishes and immediately produced the spirals (Figure 6.2b), which so closely resembled those he had observed in fungi in his own laboratory. This alone was a remarkable observation, that the inert B–Z reaction should reproduce the growth patterns of a fungus. Winfree called the pacemakers 'phase singularities' about which the wave turns into a growing spiral, in a way that proved to be characteristic of excitable media.

In the 1970s the B–Z reaction was finally accepted, and a decade later Belousov, Zhabotinsky and others who had been involved in its development were awarded the Lenin Prize for their work (although, sadly, Belousov had died in 1970). Later, Winfree and his co-workers took the behaviour of such systems much further, beginning with crude experiments to reveal the form of the reaction in a third dimension within stacks of filter paper. Using the 3-D imaging capability of powerful computers, he was able to simulate and visualise the higher levels of complexity of scroll waves created by a phase singularity that itself moves in circles to create a toroidal vortex [2].

Like other Western scientists, the Russian-born Belgian scientist Ilya Prigogine (1917–2003), first learned about the B–Z reaction at the conference in Prague. His interest was in the B–Z reaction which, like living systems, appears to contravene the Second Law of Thermodynamics. Such systems maintain conditions far from equilibrium, by transformation from disorder into order, and the dissipation of waste. For this reason he called them 'dissipative structures' [6]. They have dynamic behaviour, are adaptive and self-organising systems and create order, all features that we tend to associate with living systems. In 1977 he was awarded the Nobel Prize for chemistry.

In 1968, Prigogine and a colleague published a model that simulated the oscillatory behaviour of the B–Z reaction. The model involves four steps, one of which is autocatalytic, and simulates an oscillating reaction such that intermediate reactants fall into a limit cycle, irrespective of their initial concentrations. In 1972 an American group led by Richard Field published a similar but more complex model. They called it the 'Oregonator', echoing Prigogine's Brusselator. It requires five steps and six chemicals, is also autocatalytic and simulates oscillating concentrations of bromide in the B–Z reaction which follow limit cycles in phase space. Such models were important because they provided a basis for understanding more complex living systems. These included brain waves, waves of depolarisation of the heart and the behaviour of other excitable membranes, but one of the first examples was the similarity of the B–Z reaction to the behaviour of the slime mould *Dictyostelium* (Figure 6.2).

THE SLIME MOULD ANALOGUE

Remarkably, a number of fungi exhibit similar behaviour to the B–Z reaction, none more so than the slime mould *Dictyostelium discoideum* (Figure 6.2b). It had been used as a model by biologists since it was adopted by the American biologist J.T. Bonner in 1944. He became interested in its curious life cycle, which seemed to alternate between single-celled and multicellular life forms. 'Social' amoebae hatch from spores and consume bacteria. They multiply by fission, and wander about, feeding until they exhaust local food supplies. They then aggregate to form a 'grex' (Latin for a flock, herd or swarm), which is capable of moving in a purposeful way at a rate of 2 millimetres per hour for weeks, seeking light, warmth and humidity. Once it has found a suitable environment for the distribution of its spores, movement ceases and a vertical stem grows to a height of a millimetre or more to form a spherical mass of spores, which are then released on the wind when the sphere breaks open.

We now introduce the first of a family of biochemicals called 'nucleotides', which will become increasingly important as the chapter proceeds. One of these nucleotides, we shall just call it cAMP, is responsible for aggregation of the amoebae. Pacemaker or pioneer cells provide the phase singularity in *Dictyostelium*, and release a pulse of cAMP when they become short of food. As the radiating wavefront meets other cells, they too release a pulse of cAMP, extending their pseudopodia to begin to move towards the source. After each pulse of cAMP, cells enter a refractory period, which ensures that the cAMP signal is only propagated radially from the pioneer cells. What is so remarkable is that this slime mould, as a lawn of amoeboid cells on a petri dish, emanates waves of cAMP which create a radiating spiral pattern that is virtually indistinguishable from those of the B–Z reaction (Figure 6.2a).

Amoebae are attracted from a radius of a few millimetres and converge upon the source of the cAMP. After a few hours they constitute a grex of amoebae 100,000 strong, which forms a simple slug-like organism 1–2 millimetres long. Winfree observed that the grex has a 'pulsing wave source at its tip' that originates from the pioneer cell, and which provides the nucleus for aggregation and the origin of the spiral wave of cAMP. It seems that the coordination and movement of the grex is still controlled by cAMP. While the property of being single-celled or multicellular allows us to separate the major taxa, here we see it as a facultative change within the life cycle of a single organism.

At the subcellular level, the mechanism is important for what is to follow. The cAMP is made from ATP (adenosine triphosphate), whose transition is catalysed by an enzyme embedded in the wall of the amoeboid cells. cAMP is then released through pores in the cell wall to the exterior, where it not only acts as a pheromone, but is also received by a membrane receptor which further activates the enzyme to catalyse the production of more cAMP. In this way the production of cAMP is autocatalytic, and the positive feedback makes the process self-perpetuating for the duration of a pulse.

The waves of cAMP that guide the amoebae to the pioneer cell adopt the same form and have the same dynamics as those propagated in the B–Z reaction. The underlying processes are sufficiently similar for the B–Z reaction and the slime mould's cAMP waves to be simulated with essentially the same model. It is a remarkable example of a non-living system reproducing the complex behaviour of a living system, such that the knowledge of one can inform the other.

Such waves and patterns and the rotating spiral waves in which they grow are seen as the signature of what Art Winfree called 'excitable membranes'. Examples that behave in the same way can take the most unusual forms. In the frog egg, following fertilisation, pulses of calcium can be seen passing over the surface of the egg. The waves carry some signal through the egg, perhaps related to penetration of the sperm and to fertilisation. During evolution, excitable membranes have produced 'emergent' properties of different kinds, ranging from brain waves to gut peristalsis, which, by definition, could not have been expected from a knowledge of their cellular components alone.

The most far-reaching implications of the B–Z reaction are found in abnormalities of the mammalian heart. In the pumping of the normal heart, waves of depolarisation radiate out over and through the heart from the pacemaker node. They are relayed onward to the lower half of the heart by another node coordinating and inducing muscle contraction, and hence pumping of the heart. How these waves spread is of importance to the functioning of the heart, and the B–Z waves provide a good model. Ventricular fibrillation, otherwise known as a heart attack, occurs when the heart beats rapidly and irregularly, its contractions out of phase, as the heart becomes useless as a pump. During fibrillation, the heart squirms in an uncoordinated way because spiral waves have replaced those that progress in an orderly way from the top to the bottom of the heart. In extreme cases, the spiral waves of contraction spin like a rotor at ten revolutions per second. In the B–Z reaction, spiral waves originate from an obstacle in the path of a normal wave. In the heart tissue, it is similarly thought that some obstacle to the passage of a normal wave of depolarisation (such as damaged tissue or a clot) may induce spiral waves, which behave rather like waves at sea curling around a coastal headland.

It is clear that the work on the B–Z reaction and *Dictyostelium*, together with the simulation models that relate to them, led to significant discoveries. It also led to a collaborative interplay of ideas as experimentalists produced data, while mathematical modellers developed the dynamic hypothesis, simulating various scenarios, which fed back to inform further experimentation.

In physiology, many metabolic limit cycles are now known that oscillate indefinitely. In the 1980s, an atlas of cellular oscillators listed over 400 examples. These reflect the behaviour of metabolic processes oscillating with periods ranging from 1 second to 5 minutes. Some decay following perturbation, but others oscillate at a constant amplitude and frequency for many hours; about a quarter oscillate in stable limit cycles.

Various explanations have been put forward to account for the behaviour of these oscillators. The advancing wave fronts of the B–Z reaction, *Dictyostelium* the slime mould, and other physiological processes, indicate the outcome of cyclical processes; repeated patterns in space reflect oscillation in time, and oscillations are typically the result of a system incorporating feedback. One idea is that it may be necessary to separate antagonistic reactions catalysed by enzymes in the same cellular compartment. Their incompatibility could only be reconciled by ensuring that they do not happen at the same time, by alternating continuously between one reaction and another. It has also been suggested that metabolic oscillations have a timekeeping function, representing the ticking of a clock, such as might control circadian rhythms. Another suggestion is that the oscillations represent the consequences of control activity within a cybernetic network. This only becomes probable if we see some perturbation followed by damped oscillations. In phase space we could accept sustained oscillations as evidence of a limit cycle attractor, which could equally account for the bistable state of two simultaneous reactions. But any or all of these expectations could be true.

Perhaps the best known example is the glycolytic oscillator studied in brewer's yeast (*Saccharomyces*). In the glycolysis system, glucose is broken down to create usable energy in the form of ATP, with ethanol as a toxic byproduct, which on reaching a concentration of 15%

stops further growth by killing off the yeast. Winfree hinted that such systems, with their repeated oscillations, should provide an opportunity to study regulation. The idea was not pursued at the time and, for whatever reasons, yeast kinetics fell out of fashion. Here, yeast control system output has already been described, showing decaying oscillations from perturbation experiments as equilibrium is restored (Chapter 5). This is interesting because we can see a simple example of dual control. Rate control has already been demonstrated in yeast, but superimposed upon it is the self-limitation of cell population size or density.

SELF-REGULATION IN BIOLOGY

We find that self-regulation occurs in non-biological systems, which has informed our understanding of self-control in biological systems. We now need to look at the kinds of mechanism that might account for the control of growth processes. For reasons that will become clear, it is worth going back to the only biological paper by the mathematician Alan Turing (1912–1954) [7]. Turing provided the theoretical basis for the modern computer, but to many he is better known for his contribution to the outcome of World War II by breaking the Enigma code, working alongside a motley collection of mathematicians, linguists and chess grandmasters. Much less well known is that after the war he worked on problems of biological growth and development. His paper, published in 1952, gives one answer to the question posed by Rudyard Kipling in his *Just so Stories* on 'How the leopard got his spots'. This is not to trivialise the problem of understanding the development of biological form that remains an active field of research to this day. The key to understanding how the leopard got its spots is essentially a mechanism derived from chemical reaction-diffusion processes that establish positional information in the developing embryo, and thereby pattern.

'Morphogens' are hypothetical biomolecules involved in creating shape and form during development. Turing appreciated that diffusion could generate form by the concentration of morphogens, which would then determine the fate of the cell as it develops. He used as his conceptual model the same organism that Abraham Trembley used for his early studies on regeneration two centuries earlier. Trembley had carried out early experiments in which the head of a tentaculate polyp regenerates if cut off. Turing's thinking, and the need for simplicity, led him to look for an endless length of tissue, for which he

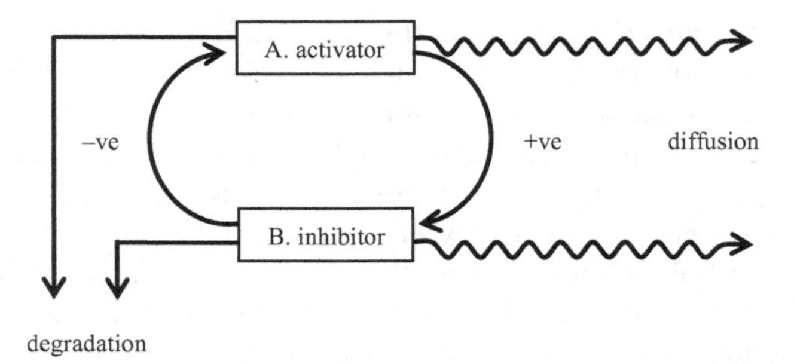

Figure 6.3. Diagram of the activator–inhibitor system discovered by Turing, and later made accessible by Meinhardt and Gierer. Redrawn with permission of Ball [4].

chose rings of cells. The open tube of a polyp, from which the head and tentacles had been removed, provided the ring of cells. Turing postulated a circular sine wave of concentrations of morphogen around the ring that would give local areas of increased concentration where development would occur. In life, five to ten darkened areas indicate the points from which tentacles regrow. Do the darkened areas, where tentacles would appear, coincide with the simulated hotspots of hypothetical morphogen?

It was not until 20 years after publication of Turing's paper that Hans Meinhardt and Alfred Gierer at the Max Planck Institute in Germany came to understand how other equations in Turing's paper make possible the creation of spots and stripes [8]. By autocatalysis, the products of a reaction stimulate its own progress to accelerate, a process whose significance in morphogenesis was first appreciated by Turing. But the greater challenge was to understand how his mechanisms could create form. Meinhardt and Gierer demonstrated the Turing model as a competition between opposing elements: an activator (morphogen A) and an inhibitor (morphogen B) (Figure 6.3). The autocatalytic production of A is localised, as this morphogen does not diffuse fast or far. Rapid diffusion of B ensures that it does not inhibit the local formation of A. Random fluctuations in the initial concentration of A trigger the self-enhancing production of A in local centres. A activates the formation of B, which is dispersed rapidly as it diffuses quickly, suppressing the formation of A in the immediate vicinity, and beyond, creating an array of spots. The inhibitor B diffuses much faster than the activator A to cause long-range inhibition. The activator has a shorter half-life, but its synthesis is promoted by autocatalytic feedback. The key outcome is that this system can produce positional information to create pattern. A and B can be produced at different rates, diffusing with a greater or lesser speed to create different kinds of patterns, such as stripes. Meinhardt later described the system with this helpful analogy. A king is ruler in his own land and suppresses contenders for his throne. This is 'long-range inhibition'. On the other hand, the king promotes those in his ambit to have power under him within his hierarchy of command. This is 'local activation' of ministers and generals.

The stimulatory effect of autocatalysis in relation to the inhibitor effect could, in principle, be responsible not just for spots and stripes, but for aspects of form such as the development of limbs. Turing's contribution is important in recognising the role of autocatalysis. But more important is the creation of a mechanism recognised by Meinhardt and Gierer as an activator/inhibitor system, which will become increasingly important as the theme of this chapter develops.

Meinhardt and Gierer also used *Hydra* to show how its amazing regenerative abilities could be explained [9]; longitudinally along the polyp, gradients of an activator and an inhibitor are produced by different types of cells, their gradients regulated autocatalytically and competitively. Later it was found that the system actually involves two sets of morphogens: an activator and an inhibitor for the head, and the same for the foot. What is important here, in relation to the control of growth, is that we have an activator/inhibitor system which, due to autocatalysis and different diffusion rates, is able to create positional information, and thereby different patterns and biological forms. While positional information is necessary in morphogenesis, what is more important here is the control of growth with respect to time. If the part of Turing's mechanism that provides positional information is removed, one is left with an autocatalytic mechanism that provides temporal control of growth (Figure 6.3).

ADENINE PHOSPHATES AND THE ORIGIN OF LIFE

As we attempt to identify the kind of system that might be responsible for Maia, we need to keep two things in mind. The first is the importance of autocatalysis, and the second is the involvement of cAMP. We do so by way of an eminent physicist's excursion to consider how life began. Freeman Dyson took what his son calls a 'mid-career detour into theoretical biology' [10] to understand the origin of life: he became fascinated by the genesis of the living from the non-living. Dyson focused on two particular nucleotides: these were AMP (adenosine monophosphate) and ATP (adenosine triphosphate). As their names make clear, the difference between them lies in whether they have one or three phosphate groups, so that it only requires the addition of two phosphate groups to make ATP from AMP, via the intermediary ADP (adenosine diphosphate), which has two phosphate groups.

Through phosphorylation (the addition of a phosphate group) ADP is converted into ATP, which has the ability to carry and donate energy within the cell. This step is reversed by hydrolysation; in the presence of an appropriate enzyme, a phosphate group is shed and stored energy is released, to leave ADP.

$$\begin{array}{c}\text{Phosphorylation} \rightarrow \\ \text{ADP + phosphate group} \leftrightarrow \text{ATP + work + heat} \\ \leftarrow \text{hydrolysation}\end{array}$$

Although this step is much less common, the cleaving of another phosphate group from ADP also releases energy, to leave AMP. The cycle from ATP to ADP and back again happens very quickly. Muscles store enough energy as ATP for less than 5 seconds' worth of activity, but it is regenerated even more rapidly within the muscle by direct phosphorylation. ATP is the universal energy carrier of the cell, providing nearly all its energy requirements, whether powering mechanical, osmotic or chemical processes.

What interested Dyson was that AMP has numerous different roles in the cell, unrelated to the energy-carrying role of ATP. Despite its role as an energy carrier, its primary role is as an information carrier, controlling a wide range of different metabolic processes. To address his question, it is necessary to step back from the detail and to consider the fundamental importance of these two roles, as well as their differences. It was Talbot H. Waterman who suggested that there are just two aspects of life that epitomise what organisms do [11]. The first is to do work and to grow: these are processes that relate to the acquisition, transfer and utilisation of *energy*. The second is to control what they do and the rate at which they do it: these are processes that relate to the detection, retention and utilisation of *information*. Energy and information are intimately involved in jointly sustaining equilibria throughout the organism in what we have already described as homeodynamic. Organisms acquire and use energy, and they accumulate and transfer information.

Dyson made the same distinction in his 'dual structure' of life, which he took as evidence for a dual origin of life of one entity that could metabolise and another that could carry information and replicate. These roles he saw as analogues of hardware and software. He proposed that in primordial cells these functions first came together, in two nucleotides: AMP molecules serve as bits of information and ATP stores and transports energy. Each is of fundamental importance in early cellular metabolism. The chemically related carriers of information and energy provide the necessary attributes for control and regulation. Dyson asked why AMP and ATP should be so chemically similar, yet end up with completely different functions in the cell. Of this he was uncertain, and the answer remains enigmatic, but he saw as a major evolutionary step that the availability of nucleotides and their components within the cell leads to the formation of chains of nucleotides or nucleic acids. In primitive cells with no genetic apparatus, many nucleotides are available as AMP and ADP, the byproducts of ATP that has donated its energy. He proposed that they coalesce to take part in the creation of the first information-carrying nucleic acid, the long single-stranded polynucleotide chain of RNA. AMP is one of the nucleotides that make up RNA.

What are the essential conditions for the formation of RNA? The German physical chemist Manfred Eigen was able to create RNA from nucleotide monomers without any template to copy from, but with a polymerase enzyme that encourages the monomers to join up and form a polymer. Leslie Orgel similarly found that nucleotide monomers would link up with RNA as a template, but without the enzyme. Normally, in life, RNA is created with both RNA as a template *and* with enzymes catalysing the process, but how might RNA have first come about without either? When it was discovered that RNA could also function as an enzyme, and catalyse its own production, it became clear that RNA might have performed all the functions of metabolism and replication alone, making possible the concept of an 'RNA world' that preceded the 'DNA world' in which we now live.

It was Francis Crick (1916–2004) who proposed that RNA was the precursor to DNA, and was responsible for heredity in many primitive life forms in an 'RNA world', before the evolution of the 'DNA world' (see [12]). This idea depended upon the fact that RNA could catalyse the reactions needed to copy itself. Ultimately RNA was replaced by DNA because its deoxyribose sugar backbone was more stable as an information store and much less prone to mutation than the sugar-phosphate backbone of RNA.

Dyson had earlier imagined that a self-sustaining network of proteins arose where the components catalyse the formation of each other from the simple chemical building blocks of polymers called monomers. Later Stuart Kauffman lent credence to the idea [13], suggesting

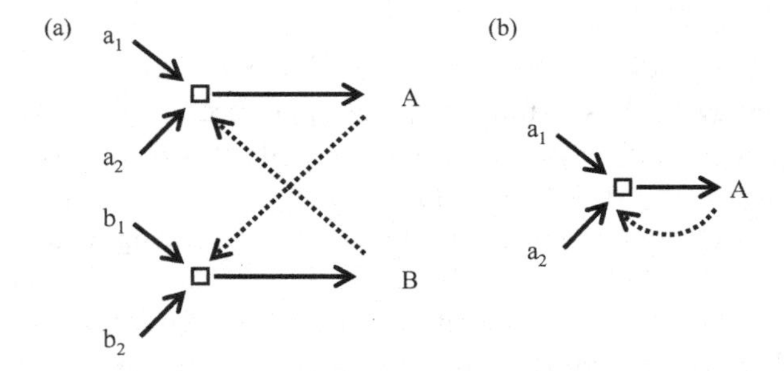

Figure 6.4. The replication of molecules: (a) molecules A and B are replicated by catalysing each other's replication, which Kauffman refers to as a collectively autocatalytic set; (b) autocatalysis of a peptide from two fragments. Redrawn with permission of Kauffman [15].

that the emergence of the self-sustaining network would be almost inevitable if a large enough assemblage of random polypeptides were brought together, as the system would catalyse its own formation. He was convinced that life is an emergent property of complex chemical networks.

The point here is that the simplest system, being the most likely to occur by chance, is that most probable to have occurred naturally. Since hardly any molecules can catalyse their own formation, it was assumed by Dyson and Kauffman that self-replication would have to arise within a soup or network of simple ingredients. It would then be straightforward to postulate a polymer that grew and then replicated, perhaps by fragmenting, as it became too long to hold together. The earliest systems were probably not based on template replication, such as evolved in RNA and then DNA, but in autocatalytic sets of molecules.

These ideas were overtaken in 1996 by two discoveries that brought such speculation to an end. The first was by David H. Lee and his colleagues, who created experimentally a self-replicating peptide [14]. This suggested to Kauffman a model much simpler than his earlier proposal [15]: a system where two molecules promote one another's production by crossover catalysis, where each promotes the generation of the other (Figure 6.4a). He described them as being 'collectively catalytic', and this led him to believe that there was no limit to the scale of such a system – incorporating ten, a hundred or a thousand molecules, or even a complete cell.

The next discovery, made in the same year, was an autocatalytic peptide, which would promote its own synthesis directly (Figure

6.4b). This seemed the simplest system and made almost anything possible. As Kauffman concluded, 'self-reproducing molecular systems are a done deal'. By extrapolation, self-replication of molecules can lead to the self-replication of modules, and thus growth, and living organisms.

We therefore reach another point where the boundaries between the living and the non-living break down, and come to realise that many characteristics of life are found in systems we know not to be living. Self-replication is no longer a property specific to life. It becomes clear that self-regulating and self-replicating systems that characterise living systems may have their origin in non-living systems.

In attempting to understand the control of growth, we must look at the control of replication, which we see at its simplest in these experiments that attempt to explain the origin of life. We have seen this in simple autocatalytic systems by which a molecule makes another like itself. Such self-referencing systems are unmistakably cybernetic in using the finished product with which to replicate itself by means of feedback. This puts the availability of the product in the position of being able to control the process of replication.

We return to Dyson's notion of two closely related nucleotides, ATP and AMP, as having different roles as energy carrier and information carrier in the simplest biological systems. Some more detail becomes necessary in order to proceed. There are two forms of AMP and the transition is important in its information-carrying role. One is called cyclic AMP (or cAMP), which is produced when a hormone from outside the cell activates the membrane-bound adenylate cyclase, the enzyme that catalyses cAMP production. cAMP, in turn, activates those enzymes that the hormone acting as first messenger targets, whereupon cAMP becomes AMP. Similarly cAMP inhibits mitotic activity and AMP does not, so the cycling of AMP between one form and the other becomes an important component of any system that controls by activation.

The purpose of information in biological systems is decision-making; cAMP is involved in the control of many biochemical processes within the cell, and is found in different cells, from bacterial to human. It is involved in the replication of cells, in immune responses and in nerve transmission, as well as gene expression. cAMP activity, and perhaps its oscillatory behaviour, also has a role in the control of growth and developmental events. Oscillations, or pulses of cAMP, incorporate information that constancy does not. cAMP is a ubiquitous message carrier involved in a multitude of functions within the cell.

THE CONTROL OF BIOSYNTHESIS

The underlying process of biosynthesis within the cell is the engine of biological growth. At a fundamental level, anabolism is biosynthesis, which creates the macromolecules necessary for growth. Catabolism is the reverse process, by which simpler molecules are made from more complex ones, due to the processes of digestion and respiration. Because anabolism and catabolism involve processes that work in opposite directions, they cannot coexist in the same cell compartment; so energy-yielding and energy-consuming processes are separated. The energy-yielding processes take place in the mitochondria, while the energy-consuming processes take place in the cytosol, the fluid matrix of the cell in which organelles are suspended.

In general, anabolism and catabolism are regulated within the cell by the same three adenosine nucleotides (AMP, ADP, ATP) acting as second messengers. Anabolism occurs under conditions that are favourable to ATP generation and consumption. Energy production is synonymous with ATP generation, and therefore with the process of phosphorylation explained earlier. In each case the nucleotides control the enzymes that catalyse the metabolic reactions.

Anabolism and catabolism are regulated by the ratios of ATP to ADP and AMP, such that anabolism is enhanced by high concentrations of ATP with low concentrations of ADP and AMP. Conversely, catabolism is enhanced by low concentrations of ATP with high concentrations of ADP and AMP. With more AMP and less ATP, growth becomes negative and catabolic, but if there is less AMP and more ATP, growth becomes positive and anabolic. Unlike anabolism, which has many specific products, catabolism produces energy in the form of ATP, but production becomes impeded by an excess of ATP, even though that inhibition is overcome by AMP. Pacemaker enzymes are inhibited by ATP and stimulated by ADP or AMP. Here, specific growth rate is a measure of the extent to which anabolism exceeds catabolism.

We can look briefly at the history of these crucial nucleotides. Earl W. Sutherland (1915–1974), an American biochemist, discovered cAMP in 1957. He was working on the way in which the hormone adrenalin brings about an increase in glucose in the blood. The discovery was doubly significant as, besides the importance of cAMP due to its many roles within the cell, it was the first reported example of hormone action by way of cAMP as a second messenger. For this work he was awarded the Nobel Prize in 1971.

What is interesting about the first and second messenger systems is that the first messenger is specific in its attachment to those

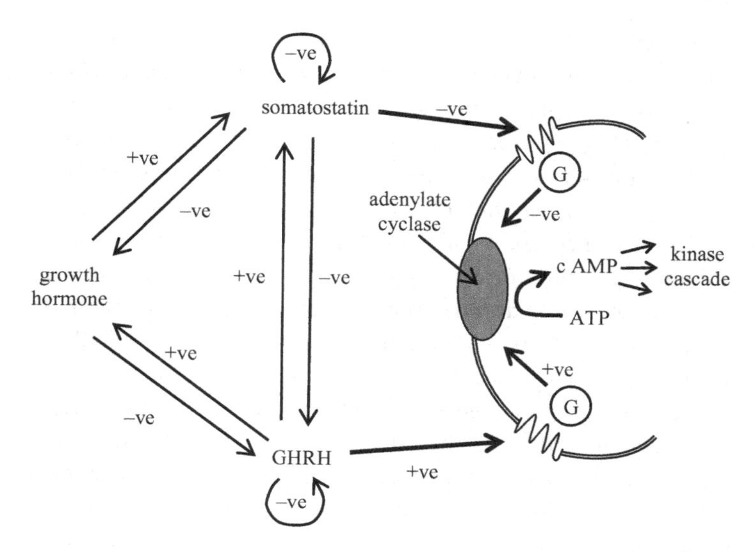

Figure 6.5. A simplified schematic diagram of the growth control mechanism from first messenger (growth hormone) to second messenger (cyclic AMP). G-proteins have the role of transferring signals from the outside to the inside of the cell. These are converted into cAMP by the enzyme adenylate cyclase, initiating various actions. Redrawn from various sources.

differentiated cells that have a matching receptor, but the response inside the cell is due to the non-specific second-messenger cAMP. The point is that while the matching of the hormone to the target cell is highly specific, the response within the cell is less so, as first messengers typically elicit their different responses by means of the same second messenger. In particular, cAMP also has a role in the control of biosynthesis, which we now look at more closely.

If we consider a diagram summarising the cell interface at which first and second messengers meet, information from a variety of sources points to a relatively simple mechanism (Figure 6.5). Growth hormone, or somatotrophin, has a key role in the stimulation via growth hormone-releasing hormone (GHRH) and inhibition by means of somatostatin (see Chapter 12). These first messengers pass signals through the cell wall to second messengers which are non-specific nucleotides, which act in the way described above. It can be seen that the positive arm (GHRH) and the negative arm (somatostatin) are linked, such that a signal by one arm tends to suppress the signal conveyed by the other, ensuring clarity in signalling.

The first messenger is a hormone which, on reaching its target cell, does not penetrate the cell membrane but fits itself to a specific receptor for that hormone. There are many different receptors, so each hormone must find and attach itself to its own specific receptor. They have complementary shapes, and the hormone must fit the receptor molecule to produce a response. Such receptors are actually long-chain proteins, which are folded back and forth between the outside and the inside of the cell membrane, crossing the membrane seven times. The receptor molecule has its receptor end outside the cell membrane, and its other end inside the cell. This allows signals to be transmitted by the first messenger into the cell, without the first messenger having to pass through the membrane. The signal arriving at the inner end of the molecule stimulates a G-protein which is specific to the receptor. The energised G-protein acts as a relay, and moves about until it finds the effector (adenylate cyclase) to which it binds. The effector is an enzyme which catalyses the conversion of ATP to cAMP, creating the second messenger which is released to move around within the cell. The regulatory role of cAMP is as a 'second messenger' for the endocrine system, since as many as eight different hormones have their effects on target cells by means of signals borne by cAMP.

cAMP can move rapidly, triggering a cascade of chemical reactions by activating protein kinases. These too are enzymes and catalysts that initiate a variety of reactions and responses within the cell. There are hundreds of kinases, which act upon small molecules for signalling purposes, priming them for some biochemical reaction involved in metabolism. Key regulatory enzymes are activated, secretion is stimulated and ion channels opened, facilitating the flux of ions into and out of the cell.

Such a path may seem complex, but some steps are necessary in order to amplify the initial signal. One hormone molecule activates many protein relays, and they are active for long enough for the effector (adenylate cyclase) to catalyse the production of many molecules of cAMP from ATP. These steps amplify the original signal, so that the effect of a single molecule of the hormone is multiplied a million times.

BIOCHEMISTRY IN CONTROL TERMS

It seems that whatever the complexities of the first and second messenger systems, and the need for amplification of signals, the summed output is probably relatively simple. The complexities of control by growth hormone originating from the anterior pituitary, which hangs beneath the brain, of signal amplification, and of distribution throughout the organism to target tissues, ultimately combine to create a simple pairing of positive and negative commands. From the outset, attention was drawn to the observation that control of any process at a minimum requires the involvement of positive and negative signals, capable of stimulating and inhibiting the controlled process. There are refinements to ensure that, whether the signal at any one time is positive or negative, it will be received 'loud and clear'; the complication of any residual and opposite signal is avoided by its suppression using cross-linking (Figure 6.5). This interpretation of the first and second messenger systems apparently delivers a relatively simple outcome, which is corroborated by the simplicity of the model that is capable of reproducing such output data. It can be seen from the control mechanism output of hydroids (Chapters 5 and 8) and yeast (Chapter 12) that the oscillatory output is relatively simple, suggesting only positive and negative controls.

A key point is that the base rate of biosynthesis is the specific growth rate, a rate that is independent of its products, reflecting the extent to which anabolism exceeds catabolism. Of course, the mean specific growth rate is typically positive and proceeds at its goal rate; that is, a rate of biosynthesis optimal for the organism. All other growing processes at higher levels relate to this fundamental rate. As has already been described (Chapter 5), a further step is required to reveal growth control with respect to this baseline. Here the rate is expressed as a ratio of the perturbed and unperturbed cultures to isolate the effect of perturbation, and to observe control behaviour that neutralises the effect of deviating growth rate from its goal setting.

Strictly speaking the control of rates is 'homeorhesis', a term coined by Conrad Waddington to differentiate the control of rates from the control of states. It is the kind of control appropriate for processes, allowing moment-to-moment control, just as travel is regulated by speed (Figure 5.2).

Growth rate control at a constant specific goal rate is important, so that the growth of populations of cells will be coordinated within a tissue and predictable to other tissues. This is the control that regulates biosynthesis as a coordinated process. A goal-specific growth rate (r) is regulated by positive and negative controls (Figure 6.6), and is maintained at a constant rate characteristic of the species.

There is one important feature of rate control that we need to be aware of, as it will feature repeatedly in what follows. Even though

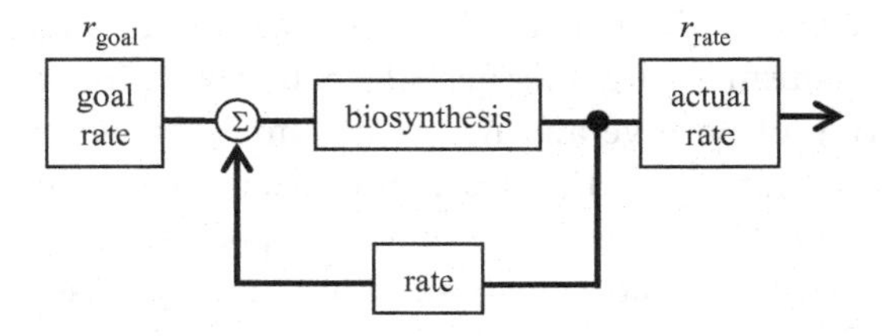

Figure 6.6. Diagram of a minimal growth control model for regulating biosynthesis. The mechanism regulates the specific growth rate (r) for biosynthesis and constitutes what will later be the inner loop of a dual control mechanism.

growth rate is controlled at a constant specific rate, integration of the products of biosynthesis results in an exponential increase in size or number. Because the products of growth also grow, growth accelerates continually. This is the primitive condition of exponential increase (Chapter 2), which must be limited if it is not to threaten the survival of the organism or population. We may have control of rate, but there is a positive feedback in size or cell number, so the system threatens to become unstable unless there is also an intrinsic limit to exponential growth and increase.

The rate-sensitive control mechanism is represented by a single loop feedback mechanism, which is the kind of primitive control mechanism likely to be found in autotrophs. Here it represents the inner loop of a dual loop control mechanism that will acquire an outer loop later that is responsible for limiting the cumulative products of growth (Chapter 9).

Even though a negative feedback loop controls growth rate, it represents a positive feedback loop in terms of size or number, which must be controlled by an additional negative feedback loop to limit size. It seems that a stable unit of control requires these two complementary control loops. The first is inward-looking to regulate the rate of biosynthesis; the second is outward-looking and limits growth in relation to the local requirements or limits for growth. Such limits might relate to the number of cells that are required for a specific tissue (Chapter 12).

In conclusion, control of the intracellular processes involved in biosynthesis is essential, and while there are various separate feedback loops involved, these are apparently synchronised by the perturbation used to reveal control behaviour in the output. We can therefore consider the intracellular part of the system (Figure 6.5) as a single output, resulting in the kind of output seen in the last chapter (Figure 5.5). But despite the constancy of specific growth rates, there is a positive feedback, such that the products of growth tend to increase exponentially. This creates potential instability, as no replicator can sustain exponential growth and its consequences for long.

SUMMARY

The specific growth rate apparently reflects the underlying control of biosynthesis. From an early stage, the properties of the control output and modelling pointed to a simple positive/negative control of growth rate. The complex first and second messenger system, linking hormonal control to the intracellular control of growth, apparently add up to a control mechanism of a simple kind. This control mechanism and its constant specific growth rate deliver a size or number of growth products that increase exponentially. In terms of the products of growth, the constant rate behaves as a positive feedback mechanism, resulting in an exponential increase in the products of growth. This is clearly unsustainable, so it is argued that rate control alone is not adequate to provide overall control of growth of the kind found in animals.

7

The three levels of adaptation

The whole organism can be seen as a coded representation of the environment.

J.Z. Young

Living things without exception are endowed with purpose.

Jacques Monod

KINDS OF ADAPTATION

Habitats fluctuate continually, and sometimes to extremes, making great demands on the organism's capacity to control its internal processes. The primary adaptation is the ability of a camel to control its own temperature, as it is the only large mammal able to survive the freezing nights and baking days of the great deserts. The most important means by which mammals keep cool is by heat loss due to sweating. For human beings, 10–12% water loss may well be fatal, but camels can lose twice as much water without harm, greatly increasing their capacity to tolerate heat. A camel can drink up to a third of its body weight in water, so can sweat copiously, and can survive for 6–8 days without drinking.

Man is a homeotherm, and has a lesser capacity to control internal temperature in extreme heat than a camel, or in extreme cold like a polar bear. Due to the work of Knut Schmidt-Nielsen and others, we know that the camel has thick fur for insulation, which not only keeps the cold out at night, but also dissipates the heat of the sun during the day [1]. To shear a camel of its fur causes its body to heat up during the day, leading to an increase in water loss by as much as 50%. While the polar bear has fur to keep it warm both by day and by night, it loses the insulating properties of its fur on entering the water, and heat loss increases by 20–25 times. Then it relies on a layer of blubber to keep it warm, otherwise it would not be able to survive in the extreme cold of the Arctic Ocean. Such evolutionary adaptations contribute to the overall capacity to keep warm or cool at extreme temperatures. The camel and the polar bear are mammals well adapted to life in the extreme heat of the desert and the cold of the Arctic. Each has evolved in different ways to thrive at temperature extremes, and these adaptations have been made possible by homeothermy and its ancillary adaptations.

There is a secondary level of adaptation superimposed upon those that are inherited. Humans have 2.5 million sweat glands, and even when you might not feel hot, sweating occurs at a rate of 0.8 litres per day. The ability to sweat creates a twentyfold increase in heat loss. In hot conditions, sweating increases to 1 or even 2 litres per hour, and following acclimation in a hot climate, sweating can increase over short periods to 3 litres per hour, or when working, to 10–12 litres a day. Since we do not usually drink such large amounts, let alone sweat that much, the ability to replace lost fluid can become the limiting factor. Acclimation is important here, and the ability acquired over weeks to increase sweating improves our capacity to thermoregulate in the tropics.

Tertiary adaptation involves behavioural traits that add greater resistance to environmental extremes through evolutionary and acquired adaptations. To restore the ability of fur to insulate requires it to trap air to act as an insulator; so on leaving the water, a polar bear, like a dog, shakes itself vigorously. Learned behaviour plays a part in where a polar bear might choose to sleep, preferably in the lee of rocks or a hummock of ice, thereby reducing the wind chill factor. When preparing to give birth to her young, the female polar bear digs a hole in deep snow for her confinement, which protects the cubs from the cold.

To make the chapter title clear at the outset, the structure of what is to follow is summarised in Table 7.1. Primary adaptation is genetic, and tertiary adaptation is familiar as adaptation due to learning, but between the two lies secondary adaptation, which is homeodynamic. Here emphasis is given to cybernetic mechanisms, which are adaptive on two levels; first, such mechanisms resist the effects of external change in relation to pre-set goals in a way that is already apparent from the preceding chapters. Second, the mechanisms are themselves adaptive, or self-adjusting, in response to patterns of external change. The reader will note that from primary through to tertiary, the time

Table 7.1. *Types of adaptation, involving three kinds of information and response systems.*

	Information	Systems	Ancillary adaptations
Primary	Genetic code	Homeodynamic mechanisms	Fur, blubber
Secondary	Cybernetic goals	Adjustments to mechanisms	Number of sweat glands
Tertiary	Cerebral	Memory and intelligence	Learned behaviour

scale of adaptive responses becomes increasingly rapid. The secondary level of adaptation is made possible by the adjustment of goal settings within the mechanism, due to adaptation of the individual and an increased capacity to resist change. Information for tertiary adaptation is stored in the brain and promotes supplementary behaviour that reduces the load on homeodynamic mechanisms. Using memory and intelligence, the organism seeks out optimal habitats and avoids climatic extremes, supplementing primary and secondary adaptation. Each kind of adaptation has been added sequentially and augments the earlier level(s) of adaptation; each also facilitates more rapid responses and adaptation than were previously possible.

A CAPACITY TO TOLERATE LOAD

It was in the 1960s that an American doctor, John H. Frenster, introduced the idea of load tolerance testing in medicine, to measure the capacity of his patients to withstand physiological challenges [2]. At a routine check-up, the doctor will take your pulse and measure blood pressure. If a problem with the heart is suspected, latent disease can be diagnosed more readily if the heart is put under load. It is much more revealing for the heart to be monitored under load, after the patient has been walking or running on an exercise machine. The surprise is that it took so long for load tolerance testing to become established as a diagnostic tool to detect latent disease. Load tolerance tests help to diagnose a patient's susceptibility to 'high output failure', such as a heart attack when running for the proverbial bus, or when the demand on the heart is several times greater than it is at rest. Nowadays, heart rate is often given on the screens of exercise machines in sports centres, so that the efficiency of the heart under load can be monitored by the runner while exercising. In this way, immediate feedback can indicate how far or how fast the runner should choose to run.

The principle of load tolerance can be applied to metabolic processes as well as those relating to physical fitness. A good example is the system for regulating insulin production to control blood glucose levels. In healthy people, the regular consumption of large amounts of glucose increases tolerance to glucose, perhaps thirtyfold, due to increased insulin production, while deprivation soon reduces tolerance for glucose. Diabetes mellitus is characterised by defective glucose utilisation due to a deficiency of insulin, such that the amount of unused glucose in the urine can be used as a measure of insulin deficiency. The glucose tolerance test is designed to elicit maximum insulin production in healthy patients. In those suspected of early insulin deficiency, excess glucose will be found in their blood. By bringing a patient to the point of high output failure, the test detects those patients whose insulin production is adequate for normal loads, but inadequate for higher loads.

It has already been established that disturbance of a controlled process, or sustained load, deviates it from a preferred state or rate, and elicits a dynamic corrective response (Chapter 5). Load provides the perturbation which is essential to elicit the action of a feedback mechanism to restore equilibrium. Here we go further in requiring the control mechanism to be put under load so that it needs to work to counteract a sustained perturbation, indicating its capacity to withstand increased load. To understand this process, we return to the marine hydroid, as this simple system provides us with a model that exhibits control and overload. How much load, provided by a toxic agent, can be neutralised before the capacity to maintain control is exceeded? We use the effect of a toxic inhibitor on the growth process as exemplary of other homeodynamic systems. In this chapter, the focus is the integrated outcome of homeodynamic systems, which relate directly to their contribution to fitness. To make comparisons with the dynamic output, the reader should refer to Chapter 5.

It is first necessary to familiarise ourselves with some basic features of toxicological relationships. The fundamental relationship in toxicology is that between the concentration of a toxic agent and the effect it has on some biological process. A simple kind of system is to use cells of an organism in culture, independently of the organism from which they originated. Such experiments typically produce linear

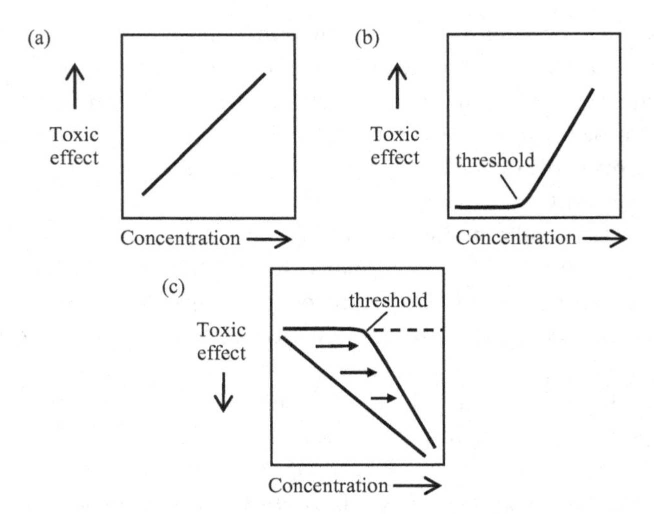

Figure 7.1. Hypothetical concentration–effect curves showing the effects of toxicity: (a) curve showing an effect that is proportional to concentration; (b) curve exhibiting a threshold; (c) the two relationships inverted, following the usual convention. The shift in the curve due to homeodynamic counteraction typically creates a threshold in the relationship.

relationships, indicating that an effect is proportional to the concentration of the toxic agent causing it (Figure 7.1a). Where the plotted line passes through the origin, it is implied that any concentration, however small, will cause some effect. When cells are isolated from organisms and kept in culture, they are removed from the intact growth control system that resists perturbation to tissues. Consequently, in cell cultures, one should not expect to observe the organism's homeodynamic capacity to counter the effects of toxic inhibition, so the effect tends to be linear. In cell cultures, the system is devoid of a cellular-level homeodynamic response to counter toxicity, as this is effectively destroyed by the removal of the cells from their normal context.

In experiments on cultures of isolated cells, using ionising radiation as the toxic agent, such graphs imply that even a single particle track will have a toxic effect and cause damage. This oversimplification has led to the mistaken belief that there is no level of radiation dose so small that it does not have an effect, and so any level of exposure must be harmful. This is called the linear no-threshold (LNT) hypothesis, and is improbable, as organisms are exposed throughout their lives to some degree of background radiation that has little or no harmful effect. Those who accept the LNT hypothesis ignore the fact that DNA repair mechanisms work continually to repair damage as it occurs, and deny the existence of other adaptive responses that influence toxicological effects (see Chapter 10).

In metazoans no effect is observed until a certain threshold dose or concentration is reached, and toxicity then increases with increasing concentration above the threshold (Figure 7.1b). In experiments that relate toxic agents to their effects, toxicologists expect to see such a concentration–effect relationship, which is typical of an intact organism. There is a range of concentrations at lower levels that have no toxicity, because inhibition is counteracted by adaptive responses due to homeodynamic control mechanisms [3]. Above a threshold of effect concentration, inhibition increases with increasing concentration. This relationship is considered the central paradigm of toxicology, as it is fundamental to understanding toxicity, regardless of the agent or organism.

While a concentration–effect relationship with a distinct threshold is typical of toxicological relationships, it is less widely accepted that the range of concentrations which have no effect is the consequence of adaptive responses that negate the effects at low levels.

The relationships shown in Figure 7.1 may seem unfamiliar to some, so the curves are brought together and inverted in the way that toxicologists typically present dose–response (or concentration–effect) relationships (Figure 7.1c). It is then apparent that the difference between the straight line and the curve with a threshold is the result of adaptive homeodynamic responses to toxicity (Figure 7.1a, b). Their response to the effects of toxicity is to neutralise the biological effect, shifting toxicity to a higher concentration, indicated by the arrows in Figure 7.1c. Where the toxic effect is on growth rate, for example, then by neutralising the inhibitory effects of lower concentrations, the onset of inhibition is deferred to a higher concentration. Depending on how toxic the agent is, the growth rate falls away steeply with increasing concentration, until the process is arrested. The first graph (Figure 7.1a) indicates a greater toxicity than the second (Figure 7.1b), due to the absence of the homeodynamic capacity, which has a capacity to neutralise the effect at lower concentrations.

This interpretation links the Maia hypothesis to the resistance of the growth control mechanism to perturbation, in the form of toxic inhibition. Inhibition is neutralised by a counter-response at low concentrations, and does not become inhibitory until the capacity of the control mechanisms is exceeded. It is necessary for the reader to remember, from the earlier analysis of growth control in hydroids and yeast (Chapter 5), that the response of a homeodynamic control mechanism to inhibition is stimulatory. Furthermore, the consequence of a stimulatory reaction in response to some inhibitory agent is the neutralisation of one by the other. As the response occurs, it is immediately taken up in neutralising the inhibition, and so both response and inhibition disappear. The response is not seen in isolation, but only by its effect in counteracting inhibition. The problem is that a counter-response cannot be induced without a load for it to work against, so the response is inevitably hidden within the final outcome.

There is one exceptional instance where the naked response is briefly revealed in a striking way. When a system is perturbed and then allowed to stabilise, the opposing factors of inhibition and a neutralising response create an equilibrium of antagonistic forces. But if the inhibitor is then suddenly removed, the naked counter-response can be briefly observed as a relaxation stimulation, which lasts for as long as the lag in the control mechanism takes to correct it. What is important here is that the peak of the stimulation indicates the magnitude of the counter-response that had been necessary to achieve the equilibrium. We can then see that the relaxation stimulation, and therefore the counter-response, increases in proportion to the concentration of the inhibitor (Figure 11.5). Relaxation stimulation, and the interpretation it provides for catch-up growth, is described in Chapter 11.

Such data enable us to construct half of the response curve, which, as the response increases to neutralise the increasing inhibition, rises in proportion to the concentration of inhibitor (Figure 7.2a). Despite increasing concentrations, control is maintained, and the controlled process is kept at the same level as in those organisms not exposed to inhibitor. The gain of the control mechanism increases in relation to the load imposed, such that, with increasing concentration, the load continues to be neutralised. The outcome is that, throughout this range, the controlled growth rate is unaffected by inhibition. The control mechanism must work harder to neutralise the load as it increases, until it reaches a maximum and saturates the control mechanism (see Chapter 5). Once the adaptive response reaches its maximum and becomes overloaded, inhibition causes the response to decline rapidly as concentration increases, giving the other half of the response curve (Figure 7.2a), which is the inverse of the effect curve over the same concentrations (Figure 7.2b). Notice that the scale of the effect curve is inverted.

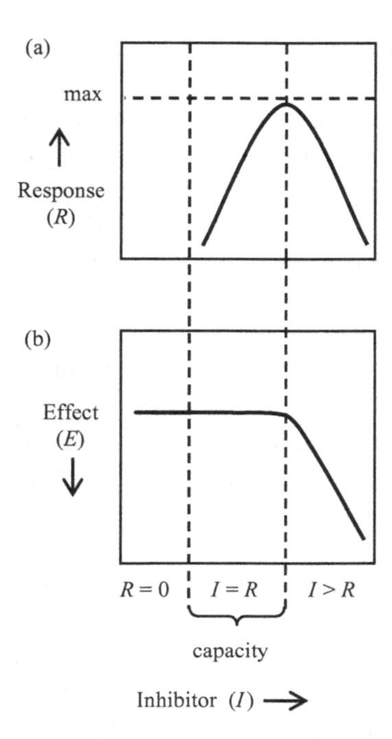

Figure 7.2. Hypothetical relationships between inhibitor (I), response (R) and effect (E): (a) the response curve is typically unseen (see text for further detail). R_{max} indicates the peak of the adaptive response before high output failure causes a decline in response; (b) the outcome of the inhibition minus the counter-response. From Stebbing [3], with permission.

We now see the response curve in its entirety as an inverted V, with the inversion point determined by the maximum counter-response, which itself determines the threshold of inhibition and the onset of an effect. The relationship between inhibitor, response and effect is interdependent and coherent. The response is stimulated by the onset of inhibition, until it can be increased no more, whereupon the response becomes progressively poisoned and overwhelmed by further inhibition as toxicity increases.

We now consider the two curves as an explanation of the relationship between a response (R) to the inhibition, the ultimate effect of inhibition (E) and the effect of the inhibitor (I). At the lower concentrations the inhibitor causes no response or effect ($R = 0$). At sub-threshold levels that elicit a response, it increases with concentration, neutralising the effect of the inhibitor ($R = I$), which consequently causes no effect ($E = 0$). When the response reaches its maximum (R_{max}), the mechanism experiences high output failure, and no further increase in inhibitor elicits any further increase in response. The limit of the response is reached, and then exceeded ($I > R$), such that the inhibitory effect increases as the response decays, and the inhibitor alone determines the outcome ($E = I$). The critical middle phase defines the capacity of the control mechanism, which extends from the lowest concentration that elicits a response to the concentration at which the counter-response is maximal (R_{max}). Thus the capacity of the mechanism to resist inhibition can be defined in terms of a range of inhibitor concentrations. Within this range, the effect is equal to the inhibition minus the counter-response ($E = I - R$). The two linked curves show how this comes about (Figure 7.2a,b), due to overload of homeodynamic mechanisms at the threshold concentration.

Most toxicology relates to the 'dose–response' curve, which typically indicates that an effect does not occur until a threshold is reached. But where there is a homeodynamic response to toxic inhibition, the onset of inhibition is the point at which the homeodynamic response fails due to overload.

A similar, but apparently unrelated, idea was put forward by R.M. Yerkes and J.D. Dodson in a paper published in 1908 [4]. It was entitled *The relation of strength of stimulus to rapidity of habit-formation*, and involved behavioural experiments with mice. The Yerkes–Dodson Law is an empirical relationship between arousal and performance. Performance increases with cognitive (i.e. knowing) arousal up to a certain point, until overstimulation occurs, inducing stress, and performance decreases. This means there is an optimal level of arousal required for maximum performance. The Yerkes–Dodson Law has become a widely respected generalisation in biopsychology. It emphasises that the stimulation of performance is arousal which, beyond some presumed optimum, becomes stressful and inhibits performance. If one takes the more strict interpretations of 'arousal' and 'performance', and instead uses terms such as 'load' and 'response', or 'input' and 'output', then one is looking at an adaptive system.

One can argue that any adaptive response to noxious agents will be stimulated on exposure to the agent, and will reach a maximum, before becoming overwhelmed with increasing concentrations that become toxic and finally lethal. Thus all adaptive responses to noxious or toxic agents may be expected to take the form of the V-shaped curve.

W. ROSS ASHBY: AN EARLY PIONEER IN CYBERNETICS

For years I had not looked at W. Ross Ashby's work in depth, despite being aware of his important contribution to cybernetics. I was not alone in finding the unfamiliarity of set theory an obstacle to understanding. The consequence was that it was a long time after formulating a conceptual model linking inhibition and response to the overall effect (Figure 7.2) that I realised that Ashby had provided the theoretical basis for the central idea of homeodynamic counteraction 50 years earlier.

W. Ross Ashby (1903–1972) was trained in medicine and psychiatry and, after service in the Royal Medical Corps during World War II, became director of Barnwood House Hospital, a private mental hospital near Bristol. It is said that, for peace and quiet, he sometimes worked in a padded cell. During the 12 years he spent there, he developed his ideas on brain activity, using a simple analogue simulator, which he called the Homeostat. Ashby worked alone in the early years, and his thinking was so far ahead of others in the field that the true worth of his research was not acknowledged until much later.

Operating in a similar way to other scientists in the UK during the post-war era, he built the Homeostat on his kitchen table from war-surplus electronic components. It was designed to demonstrate adaptive behaviour, as a vehicle for experiment, and helped him to begin to provide a mechanistic explanation for adaptation. During this period, he wrote two books. The first was *Design for a Brain: The Origin of Adaptive Behaviour* (1952), which includes a description of the Homeostat and the behaviour it was able to demonstrate. Later he wrote *An Introduction to Cybernetics* (1957), which had enormous influence on the early development of the field.

In the USA, Ashby was seen as one of the founders of cybernetics. In an appreciation of Ashby's work, George J. Klir wrote that a survey conducted in 1977 found that 'Ashby was at that time by far the most influential person in the whole systems movement'. This dated back to an invitation to attend the 1952 Macy Conference. His growing reputation in the USA led to an appointment in the Department of Electronic and Electrical Engineering at the University of Illinois. Ashby's analysis of information theory, and in particular his demonstration of the importance of Claude Shannon's work on communication to cybernetics, is seen as a major contribution. He later said that this was the most productive period of his life, but despite his reputation in the USA, his contribution remained undervalued in Britain and elsewhere.

Ashby's Law of Requisite Variety is the concept of importance here. It refers to the resistance of disturbance by a regulator. As we shall see, the law is important to the general theory of regulation, yet has been largely ignored. In recent years it has attracted increasing interest, but even now is sometimes misunderstood. It is therefore worth trying to recognise the problems with its acceptance, before explaining how the law relates to homeodynamic capacity. Set theory, although accessible, is more appropriate for dealing with the logic of ordinary experience, and is much less suited to complex mathematical concepts.

In his Law of Requisite Variety, it was Ashby's use of the word 'variety' that introduced ambiguity [5]. It seemed that he was referring to the variety of different kinds of responses to disturbance. He therefore continued to use the term 'variety' as a collective term for disturbances, responses and outcomes, set in a matrix. This tends to obscure the fact that he also meant the continuous variation of each of these factors. He wrote:

> if the varieties of D (disturbance), R (response) and the actual outcomes are respectively V_d, V_r and V_o, then the minimal value of V_o is $V_d - V_r$. If now V_d is given, V_o's minimum can be lessened *only by a corresponding increase in* V_r. [italics in original].

It now becomes clear that the *capacity* to regulate is what Ashby meant by 'variety', and the outcome (o) is the sum of the disturbance (d) minus the response (r), all terms being given as subscripts of V, meaning variety. For a given disturbance, the outcome can only be reduced by increasing the response that counters it. When Ashby wrote 'This is the law of requisite variety', he was referring to sets, meaning different levels of disturbances, rather than a continuum of a single factor. Set theory was an inappropriate way to describe the dynamic behaviour of control mechanisms, as subsequent work has shown. Its use explains the ambiguity of what he meant by 'variety'.

The significance of the misunderstanding is clear. An earlier definition, published in 1956, said that 'only variety in R (response) can force down the variety due to D (disturbance); variety can destroy variety'. By the second edition of *Design for a Brain*, published in 1960, the problem had been recognised and much of the text had been rewritten. Ashby writes in the later and revised edition of *Design for a Brain*,

> all the processes of regulation are dominated by the law of requisite variety. This law says that if a certain *quantity* of disturbance is prevented by a regulator from reaching some essential variables, then that regulator must be capable of exerting at least that quantity of selection.

The overall effect or outcome (*o*) of Ashby's Law is the product of the disturbance (*d*) minus the response (*r*) to it, or $o = d - r$. This is the same as my expression $E = I - R$. Both can be better expressed in terms of the concentration (*C*) of some inhibitor, such that $C_{\text{toxic effect}} = C_{\text{inhibition}} - C_{\text{response}}$. So the ultimate toxic effect is the inhibitory effect minus the capacity of the organism to neutralise the toxic inhibition.

There remain misunderstandings as to the meaning of the law, even during the last decade. The most important of these is the expected misconception that 'variety' means the number of different *kinds* of regulatory responses, rather than the *capacity* of a single system. We now also recognise that the Law of Requisite Variety has obscured the concept that, as Ashby claimed, 'is of fundamental importance to the general theory of regulation'. His law captures the essence of control as the capacity to neutralise disturbance. It is important to recognise that all homeodynamic mechanisms have a capacity to neutralise load that can be defined in terms of levels of load. Their capacity is marked by the lowest level at which a control response is initiated, extending to the level at which overload occurs (Figure 7.2b).

It should be noted that the response in normal circumstances is not apparent, because it is only elicited in response to some perturbation. When an inhibitor and a neutralising response come together, the inhibition and the stimulatory response are dissipated in neutralising one another.

PRACTICAL IMPLICATIONS OF ASHBY'S LAW

We can now look at some implications of Ashby's Law. First it provides an interpretation of the toxicologists' concentration–effect curve[1] as it relates to sub-lethal effects, but here to homeodynamically controlled processes. Homeodynamic mechanisms are so ubiquitous in the physiology of organisms that any sub-lethal response may well be determined by such mechanisms. Key to this is a capacity to control a process, maintaining constancy and equilibrium by countering any deviation of the controlled process from its pre-set goal.

[1] The use of the 'concentration–effect' curve is a deliberate renaming of the toxicologists 'dose–response' curve. 'Dose' is an inappropriate term for those biological systems in aquatic culture, whether a hydroid, fish or cell suspension. 'Dose' relates more specifically to mammalian toxicology where animals are dosed by mouth. 'Response' is also inappropriate, since it actually refers to the organism's response to inhibition, rather than the outcome.

The capacity of the control mechanism can be estimated in terms of the range of concentrations of some inhibitor over which it maintains control. The more capacity that is committed to counteracting inhibition, the less remains to counter inhibition, and the closer towards a threshold the system moves. This is the point at which some further small addition to the total load exceeds the threshold and inhibition occurs. To use the original form of this idiom, 'It is the last feather that breaks the horse's back'.

The capacity to neutralise inhibition confers fitness upon an organism, as it is thereby able to offer resistance to the effects of any environmental agents that inhibit homeodynamic processes. Using the hydroid, we can quantify the capacity of the growth control mechanism in terms of the concentrations of inhibitor over which control can be maintained (Figure 7.2). For example, copper induces a response at 1 μg per litre and become overloaded at 10 μg per litre, so control is effectively maintained over a tenfold range of concentrations. These concentrations define the range over which copper is effectively neutralised, so the capacity of the organism to resist copper is an order of magnitude from 1 to 10 μg ℓ^{-1}. It was interesting to discover that this capacity was adequate to resist the range of variation of copper in the hydroids' estuarine habitat. The toxicity of metals in fresh water is greater than that in salty water, and the capacity of the freshwater hydra to resist copper ranges from 0.5 to 5.0 μg ℓ^{-1}. Different metals and other agents vary in their toxicity so, for each, these pairs of concentrations are likely to be different.

Ashby's Law helps interpret the combined effects of mixtures of toxic agents; the more loaded by one inhibitor, the less capacity remains to deal with another, and the more vulnerable the organism becomes to any further load. Then finally, as the threshold is approached, the smallest additional load precipitates inhibition. For a mixture of toxic inhibitors, their combined inhibition will be greater than the sum of their effects when each is tested separately. This is because in the first case the capacity to neutralise inhibition is divided between them, but in the second each is tested alone and the whole capacity to neutralise inhibition is directed at only one inhibitor. Consequently, for each inhibitor the toxicity of mixtures will tend to appear greater in mixtures than individually; their effects become greater than simply additive. This suggests that synergistic effects are to be expected whenever homeodynamic responses are involved, due to the partitioning of the capacity to counteract between more than one toxic inhibitor.

As the control mechanism responds to inhibition as a deviation of the process from some preferred condition, it is to be expected that any cause of inhibition will elicit a similar response. The counter-response due to growth control mechanisms has been shown to be the same for a wide range of agents, such as salinity for a marine organism, temperature change in yeast, as well as toxic agents. We can assume therefore that any inhibitor will be neutralised in the same way, until the control mechanism is saturated and becomes overloaded.

The generality of homeodynamic systems is important, in that their response is to the deviation of the processes they control, and not to the specific agent or toxin causing the deviation. Such generality of response is a powerful evolutionary adaptation, in that any organism so endowed has the means of responding in this way to neutralise perturbation, irrespective of its cause. In effect, the system is pre-adapted to counter the effects of any inhibition, including toxic agents synthesised by mankind.

SECONDARY ADAPTATION

Discussion of Ashby's Law and its application to homeodynamic mechanisms relates to primary adaptations, which are coded for within the genome and are lifelong. During evolution, animals have come to grow larger and live longer than before; so individuals live with an unchanging genotype, and there is only primary adaptation to change for the individual. However, a range of secondary adaptations have evolved – mechanisms that provide adaptation for an individual during its lifetime. Secondary adaptation is concerned with short-term adaptation, and here the focus is primarily on adjustments to homeodynamic mechanisms. Such adaptation involves acquiring greater capacity through the adjustment of goal settings in relation to recent changes or trends. Such adaptations are phenotypic and specific to the individual organism, and so vary throughout populations of a species in response to the pressures upon them to adapt to environmental change.

'Health' is a concept that we use daily, often without being quite sure what we mean by it. It can mean freedom from disease, but is also a term for well-being, and for the ability to adapt to environmental change, and the stresses and strains that challenge all organisms. Health can be measured in terms of the capacity of a person to resist physiological load, as in load tolerance testing, or in the capacity of a hydroid to resist the effects of a toxic inhibitor.

Clearly secondary adaptation can take many forms, whether to tolerate a high sugar intake, to increase athletic fitness or even to prepare for a beer-drinking competition. The homeodynamic paradigm is often expressed as: 'processes hypertrophy with use and atrophy without', referring to the response by enlargement or shrinkage of tissues and organs. The point is that homeodynamic systems are constantly adjusting to the most recent loading due to environmental change. The capacity of systems increases in response to sustained demand and decreases with reduced demand. Regular exercise increases the ability to take more exercise, while the absence of exercise reduces it. Loss in fitness of one function studied by physiologists occurred more rapidly due to inactivity than it was gained by training. So we accept the training paradigm with respect to sport, and we have already considered the way in which fitness leads to the extreme development of the heart, for example, in endurance in human and equine athletes (Chapter 4). 'Training' also occurs as one of the consequences of disease. In people with emphysema, which causes congestion of the lungs, enlargement of the heart comes about due to its additional work load as the diseased lungs become less efficient. Technology has also come up with self-adapting control mechanisms that are designed to sense ambient conditions and adjust themselves to respond more effectively to the new conditions; for example, the automatic adjustment of fuel consumption in jet engines for maximum efficiency with increased altitude. With height, the air gets thinner and offers less resistance but provides less oxygen so, with a change in height, adjustments are made automatically and continuously to the fuel flow in order to maximise efficiency.

A return to toxicology will show that 'training' is a term that relates to negative as well as positive effects. Some interesting examples of secondary adaptations relate to adaptation to potentially toxic substances. We become aware of our own adaptation to toxic agents through training, by the use of caffeine, nicotine or alcohol. First experiences of smoking are sometimes enough to put the young off, but some persist with the unpalatable. In time we not only tolerate the taste, but learn to enjoy the sensations that smoking gives. Various processes are involved, including suppression of the choking response and adaptation to the toxins in the smoke, as well as to the intake of nicotine. Nicotine is rated as highly toxic; in fact, eating the contents of a packet of cigarettes could be lethal. Yet, in low doses, its uptake is part of the daily lives of millions. The physiological effects of nicotine are complex, with various effects, but essentially it is a toxin that stimulates the nervous system, having relaxing and stimulatory effects at different doses.

Caffeine is also a stimulant of the nervous system that increases alertness, staving off fatigue when sustained concentration is required. It is also known to increase physical work output by increasing the mobilisation of glucose from the liver. As with nicotine, tolerance is acquired over time, and the effect of caffeine depends upon the amount consumed. Regular users with high intakes are more responsive to its effect if they abstain from caffeine for a few days. Similarly with alcohol intake, there is adaptation, and regular intakes of alcohol increase the capacity to metabolise it. The point here is that with adaptation to high intakes, whether due to tolerance or improved excretion, habitual users must consume greater amounts in order to experience the same effects, due to the adaptation of cells and tissues to metabolise the drug, such that addiction increases the craving.

'Social toxins' may initially be distasteful, but this is often followed by tolerance, dependence and addiction. In the case of caffeine, nicotine and alcohol, we learn that they have a useful social effect in helping us to keep awake, as a stimulant focusing concentration or by having a relaxing effect but, in the long term, conditions that damage health await heavy users.

Mithradates the Great (d. 63 BC) was the King of Pontus, on the southern shores of the Black Sea. He trusted no-one and feared assassination by poisoning, so he saturated his body with a cocktail of poisons, and therefore believed he was immune. A.E. Housman captured his efforts in verse:

> He gathered all that springs to birth
> From the many-venomed earth;
> First a little, thence to more,
> He sampled all her killing store.

When his own troops, led by his son, revolted against him, he tried to commit suicide by taking poison, but – ironically – this failed and he had to order a Gallic mercenary to kill him.

Similarly, we know from experience that a medicinal drug taken over months or years can become less effective. Long-term use and acquired tolerance means that progressively larger doses are required to achieve the same effectiveness. The potency of morphine in relieving pain becomes less with repeated injections in the presence of intense pain. Similarly, heroin gradually becomes less effective in achieving the same heightened state of euphoria. Heroin was originally advertised and marketed as a non-addictive alternative to morphine and a safer, less toxic, alternative to codeine. Unfortunately, and paradoxically, it was later found to be four times more addictive than morphine, but not until it had been commercially available for more than 20 years. Laudanum, a solution of opium and alcohol, was created by the Swiss physician Paracelsus in the sixteenth century, who pioneered the use of chemicals and minerals in medicine, in place of traditional herbal remedies. Laudanum was a fashionable medication for Victorian authors seeking inspiration, but they risked the reverse effect, and the possibility of addiction. Heroin taken with alcohol is a potent combination, often taken by drug addicts. They are metabolised in the same way, each competing for the capacity of the same biochemical pathways to metabolise them. Each makes the other more effective, due to overloading the capacity to neutralise toxicity by the effect of the combination.

Whether used for recreational or medicinal purposes, the vicious cycle of addiction towards overdose is accelerated by the body's remarkable capacity to adapt, or be trained, to tolerate the uptake of toxic chemicals. In response to repeated doses of the drug, the capacity of biochemical systems grows in order to metabolise greater amounts. As tolerance and dependence take hold, metabolic pathways to metabolise the drug take over the normal functioning of cells, as massive adaptation of cells becomes necessary. Their purpose is subordinated to drug metabolism. So the cycle of adaptation and dependence deepens, with progressive increases in dose required to recreate the same euphoric state.

Cravings of need are best understood in the context of the transformation of cell metabolism that accompanies addiction, as the role of cells and then the addict's metabolism become subverted to the drug. Withdrawal is difficult because cell metabolism not only depends upon the drug, but its very purpose has been taken over by the need to break it down. Inevitably a major re-adaptation is necessary before normality can be restored, explaining the difficulty of withdrawal from dependency. Symptoms may include diarrhoea, vomiting, dilated pupils, loss of appetite, tremors and gooseflesh – which explains the origin of the expression 'cold turkey'.

A MECHANISM FOR ACQUIRED TOLERANCE

The hydroid model provides a simple system to understand how homeodynamic mechanisms may account for adaptation to toxic

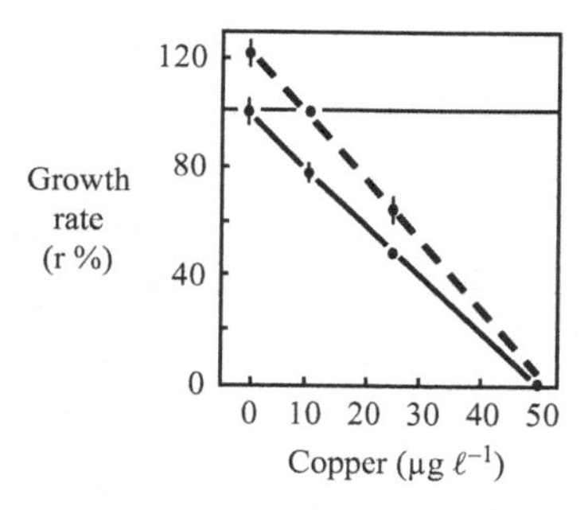

Figure 7.3. The effects of pre-exposure to a toxic inhibitor (copper) on the hydroid *Laomedea*. Colonies indicated by the *dashed line* had been pre-exposed to 10 μg ℓ^{-1} for 3 weeks and are compared with others that had not been pre-exposed (*solid line*). Redrawn from Stebbing [3].

agents [3]. 'Acquired tolerance' is not really the right phrase, since what is observed in the drug addict, and the hydroid, is active adaptation. It has two facets. The first relates to the biochemical responses that metabolise, or otherwise detoxify, poisonous substances. Organic toxins tend to be metabolised and broken down into simpler water-soluble molecules that can be easily excreted, but metals are elements that cannot be further degraded. These are tied up by metal-binding proteins which render them non-toxic (see Chapter 4). The second kind of adaptation relates to the adaptation of homeodynamic processes, increasing their capacity by self-adjustment in response to sustained exposure to toxicity.

Adaptation to toxic metals can be observed by pre-exposing hydroids to a toxic metal for a period of time (Figure 7.3). In this case copper was used, but the response to other toxic metals would be much the same, as the adaptation of the control mechanism is to inhibition, and not to the specific properties of the toxic agent. The lowest concentration of copper that inhibits the growth rate of hydroids is 14 μg per litre, so a lower concentration of 10 μg per litre was used to train them by pre-exposure, so as not to poison the organism which we expect to show adaptation. Here the hydroids were exposed to copper for 3 weeks. If we compare hydroids that have been pre-exposed with those that have not, we see that the pre-exposed hydroids become much less susceptible (Figure 7.3). That is to say, following pre-exposure, it takes more copper to have the same effect or, conversely, at the same concentration there is less inhibition. They have become more tolerant; in fact, the benefit of pre-exposure at one concentration is sufficient to confer resistance that extends across a wide range of concentrations.

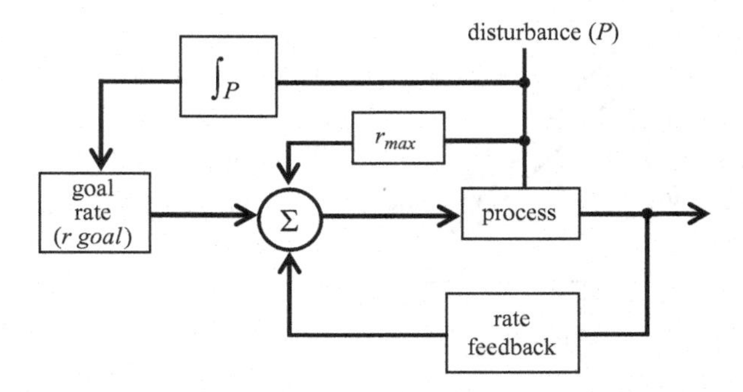

Figure 7.4. Diagram of a growth control mechanism incorporating two feedforward loops that account for adaptation of two kinds. It can be deduced that the inner feedforward loop is responsible for the maximal overcorrection in response (r_{max}) to the initial inhibition. The outer feedforward loop accounts for adjustments in the goal rate setting (*r goal*) that are responsible for acquired resistance to inhibition.

How does one explain secondary adaptation from a cybernetic perspective? It has already been mentioned that, whether a technological or a biological control mechanism, the underlying system is likely to be fundamentally the same. Essentially these are systems that have an ability to adjust their own goal settings in relation to external loading. After pre-exposure to copper for three weeks, adaptation is so complete that those colonies in the experiment exposed to 10 µg ℓ^{-1} grew at exactly the same rate as those colonies not pre-exposed to copper. Adaptation is so complete that the pre-exposure level has no inhibitory effect.

How has this come about? It seems that acclimation to the inhibitor has occurred due to self-adjustment of the goal setting to a higher rate than before, raising the baseline from which a normal response arises. The response to neutralise inhibition is necessarily stimulatory, so any increase in the pre-set growth rate has the effect of increasing the capacity to resist toxic inhibition. This is sufficient to neutralise the pre-exposure level, so that the colony grows at its previous normal rate (Figure 7.4). Without pre-exposure, the hydroids exposed to 10 µg ℓ^{-1} were inhibited by 20%. With pre-exposure the adjustment neutralised the inhibition so completely that the hydroids grew at the same rate as those not exposed to copper. But pre-exposed hydroids then exposed to zero copper in the experiment were stimulated by 20%. The reduction in load had caused an equivalent relaxation stimulation. The adjustment is revealed for what it is: an adjustment in the pre-set growth rate that precisely neutralises the effect of the inhibitor.

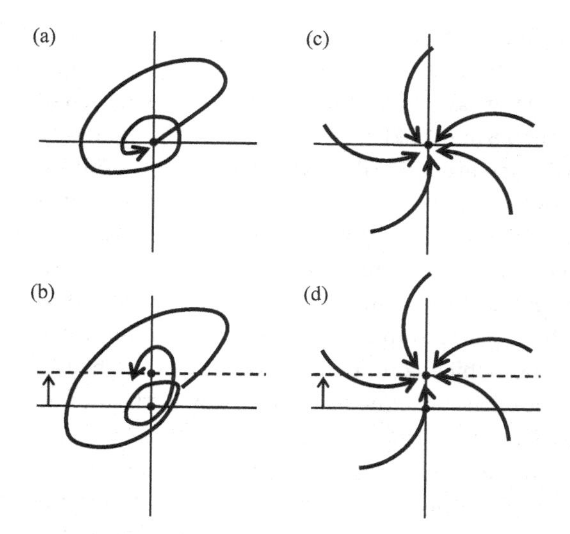

Figure 7.5. Adaptation of control behaviour in phase space following perturbation. For clarity, the magnitude of adaptation is accentuated. Diagrams (a) and (b) indicate typical results (see Figure 5.8) showing equilibration (a), and equilibration following adjustment of the goal setting (b); diagrams (c) and (d) indicate attractor fields. Diagrams (a) and (c) show attractors the intercept; (b) and (d) show attractors to that have moved upward with adjustment of the goal setting indicated by the vertical arrows.

The adaptive properties of feedforward control have already been alluded to (Chapter 5). Such control is of two kinds (Figure 7.4) and is generally associated with more rapid responses to disturbance achieved by shortening the delay in the response time. We can identify feedforward control of two kinds: the first is that which causes a maximum counter-response irrespective of load. This response causes the spectacular over-corrections in hydroids when they first detect toxic inhibition (Figure 5.5). It is the second feedforward loop that interests us here, which feeds forward information that is some integral of disturbance, which causes an increase in the goal setting, and effectively increases the stimulatory counter-response in relation to toxic inhibition.

A better way of depicting such adaptive behaviour is to use phase space diagrams. The pre-set rate is indicated by the intercept on axes giving growth rate and rate of change of growth rate (Figure 7.5). The intercept is the attractor and the goal setting of the control mechanism. Following perturbation, the closer the system approaches to the attractor, the stronger is the draw for equilibrium to be restored. The line of attraction indicates how, within its field, the controlled process is drawn to the attractor (Figure 7.5c). The course of the controlled process to the attractor is indicated in Figure 7.5a, when behaving as hydroids do (Figures 5.8 and 11.2). With an upward shift in the attractor to a level that will increase its goal growth rate (Figure 7.5d), the process is shown to stabilise at a new and higher level (Figure 7.5b).

This application of some integral of load causing adjustment of the goal setting provides a mechanism for adaptation that confers greater resistance to inhibition by increasing the goal growth rate. In effect, the adjustment has the effect of amplifying resistance to inhibition by raising the baseline from which counteraction occurs. In the hydroid, the upward adjustment of the baseline about which growth rates oscillate amplifies its capacity to resist future challenges by 10–20%.

We will return to such adaptation when explaining hormesis (Chapter 11). It will be appreciated that any increase in growth rate, such as would be caused by this adaptation, must also increase growth and cause hormesis. A more detailed examination of hormesis will provide further evidence for such adaptation, as acquired tolerance and hormesis are two facets of the same adaptive process.

It is important to recognise that acquired tolerance to toxic loads is a form of memory (Figure 7.5). Sustained exposure to such agents creates a memory of past exposure, revealed as a response involving goal adjustment. The memory is adaptive in that later exposure to the same agent is more effectively neutralised by the supplemented response. The adjustment to the goal growth rate is not immediately lost, as we shall see in later chapters on hormesis (Chapters 10 and 11). The memory of prior exposure to a toxic inhibitor is retained initially, to be lost slowly over time, unless reinforced by further exposure to toxic inhibition. It is interesting to add that such behaviour is observed here in coelenterates, with the simple nervous system of a distributed nerve net, found also in echinoderms and hemichordates, from which all more complex systems evolved. The fundamental properties of biological systems are sometimes more accessible when using simpler biological models.

SPEED OF ADAPTATION

The three kinds of homeodynamic adaptation relate to three types of adaptation that apparently evolved sequentially; they depend upon three different kinds of information, and are used in different systems (Table 7.1). Information is acquired, stored and renewed on different time scales and by different means. The three levels of information coexist in animals, including humans, to this day. During evolution, the effects of each adaptive capability have been cumulative, supplementing previous functions with new ones, while storing information in ever more accessible and flexible ways.

During evolution the amount of information vested in organisms has increased, as they became more complex. Much of that information relates to coding for the organism and its development, but much is used continuously in relation to life and the organism's habitat. It constitutes a template of the environment (Chapter 15) in which the animal lives.

Information for *primary adaptation* is genetic, and codes for the control mechanism itself. Information is stored as goal settings, which are the preferenda for each process. Initially, at least, they are genetically predetermined states or rates, the 'self' to which the feedback mechanism repeatedly refers, in these self-referencing systems. Goal settings are relatively few and may be one or two for each control mechanism, but the information vested in them is referred to continuously. The settings of such goals as optima are the purpose of this mechanism and, collectively, homeodynamic goals set the physiological purpose of the organism.

Secondary adaptation of homeodynamic mechanisms is represented by the adaptive adjustment of goal settings to better provide levels of control tuned to the kinds of challenges the environment presents, increasing the organism's capacity to withstand the effect of some change threatening the stability of internal processes. Adjustable goal settings are a simple form of memory, in that some information of past exposure to inhibition is integrated over time and then used to reset goal settings, adaptations that we have already considered in some detail. This short-term memory can be termed 'cybernetic', since it continuously modulates the set points of homeodynamic mechanisms. This information is extra-genetic, and is lost when the organism dies, relating only to the life and experience of the individual.

Tertiary adaptation provides information that is stored in the cerebral cortex of more complex organisms, and informs the organism in a way that guides its behaviour to protect homeodynamic control from overload, for example by avoiding extreme conditions. At this level, the turnover of information is adaptive to life in the present time; it is continually amended in relation to changing circumstances. Memory of this kind is used to avoid environmental extremes that might overload the organism's capacity to maintain control of its physiology.

Each kind of information has been added sequentially during evolution, each contributing to the quality of homeodynamic control in relation to external change. The amount of information available to organisms has increased during evolution, and their fitness is related to the amount of information at their command that relates to their life and habitat.

THE SHORTENING OF RESPONSE TIMES

Each kind of adaptation is accessed on different time scales, in that genetic memory is only changed with each generation, goal settings are adjusted continually but retrospectively, while cerebral memory is turned over continually. Each kind of information was added sequentially during evolution, supplementing that previously in place.

We are aware that, throughout evolution, functionality and complexity increase steadily, and along with this we see a trend to reduce the response time of homeodynamic adaptation that has required some of that complexity. Some have referred to the delay in feedback as a 'design flaw'. The yeast control mechanism is crude and the consequences of the initial dead time delay in responding to inhibition take some time to overcome (see Figure 12.2). The multicellular hydroid has apparently evolved a feedforward loop (Figure 7.4) that ensures more rapid reactions by providing an initial counter-response to any inhibition that is immediate and maximal (Figure 5.5). Such adaptations reduce reaction times, and minimise the time delays typical of all feedback mechanisms. Clearly there are great advantages in increasing the speed of reaction and adaptation, which contribute to fitness in a variety of contexts from survival to competition.

The mammalian body carries with it vestiges of its evolutionary past, particularly with respect to the central nervous system, where new levels of complexity are superimposed upon the ancestral vertebrate nervous system. The autonomic nervous system is the primitive mammalian nervous system, controlled by the hypothalamus, which has two elements: the *sympathetic* nervous system, which stimulates homeodynamic processes, and the *parasympathetic* system, which inhibits the same processes. This dual system provides nervous control of the internal organs. We see here again the same minimal elements of control provided by opposing positive and negative signals (Chapter 6). Such processes remain in an earlier evolutionary state, where control mechanisms operate in arrears of actuality and stabilise following perturbation in an oscillatory manner not unlike that of hydroids

(Figure 5.5). The control of our internal organs is by the autonomic nervous system, implying that more rapid control would not contribute significantly to fitness. Such systems observe themselves retrospectively, responding to errors that relate to the past, such that by the time some corrective action has been taken, the moment in time to which it applied has passed. Such delays do not matter with respect to many processes, such as the control of digestion or, more obviously, of calcification (Figure 4.3).

During evolution, the central nervous system was superimposed upon the autonomic nervous system. Much faster coordination and response times were required for motility in vertebrates, while the slower and more ancient autonomic nervous system continued to be responsible for the control of homeodynamic systems.

In active animals, the cerebellum, or small brain of aeons ago, became the cerebrum, with a development that relates directly to the motility of animals. So it is of a moderate size in fish, a much smaller part of the brain in reptilians, and a much larger part of the bird brain. This involves the coordination of limb positions with the stimulation of muscles to effect movement. Coordination of the movement of complex organisms requires that all the nerves and muscles function together at the same speed. The cerebrum integrates sensory information from the exterior with motor functions, bringing contextual sensation together with the control of motility. The speed of animals requires the development of the central nervous system for the rapid conduction of nerve impulses.

This is most obvious in the success of the predator, or the survival of prey, whether on land, sea or air. Speed is crucial in competition too, because to be first, an animal must be faster. Given such selective pressures, rapid responses to stimuli and speed of movement became a priority. The central nervous system has myelinated nerves that pass impulses 150 times more rapidly than those of the autonomic nervous system, which are unmyelinated. The fast nerves of the central nervous system mean that its responses occur so rapidly that sensation and reaction fall almost within the same time frame.

Mobility requires a nervous system that operates quickly, and the speed of nervous conduction imposes a limit on the speed of mobility. Animals would simply run into trees without nervous conduction that is as fast the approach of obstacles, in order for them to sense, react and divert their path. So it is that invertebrates, without rapid myelinated nerves, tend to move more slowly than vertebrates do. Consciousness has evolved in a way that allows action to be controlled due to the speed of reaction. At the speed of a running cheetah, its reaction to the jinking gemsbok or impala is rapid enough to bring it down, as its action, sensation and reaction are close enough to match the actions of its prey. The disadvantages of slow nervous conduction can be envisaged, if we imagine walking down a busy street wearing spectacles that delay the received image by a second or so.

Such capabilities evolved in the cerebrum, and before that in the cerebellum, the site of visual processing, movement, orientation, recognition and calculation, which to control mobility must provide awareness in the same time frame. It was the minimisation of lag times to facilitate rapid motility which provided the real-time responses that made consciousness possible for other purposes. What is interesting here is that the evolution of adaptations for speed of reaction were already taking important steps forward in the evolution of the simplest multicellular organisms from unicells.

As animals evolve to minimise delays in homeodynamic systems, and nervous conduction allows rapid movement with real-time responses to obstacles or approaching predators, there remains a further refinement that adds to fitness. François Jacob wrote that 'one of the deepest, one of the most general functions of living organisms is to look ahead', adding that 'the need to imagine the future and the impossibility of knowing it are woven into the very fabric of life' [6].

Any adaptation that provides the ability to anticipate has great survival value. Of course, nature cannot predict a specific event, but adaptations to the lifelong rhythms of change, like the rising and setting of the sun and the ebb and flow of the tides, become incorporated into the genome after many generations. So too does the change of the seasons. Any event that is repeated or cyclical becomes predictable by possessing pattern; in this way swallows and other migratory birds 'predict' the changing seasons as they prepare to fly south on their long migration. But far greater challenges are random changes, such as a sudden storm, wind and rain, heatwave or drought, or the intermittent biological challenges of disease, competition and predation.

The survival value in being able to anticipate such challenges is so great that a simple strategy has evolved. A greater frequency of some kinds of changes in the recent past is a statistical indicator that similar changes may occur again in the future. The simplest indicator of the near future is the recent past. For example, six days out of ten, tomorrow's weather will be like today's. The odds are sufficiently better than even for natural selection to exploit such opportunities to increase fitness. However, predictions become more reliable when they are integrations of some variable, over say a week or more, because the change detected will be smaller. An investment in being ready for some future rare event is worthwhile if the cost of not being prepared is reduced survival, such as being alert for predators. Similarly, we have seen hydroids adapting to toxicity; to act in anticipation is worthwhile so long as the action is not too costly, and there is an occasional reward for correctly anticipating some challenge. A 'memory' of past events, or loadings on control mechanisms, integrated over time, increases future capacity and adds to fitness by a 'probabilistic anticipation'. In hydroids, secondary adaptation works by adapting capacity to changes in sustained load by adjusting the set point. As we have seen, such an adjustment provides an enhanced resistance to meet a similar future challenge. Adjustments to the recent past prepare the organism for the near future.

Natural selection has come up with probabilistic anticipation by reading the subtle signs of future change, making it possible for fast-moving creatures to instinctively avoid obstacles, or warning prey of the rapid approach of a predator. Alarm reactions and calls then warn others of an impending threat. Any knowledge of some future event, whether positive or negative, has advantage and improves fitness. The successors of these biological early warning adaptations are the long-term data sets and computer models that allow mankind to anticipate change and react accordingly. At the same time, any response that could accrue advantage by acting prospectively, so as to anticipate change, often implies great advantage, whether the change is a threat or a benefit, in the short or the long term.

Over time, organisms have evolved to overcome delays or tardiness in their reactions to life's vicissitudes. Delays have been shortened, responses have been sharpened and probabilistic anticipation is commonplace. Higher mammals anticipate the movements and habits of their prey. If we consider an arrow representing time (Figure 7.6), it is apparent that it has been an important factor in the adaptation of life to reduce the time delay inherent in the responses to any challenge. Any organism that has a short response time has an advantage over another that is slower. This matters with respect to competition and predator/prey relationships in many obvious ways. It is clear that when the outputs of perturbed yeast and hydroids are compared, the reduction in the lag time of the hydroid (Figure 7.4) by means of its immediate maximal response to perturbation is advantageous compared with the yeast's delayed response (see Figure 12.2).

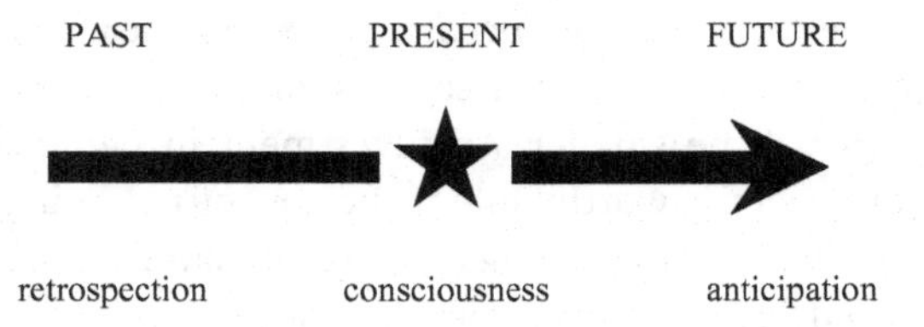

Figure 7.6. Time's arrow through past, present and future indicates that while life is tied to the present (*star*), the time frame of adaptation has shifted during evolution, from being retrospective to real time and consciousness to one that looks into the future.

Driven by natural selection, the evolution of rapid nervous conduction is made possible by myelinated nerve fibres; stimulus and response become almost instantaneous and fall into the same time frame. This has made the knee-jerk response possible, and the flight of a bird through hedging, as though it was not there. Such rapid reactions make it possible for sensation, decision and reaction to occur almost instantaneously.

From an initial perspective of life lived retrospectively, evolution has caught up to the extent that higher mammals appear to have the advantage of consciousness, made possible by living in real time (Figure 7.6). Animals have evolved intelligence that enables them to predict the habits of their prey, and human beings are still working to improve prediction of the weather because it impinges on so much that we do. Any organism that can live a life that is informed more prospectively than retrospectively will have advantages and greater fitness over those that cannot. The contest between animals to eliminate delays and to learn to anticipate has been ongoing throughout evolution.

EVOLUTION OF CONSCIOUSNESS

The brain is the seat of memory, decision-making, consciousness and control. It is also where the sense of autonomy, self-control and ownership resides. The brain is the principal cybernetic organ.

When Heinz von Foerster edited the Macy Conference proceedings during the 1950s, the conference title became *Cybernetics: Circular Causal and Feedback Mechanisms in Biological and Social Systems* and, as Wiener's concept took hold, these were the two ideas that became established in people's minds. J.Z. Young returned from the ninth Macy Conference in 1952 convinced of the importance of cybernetics. No published book has imparted the sense of pervasiveness of cybernetics in biology more clearly than his classic work on *The Life of Mammals* (1957). The first chapter, written as an introduction to the book, was entitled *The control of living systems*, and leaves the reader in no doubt as to the importance of control theory as a unifying concept. Although a great generalist, the strength of Young's thinking came from his experimental work on octopus. He wrote about the frequency of circular connections of nerves, which were identified as self-exciting circuits, with nerve impulses cycling around them. The ubiquity of feedback was plain to see. He was interested in circular self-exciting circuits, which live for as long as there are impulses going round them. Was this how memories were stored, he wondered.

The location of memory has been a key question in understanding the brain, but so also must be the site of decision-making and control. It seems that such functions are hidden within the network of nerves. Young noted that 'circular paths of action' were common, from octopus to human, and such circuits of nerve cells became larger with use by cycling impulses, and smaller if disconnected, reflecting the ubiquitous attribute of homeodynamic mechanisms to hypertrophy with use and atrophy without.

J.Z. Young was Peter Medawar's tutor at Oxford, and Medawar contributed to a programme on the healing of peripheral nerves. Young remained an influence throughout Medawar's formative years at Oxford, and he must have been aware of Young's post-war interest in cybernetics. Much later, Medawar used the principles of cybernetics to provide insight to his analysis of research as a process of recurrent hypothesis testing with feedback that modulates the hypothesis to conform to the data. His involvement in research had ended when he wrote: 'Among the most fertile ideas introduced into biology in recent years are those of cybernetics ... control theory obtrudes everywhere into biology'.

In the 1950s and 1960s, neurophysiologists in Europe began to study the sea slug *Aplysia*. The large and intelligent octopus had been difficult to contain in the laboratory, and facilities had to be scaled accordingly. The attraction of *Aplysia*, apart from its size and availability, was its specialised large nerve fibres, typical of some invertebrates, for speed of conduction. Eric R. Kandel collaborated in Paris with Ladislav Tauc, and later took his study on *Aplysia* to New York [7]. There he searched for the seat of memory for the duration of his career, sharing the Nobel Prize for physiology and medicine in 2000. But the emphasis of work on the nervous system had shifted by then to the use of computer modelling of these systems, and attracted funds from the US 'Star Wars' Program.

Douglas Hofstadter came from a background of maths and physics; he lists his research interests as consciousness, thinking and creativity. In his book, *I Am a Strange Loop* (2007), he includes a chapter about consciousness. Self-referencing has long been recognised as a unique property of feedback mechanisms, but Hofstadter's insight goes further. He wrote, 'However simple a feedback mechanism - its reference to itself is the germ of self-awareness'. Self-referencing and self-adjustment lead inevitably to self-awareness and consciousness.

In simple feedback mechanisms, decision-making is automated. But sensory inputs to the process have increased during evolution. In more complex organisms, with more sensors and more memory, decision-making become multifactorial. In principle, better decisions can be made, and greater fitness accrues, when more relevant information can be brought to bear. The gathering of information and its use in real time has therefore become a priority for more complex organisms. This is where a more advanced form of overall control becomes necessary for decision-making, environmental information and memory.

A control mechanism is a miniature organ of decision-making, with inputs of the goal setting from memory, and the sensed feedback giving the performance of the process at the output. Information flows converge on the comparator, which literally decides on the action required. Alluding to this mechanism, Young wrote, 'We must compare things, because that is the way our brains are constituted'. The comparator is the focal point of decision-making, because in the simplest feedback mechanism, this is where sensory and stored information converge and determine the difference between actuality and the goal in order to initiate corrective action. Decision-making is the function of the simplest feedback mechanism, which could be visualised in terms of nerve circuitry, and decision-making is the function of the brain; understanding how one informs us about the other is mainly a matter of scaling.

If one were to imagine improvements to a simple control feedback mechanism, what might they be? To reduce the lag time? The experimental data show that hydroids respond rapidly to any perturbation with an immediate maximal response, irrespective of load. Increase in the capacity of control mechanisms to withstand greater load? We see this in the adjustment of goal settings in response to sustained load (Figures 7.3 and 7.5). Both are apparently due to feed-forward loops (Figure 7.4). Similarly in scaling-up a simple control mechanism for decision-making, one would plug into it experiential memory to inform decision-making and prediction. Previous experience is valuable in making decisions. In some respects the brain is like a security videotape, in that it provides a continuous record of experience. Recent experience is of the most relevance and worth, but as it ages it becomes redundant, and so the tape is continually erased by forgetting. The value of information is ephemeral, so a rolling memory ensures that only information of the greatest worth is stored. Apart from the addition to performance feedback of the process itself, there is the matter of sensory information from the environment. If assimilated quickly, more information makes for better decisions. The massive flux of sensory information about the immediate surroundings of a human being amounts to 10^7 bits of information per second, according to an early estimate. The issue is one of filtering the information relevant to a decision and making it available quickly. Sensory and stored information are both required.

'Real time' means the time frame that encompasses the factors that impinge upon an animal within a period of time, and which it must relate to and react to simultaneously. This includes the pace of natural events, such as the flow of a river, a falling branch or wind speed. Then there is the speed of other animals, such as predators they need to avoid and the prey they must catch. To work best, consciousness must provide external sensory inputs and relevant experiential memories to the point of decision-making, which for convenience can be termed the comparator. These inputs are related in the comparator to the goal setting, and a decision for action is made.

It seems that consciousness developed along with the faster nervous systems that evolved with faster-moving animals. For man, continuous sensory input allows decision making and reaction to fall into the same time frame of consciousness. We can switch consciousness from thinking as we read or write, and being distracted by a fire that needs stoking or someone who enters the room – all in the space of seconds. Perhaps, since consciousness can obviously be redirected, it is like the 'read/write head' of a disk drive. It reads by focusing on, or switching to, different streams of information that flow in parallel. If one took a number of feedback mechanisms and laid their main east-west axes close together, such a read/write head could read the flow of information in any control mechanism, or switch between channels. But this could not replicate consciousness.

A read/write head would be necessary to contribute to the flow of information, and inform or overrule the decision being made. The reading head could listen in to the comparator, and then intervene in the decision-making process. The reading head could provide selected sensory information and relevant memories, drawn from sense-organ input and memory stores. Higher-level information that is relevant is brought to the individual channel in which the read/write head is temporarily exchanging information, and the consciousness of the organism is focused. It could impose a higher level of control on the mechanism. While it is obvious that the filtering of information is important, decision-making requires other subtleties such as priority setting, making trade-offs, weighting of decision-making factors and so on.

Francis Crick (1916–2004), in his later years, worked on the question of consciousness. He is quoted as saying, 'I think the secret of consciousness lies in the claustrum – don't you? Why else should this tiny structure be connected to so many areas of the brain'. Crick thought that the seat of consciousness must be at the point in the brain of greatest connectivity. This implies that consciousness can only happen where lots of kinds of information can be brought to one place.

If we return to the earlier speculation, the simplest form of decision-making occurs in a single loop control mechanism. Performance feedback on the controlled process is related to a goal setting, and the error between them determines the sign and magnitude of the action to correct a deviation in the process. If we imagine many such control mechanisms, with the comparator of each lying within the claustrum, a mechanism analogous to a read/write head could provide the means of switching consciousness between one channel of control and another. The claustrum has two-way connections to most of the regions of the cortex, including sensory areas, the thalamus and hypothalamus.

SUMMARY

Control mechanisms counteract any deviation of a rate process from its goal setting. They have the capacity to neutralise the forces of deviation. Here, where toxic agents are used to perturb growth, the implication is that the organism has a capacity to counteract toxic inhibition. Such ideas are embedded in the hypothesis that W. Ross Ashby proposed explicitly with his Law of Requisite Variety. Three levels of adaptation of homeodynamic mechanisms are identified: primary adaptation is due to the homeodynamic mechanism itself, as genetically prescribed; secondary adaptation is due to the modulation of the goal settings on homeodynamic mechanisms in response to sustained inhibition, such that its capacity increases with use and decreases without use. It is at the tertiary, and highest, level that the brain and memory influence the behaviour of the organism in avoiding extreme conditions. Each has been added sequentially during evolution, and each provides for more rapid responses to perturbation. The overall system has shifted during evolution from operating retrospectively to prospectively. This has been achieved using memory systems and the organism's complete template on the environment, changing its information with increasing speed, and ultimately basing behaviour upon anticipation.

8

Population growth and its control

Population, when unchecked, increases in exponential ratio.

T.R. Malthus

Nobody foresaw how the apparently simple Logistic Map would yield this graphic beauty.

Robert May

A RATIONALE FOR SELF-LIMITING POPULATION GROWTH

The simplest, and probably the earliest, way in which population growth was regulated is by the limitation of raw materials necessary for further growth. In Chapter 1, the blooming of huge populations of the single-celled alga *Emiliania huxleyi* was used to illustrate the exponential growth of populations. Each year in the ocean, *Emiliania* numbers grow rapidly following the peak in diatom numbers. Blooms appear, preferring the lower concentrations of phosphates and the high light intensities found in mid-summer. The growth of other micro-algae is similar, in being limited by the availability of the essentials for life: space, nutrients, trace elements, vitamins and so on. When any essential nutrient becomes less than optimal, the growth rate slows and populations decline.

Such a cycle of events is inevitable for those organisms that synthesise their own organic matter from inorganic sources. Plants and algae in the sea are autotrophs and manufacture what they need from recycled nutrients, carbon dioxide and water, with energy from the sun. When the resources essential for their growth are available, and conditions are optimal, densities grow to millions of cells per litre, and blooms cover vast areas of the world's oceans (Figure 2.1b). Their exponential growth rapidly strips the surface waters of available nutrients, which ultimately causes a decline in numbers as rapid as their appearance. The blooms die back, but extinction is avoided by the creation of dormant spores which descend into the deeper, cooler water. The following spring, when conditions for growth and multiplication return, the equinoctial gales stir up the oceans and bring nutrients and algal spores back to the surface waters. With warming water temperatures, the spring phytoplankton outburst recurs in the month of May, named after the Roman goddess Maia.

It seems probable that phytoplankton microalgae possess growth control of the kind described in Chapter 5, which has a simple feedback loop that controls specific growth rate (Figure 6.6). Its goal setting is termed the 'intrinsic rate of natural increase' ('*r*'), which is specific to each species. Exponential growth is the result, which continues to accelerate until the resources needed for growth become exhausted, whereupon the population crashes. When autotrophs exhaust the last vestiges of nutrients from seawater, these nutrients will be recycled in due course by their own death, and the cycle of 'boom and bust' begins again. In a world where survival is all, autotrophs do not risk driving any resource to extinction by stripping the ocean of nutrients; they only have to die for the resource to be recycled.

Animals are 'organotrophs', which depend on acquiring organic sources of carbon from other organisms; they are consumers and do not manufacture organic matter as autotrophs do. Instead they consume other organisms, or their products. If organotrophs behaved as autotrophs do, they would exhaust their sources of organic matter, and drive the living species on which they depend to extinction. Inevitably, being left without their food source, such a strategy would lead to their own demise. In evolution, survival is all; so what prevents organotrophs from causing their own extinction? To survive, they must ensure the survival of the organisms they use as food by limiting their own population growth. The evolutionary solution was to superimpose another feedback loop on top of the first, with a goal setting of a population limit, which imposes limitation on the inner loop and specific growth rate. The particular level at which population growth is limited relates to the resources that the organotroph needs in order to survive, and the minimal viable population size for the food species. It is proposed that animals have a mechanism that limits their own population density in relation to their prey, thus preventing overexploitation of their living resource. This level is termed the maximum sustainable carrying capacity or '*K*', and guarantees a sustainable food resource, and a sustainable limit to population growth for consumers, which would otherwise increase exponentially as autotrophs do. Although the controlled populations and their prey species may fluctuate, and goal settings vary, this mechanism has evolved to maintain stable relationships between animals and their food species. The problem of overexploiting food resources is avoided by the simple expedient of the consumer limiting its population size to a level of resource exploitation that the food species can withstand.

What follows is an interpretation of population self-limitation using the logistic equation as a model control mechanism, beginning with a brief history of the development of the basic equation.

VERHULST'S LAW OF POPULATION GROWTH

In Chapter 2, the question of exponential growth was explored, as was the paradoxical part it plays in ecology. It can be expressed by the Malthusian parameter rN, which allows organisms to grow at their own intrinsic rate of natural increase. As we have seen, even a constant specific growth rate results in exponential growth (Figure 8.1a). But how do animals, when driven by the positive feedback of exponential growth, limit their own population growth? This idea is considered by examining the logistic equation, and its key developments over the last 150 years. It was devised by the Belgian, Pierre F. Verhulst (1804–1849). He had been inspired by reading Malthus' *Essay on Population*, just as Charles Darwin and Alfred Wallace had been in arriving at the idea of evolution by natural selection. Verhulst was searching for a more realistic law for population growth, believing that there were natural forces at work that prevented populations from growing exponentially. He showed that while the population of the USA behaved in the way that Malthus would have predicted, doubling every 25 years between 1790 and 1845 (and continuing to do so for another 50 years); older European countries such as France were growing much more slowly, doubling every 400 years. He looked for an equation which would provide a more realistic law of population growth that incorporated Malthusian growth, as well as growth that became self-limiting with time. It would need to emulate exponential growth in its early stages, and then growth that slowed in more mature countries as population saturation was reached. We have already established that it is the potential of all organisms to increase exponentially (Chapter 2). It follows that the positive feedback of exponential growth will continue to increase explosively and unrealistically, unless coupled to a negative feedback mechanism.

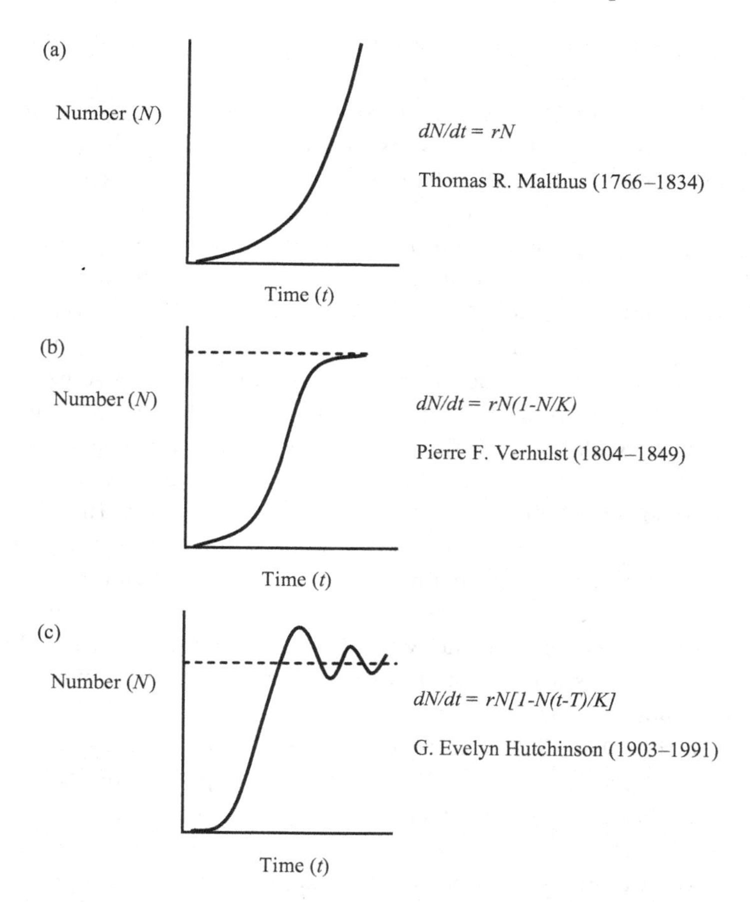

Figure 8.1. The origins and development of the logistic equation as a model of population growth indicated in three steps: (a) the Malthusian parameter creates exponential growth; (b) the logistic equation of Verhulst provides self-limitation; (c) Hutchinson adds a time lag which creates oscillations typical of biological systems.

To achieve this, Verhulst added a self-limiting term to the function for exponential growth, to create the logistic equation. He took the Malthusian parameter for exponential growth $dN/dt = rN$ (see Chapter 2) and added to it a self-correcting term to give $dN/dt = rN (1 - N/K)$. As before, N represents the population size, r is the intrinsic rate of natural increase and K is the carrying capacity of the system, or rather the maximum population that the capacity of the ecosystem can sustain. Normalising the equation to unity was an important step, as it simplifies the problem of defining a population of an indeterminate size in an undefined space. In the logistic equation, population sizes are represented as ranging from zero to one, such that 0.5 represents 50% of the population size. This allows populations to be handled within the equation as an abstraction that can be quickly converted into real numbers when they are known or required.

Overall, the logistic equation has two competing factors: rN causes exponential growth and $1 - N/K$ limits growth to unity. For all values of N, the logistic equation gives a symmetrical sigmoid curve of rising values of N over time (Figure 8.1b). When $N < K$ the term is positive and when $N > K$ it is negative, creating a self-stabilising population with K as its point of equilibrium. In the early stages of growth of a population, N is small and $1 - N/K$ is close to unity, and the rate of population growth is close to the organism's normal rate of reproduction given by rN. In this way, the population initially grows exponentially in a Malthusian manner. With time, the term $1 - N/K$ becomes close to zero, and the population growth rate (dN/dt) slows and stops, as the system reaches equilibrium. When the population has a time delay (see below), it will overshoot K, whereupon $1 - N/K$ becomes negative and will return to an equilibrium at K. In this way the logistic equation mimics the essential features of a self-limiting population.

In the 1920s, Raymond Pearl and Lowell Reed, working at the Johns Hopkins University in Baltimore, USA, hoped to identify an equation that could be used as a law of population growth. They found Pierre Verhulst's equation, which had been devised with the same aim in mind. In their paper published in 1920, Pearl and Reed used the equation, echoing Verhulst's name for it in recognition of his original discovery. They used the logistic curve to describe human population growth, believing the equation to be a universal law. However, Pearl's efforts to convince others of its generality, and his belief that it was capable of predicting growth beyond the data, turned people away from his work and the equation.

Explicit experimental corroboration of the logistic equation and its sigmoid curve came from an unlikely source. This was the Russian biologist Georgii F. Gause (1910–1986), who in the 1930s, through the darkest days of Stalin's rule, worked at Moscow University studying competition between populations of protozoans in test tubes. His work led to what became known as Gause's Principle, or the 'competitive exclusion principle', interpreted as meaning that two separate species cannot coexist in the same 'ecological niche'. A niche, in this sense, is analogous to the way in which an organism makes its living. If two species do coexist in the same space, it is likely that they are actually occupying slightly different niches. So 'one species: one niche' is the rule of thumb for ecologists to this day.

Gause provided experimental data in support of the logistic equation with his experiments on micro-organisms [1]. He worked with fast-growing microbial systems in culture, such as the yeast *Saccharomyces*, with a doubling time of ~30 minutes, and the ciliate *Paramecium*, which doubles in ~12 hours. His experimental method provided data showing that their growth can be described and defined by the sigmoid curve of the logistic equation. An exponential phase is followed by slowing of growth, as limitation is achieved by metabolic inhibitors, such as alcohol in yeast cultures, that slow growth to a tolerable density of cells. Given that the logistic equation was intended to describe human population data spanning centuries, it was useful to have experimental data that could provide corroboration of the sigmoid growth curve in just a few days (Figure 8.1b). Gause's results not only concurred with the logistic equation, but also testified to its generality.

HUTCHINSON'S INSIGHT

In 1948 the American ecologist G. Evelyn Hutchinson (1903–1991) presented his classical paper on *Circular causal systems in ecology* at a meeting on *Teleological Mechanisms* at the New York Academy of Sciences [2]. Norbert Wiener, who had not long published his seminal work on cybernetics, also gave a paper. Hutchinson paid respect to the scope of Wiener's ideas, noting that 'it is well known from mathematical theory ... that circular paths often exist which tend to be self-correcting'. He recognised that the last term of the logistic equation $(1 - N/K)$ was a 'formal description of the regulatory system'.

Hutchinson realised that, as a control mechanism, the logistic equation as a model would provide an unrealistically instantaneous response; there was no time lag within the logistic equation, if it was to be used as a population model. The sigmoid growth curve converges on its goal level as a smooth curve at a steadily declining rate (Figure 8.1b). Actuality is different, because there are delays in the biological feedback loop. It takes time for the consequences of exceeding the carrying capacity (K) to become apparent, for a corrective response to take effect and for the population to decline. Hutchinson had seen such oscillations in populations as diverse as planktonic crustaceans and field mice, and postulated that fluctuations in numbers seemed to characterise most ecological systems. He drew the conclusion that oscillations in population data implied the action of feedback mechanisms, and recognised that the logistic equation required a lag term if it was to behave in a realistic way. Given that populations cannot respond to change instantaneously, some time delay is inevitable and the addition of a delay term $(t - T)$ added realism to the logistic equation. Following perturbation, a time lag is necessary for the logistic equation to have an oscillatory output and behave as natural populations do (Figure 8.1c).

Hutchinson made this important link between ecology and cybernetics, recognising the logistic equation as a cybernetic mechanism which could be used to describe the behaviour of populations. As such, mechanisms behave with apparent direction and purpose by leading a population to K as if it were a goal. They are sometimes called teleological mechanisms, in that - unlike evolution itself - they recognise a purposive goal. It is a remarkable consequence of evolution that a process that has no goal or purpose can produce mechanisms that do. A system that behaves with foresight is an extraordinary product of a process that only has hindsight.

Hutchinson also realised that the logistic equation was not only applicable to populations, but that it describes equally well the growth of many individual organisms. So it is usual to regard multicellular organisms as growing like populations of protozoa, by cell division. The point is that the major growth processes, whether population number or the growth of individual organisms, are primarily due to the multiplication of self-replicating modules. It follows that the growth of individuals in a population and the growth of cells in an organisms have similar mathematical properties. The products of growth also grow, just as the products of reproduction also reproduce. In each case we see the outcome of multiplicative processes, which can be described in the same way. An exponential phase of growth is followed by a gradual slowing down as growth approaches an asymptote at a population equilibrium (K), or the size of an organism (Chapter 12). Rapid growth is gradually slowed and limited as the goal size or density is reached.

Of course ecologists were not unaware of the control properties of the logistic equation. Robert May, and more recently Peter Turchin (2003), along with other population ecologists, have appreciated that the logistic equation mimics the self-limitation of population numbers. It is just that they did not see it as a dual loop feedback mechanism, or exploit the insights provided by a cybernetic interpretation.

THE LOGISTIC EQUATION AS A CONTROL MECHANISM

To help with our attempts to understand control and limitation, the reader is asked to consider the logistic equation as a mathematical metaphor. Historically, it tends to be seen as a growth equation that describes a sigmoid curve. My purpose is to emphasise the properties of the logistic equation as a regulatory mechanism that mimics the control of population growth. Here the intention is to explore how the logistic equation might serve as a hypothetical control system that could represent the regulation of population growth and numbers. Before going into detail, it should be mentioned that the original equation is but one of a large family of population equations derived from the logistic equation; the original form is now seen as an oversimplification of reality.

Given Hutchinson's insights, we can now represent the logistic equation as a control mechanism, and use a control diagram as an aid to understanding its operation (Figure 8.2). It consists of two nested loops, which represent the two terms of the logistic equation. The inner loop is responsible for the first term of the equation, and controls the rate of reproduction of individual organisms. It has a goal setting (r), and the output is sensed to provide rate feedback. The outer loop is responsible for the second term, and limits population size by some measure of the density of individuals. Both loops incorporate time delays which are different, as they control processes on different time scales. The two loops are of necessity closely coupled, with a goal population density represented by K, and a goal growth rate represented by r. The

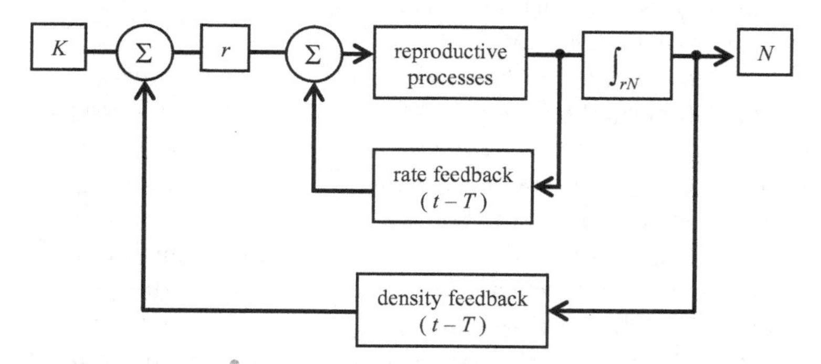

Figure 8.2. The logistic equation depicted diagrammatically as a dual loop control mechanism. Rate control is represented by the inner loop and the goal population growth as a specific rate (r). The outer loop can also limit growth rate, and so limits population density at a goal density (K). Time lags are indicated for each feedback loop, where t is time from which the lag time T is subtracted.

outer loop that limits population size exerts constraint by slowing the growth rate, which is controlled by the inner loop.

The control of population growth by the logistic control mechanism is effective in terms of both growth rate and population size. In fact it was introduced earlier in describing the control of growth at lower levels of organisation (Chapter 6), and a dual control was mentioned, but only the inner loop was described in detail, as the chapter was essentially a search for the rate control mechanism first described in Chapter 5. But even then it was realised that a dual control mechanism was the simplest control module, which is likely to have evolved on more than one occasion as it provides the minimal properties to control the growth of any replicator. Such ideas will be taken further in the next chapter on the hierarchical organisation of growth control.

Control of any process is much more sensitive if controlled by regulating its rate, but control of the specific rate is independent of population size, providing a growth rate per individual per unit time. The constancy of such a rate is likely to be important in the metabolism of the organism, but provides no control over numbers, which accelerate exponentially in a Malthusian way. One effect of control of the specific rate ($\log_e N.dN/dt$) is that, in terms of numbers, the inner loop is a positive feedback mechanism, because the growth rate (dN/dt) accelerates as the numbers grow. The inner loop, by maintaining a constant specific growth rate, creates exponential growth. Clearly such growth is unsustainable, and at some time requires a means by which it can be limited. This is provided by the second loop, which imposes negative feedback control on numbers before the exponential increase becomes maladaptive.

A sigmoid curve of the logistic equation ends by converging at the population equilibrium (K), representing the carrying capacity of the system. While this is the theoretical expectation, the curve here also represents the output of a control mechanism. If time delays are incorporated, it is to be expected that the output takes the form of a series of oscillations, decaying in amplitude as equilibrium is restored. The frequency of the oscillations is related to the lag in the system in a manner already described. The oscillations could be presented as oscillating about the sigmoid curve (see Figure 8.6b), but for simplicity the curve is given as a baseline for the oscillatory output (Figure 8.1c).

Before going further, we need to look at a single cycle of an oscillatory system to explain the relationship between its oscillation and the time delay that causes it. A complete cycle, conventionally represented by λ, is shown that over- and undercorrects about its

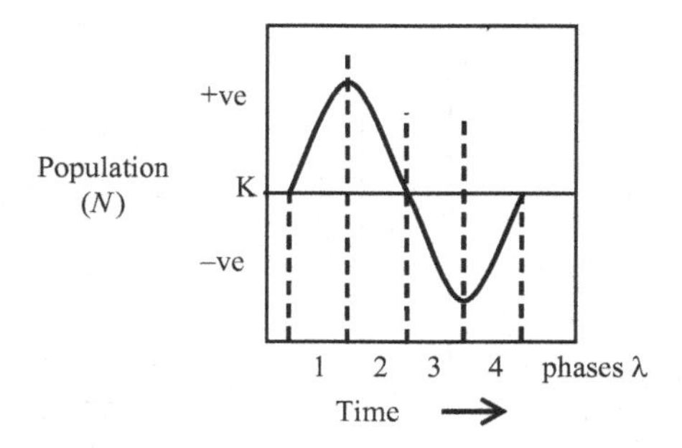

Figure 8.3. The cycle (λ), shows how a time delay in a feedback loop causes oscillation in the different phases of a cycle (see text for further explanation).

goal level (K), to which it stabilises at some future point (Figure 8.3). Phase 1 gives the time required to respond, due to the lag in the system ($\lambda/4$), to effect a correction that returns the population size to K (phase 2). But, due to the time lag, the curve passes through its goal level and undercorrects, and a similar delay passes (phase 3), before there is a corrective response. Oscillation is due to the delay between the control mechanism embarking on a corrective response, and detection after the time delay of a new error.

It some senses, time lags in biological control mechanisms are akin to a 'design flaw', in that the mechanism can only look backwards at the process it controls. The process cannot sense itself in real time, so it has a retrospective view of reality. In nature there is much to be lost in a delay of control responses, and nothing to be gained. As we shall see, adaptations have evolved to help organisms overcome this problem of delayed feedback (Chapter 11). With this understanding, delays are inevitable in the operation of control mechanisms, but the oscillatory output tells much about the system's behaviour.

The behaviour of the logistic equation as a control model, with a time delay, is likely to behave in a similar way to population growth of the freshwater *Hydra littoralis* (Figure 8.4). Here the baseline or goal setting given by the unperturbed controls (100%) is their relative growth rate, and results in an exponentially increasing population size. Hydra is not colonial but reproduces by asexual budding so, unlike the colonial hydroid (Chapter 5), these data represent the rate of asexual replication of a population. When growth is perturbed, the results reveal the characteristic oscillation of the output of a feedback mechanism [3]. Like the hydroid *Laomedea* (Figure 5.5), rates overcorrect initially due to the maximal response to inhibition. At lower concentrations of

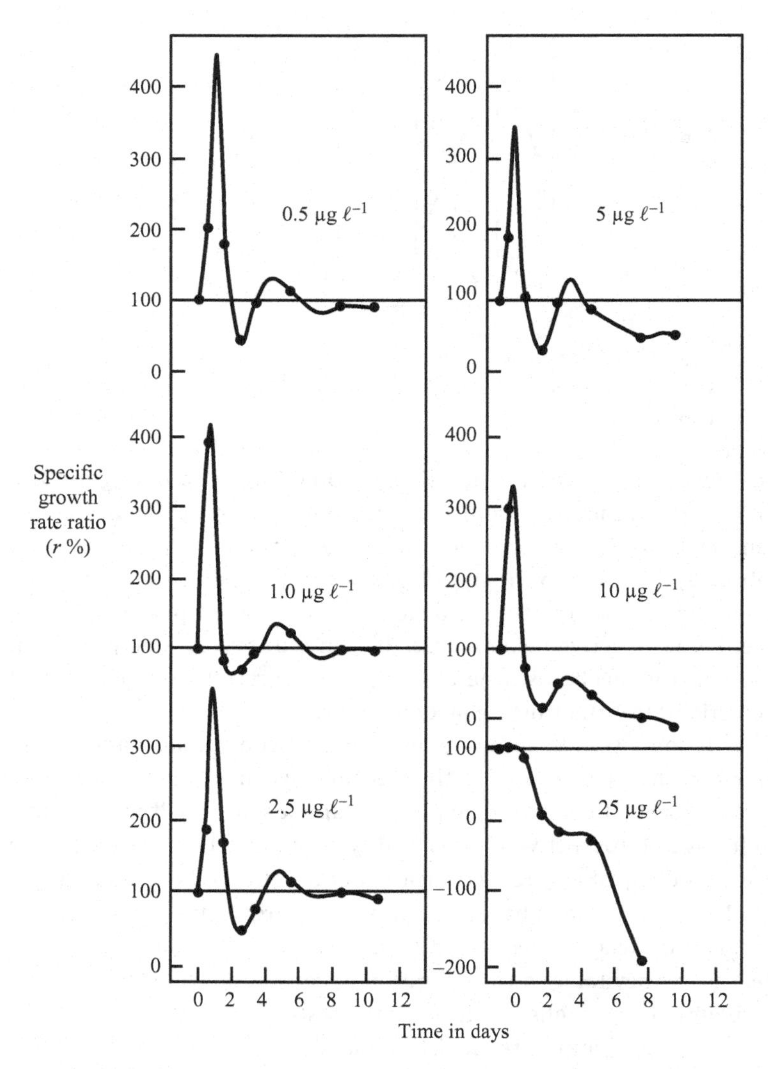

Figure 8.4. Experimental data for *Hydra littoralis* giving the replication rates of populations exposed to different concentrations of copper as a toxic perturbation of growth rate. Redrawn from Stebbing and Pomroy [3].

copper, an equilibrium is then restored. At higher concentrations of copper, overcorrection cannot be sustained and leads to a decline in replication rate after a single cycle, as the control mechanism becomes overloaded. At the highest level of inhibition, the growth rate immediately declines, at a concentration that quickly proves to be lethal.

Control of specific growth rate (*r*) by the inner loop, probably representing the rate of underlying metabolic processes and biosynthesis, is synchronised and loaded by the toxic perturbation, so even though evidence of control is provided by a population of hydra, it also represents control activity at a lower level indicative of the rate of biosynthesis. Despite evidence of control over the specific growth rate, the outcome is an exponentially growing population.

It has long been known, initially from the work of L.V. Davis in 1966, that *Hydra* also limits its own population density using a specific inhibitor [4]. As density reaches some optimum level, the concentration of inhibitor increases and begins to slow down budding and asexual reproduction, ultimately imposing a ceiling on numbers. A water-borne pheromone is responsible, which can be demonstrated by transferring water from a dense culture of hydra to a sparse culture, which then inhibits budding and further population growth. This limitation is represented by the outer loop of the control mechanism and provides an equilibrium population density (*K*). Here then we see dual control of replication and population size, which follows control of the kind that would be expected by the logistic control mechanism (Figure 8.2).

r AND *K* AS GOALS

If we accept the logistic control mechanism as offering a hypothesis as to how populations might be regulated, we must look more closely at how they might operate in reality. Within the logistic equation, *r* and *K* are positive constants, but in the logistic control mechanism they are its goal settings. We have already considered the specific growth rate *r* as the rate at which biosynthetic processes are controlled, as reflected in the earlier analysis of growth control (Chapter 5). That the carrying capacity *K* should also represent a goal setting for population numbers in relation to their environmental setting is appropriate here, as it was Verhulst's intention to calculate such a limit to population growth in his equation.

Feedback mechanisms have pre-set goals, so it is appropriate that *r* and *K* become goal settings for the inner and outer loops, respectively, but they also identify the purpose of the hypothetical mechanism as the targets for its control activity. We assume *r*, as a goal setting, to be the intrinsic rate of natural increase, and *K* to be the carrying capacity. That is to say, the organism has an endogenous carrying capacity, which gives the maximum number of individuals the habitat can sustain. This goal setting is likely to fluctuate for a carnivore, for example, with the number of available prey, as in the classical data sets for the lynx and the hare. Experimentally, the goal setting r emerges rather differently, representing a state rather than a rate. Conditions under which the hydra flourish in culture, and have a high growth rate, are assumed to be, and are expressed as, the baseline (r = 100% in Figure 8.4). Fluctuations with respect to the baseline provide control mechanism output and reflect the process of the recovery of equilibrium over time. Further details of the experimental design and analysis of the raw data required to provide such data are given in Chapter 5.

At first sight, it is apparent from the output of the control mechanism that oscillating growth rates exceed the baseline by as much as a factor of four in overcorrecting in response to exposure to the inhibitor (Figure 8.4). The ability to provide such high rates of growth in response to inhibition is analogous to using the power of a lorry to draw heavy loads uphill, rather than to provide raw speed. In hydra, the ability to grow much faster is a latent capability, used to counter the effects of disturbance or noxious agents that deviate growth from the baseline. Such effects are due to the homeodynamic properties of the system, and its innate resistance to inhibition (see Chapter 7). The counter-response to inhibition is stimulatory such that, in neutralising growth inhibition, the response has the effect that they cancel one another out. This stimulatory capability is not sustained, but is a maximal initial response that fills dead time with a positive response due to the lag following perturbation before the initial response occurs. This is observed in yeast (Figure 12.2) but not in hydroids.

Within the logistic equation, the hypothetical self-limitation of population size is imposed by K for each species, which represents a goal limit to numbers and prevents density from exceeding the carrying capacity. The population goal setting (K) for a species is the carrying capacity expressed as the maximum number of individuals that can be sustained. In actuality, such limitation in the case of hydra, and many other organisms, is provided by mechanisms by which population density is limited by some agent, such as a pheromone which acts as a density-dependent inhibitor. Here a measure of population size is given by the concentration of a pheromone to which all members of the population contribute and respond. The implication of such an adaptation is that there is fitness and advantage for the organism in maintaining its population numbers at some lesser density than would be limited by resources alone. The very existence of such mechanisms confirms the concept of self-limitation, and we will return to the question of such self-limiting mechanisms in the next section.

Ecologists believe that organisms are either r-selected or K-selected. That these are also the goals of the logistic control mechanism is of interest to the interpretation of r/K selection, because it implies that r and K are goal settings for a dual loop control mechanism and define its purpose. For the purposes of discussion, they are taken as the extremities of an evolutionary continuum by the graph of r versus T (generation time of a species). This well-known graph, first published by the American ecologist Frederick E. Smith in 1954, has been elaborated by ecologists over the last 50 years (Figure 8.5). It provides important insights, by relating population growth rates (r) and generation times (T) for organisms of various growth rates and generation times over many orders of magnitude. The inverse linear relationship is striking, as it also indicates an evolutionary trend from the simplest and smallest, which multiply fastest, to the most complex and largest, which reproduce the most slowly.

Smith's curve incorporates some other less obvious generalisations. Two significant correlations make the implications far-reaching. Population growth rates are proportional to the rate of metabolism (m) of the organism ($r \propto m$), so growth rate tends to reflect the operation of the underlying biosynthetic machinery of the organism. Second, generation time is also proportional to the size (W) of the organism ($T \propto W$), which has tended to increase steadily through evolution.

The fastest-growing organisms are autotrophs, for which rapid multiplication and a boom and bust lifestyle are dominant traits. It seems from their behaviour that autotrophs may be too primitive to have an outer loop for K-selection (Figure 8.2), as ultimately their numbers are only limited by the available nutrients. Their growth in spring and summer is huge on a global scale, covering vast areas of the world's oceans. The scale of the blooms turns the oceans milky, and their exponential growth quickly creates a feature of our global oceans that is visible from space. As the smallest organisms also have the highest metabolic rates, this confirms that the unicellular organisms, with their spectacular growth rates, are r - selected. Phytoplankton species do not limit their own numbers, and their explosive multiplication is ended by using all the nutrients that are present. In exhausting the available nutrients, they do not cause the extinction of their food species, as is the analogous case for organotrophs. As nutrients become exhausted, they die back, forming dormant spores until nutrients are recycled and conditions favourable for growth return. With their ability to rapidly

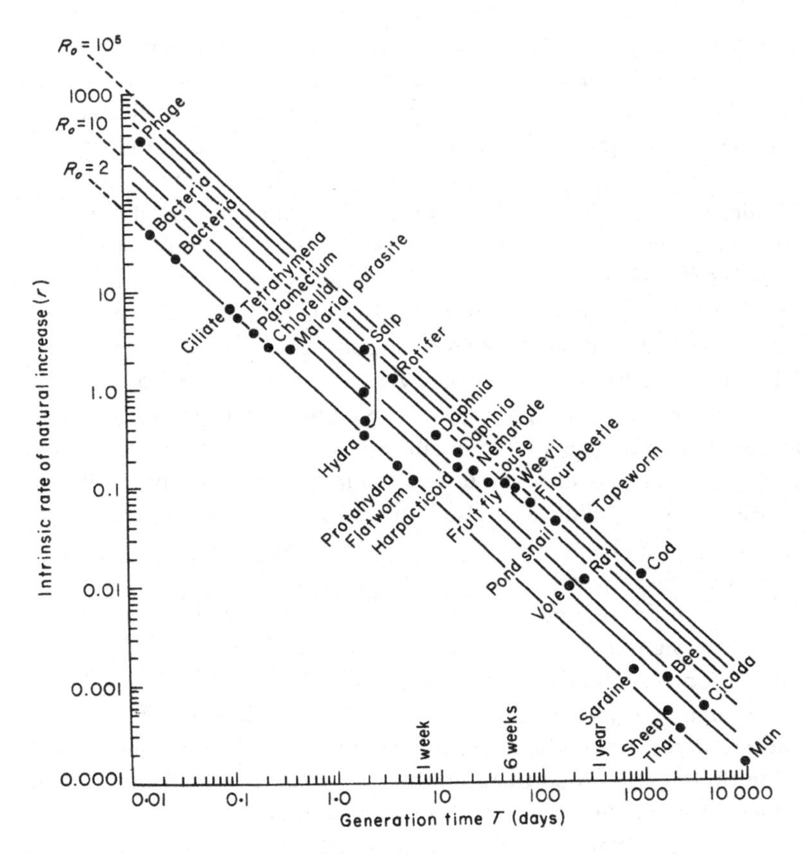

Figure 8.5. Graph showing the relationship between relative or specific growth rate (r) and generation time (T, days) for a wide range of organisms. The parallel lines indicate the fecundity of organisms from doubling ($R_0 = 2$) to tens of thousands in the case of organisms producing large numbers of eggs ($R_0 = 10^5$). The figure is taken from May [14], courtesy of Wiley-Blackwell.

colonise an ephemeral habitat, these autotrophs exploit temporary niches, utilising their rapid transition to a dormant phase, when the niche is exhausted. They are extreme r-strategists, and are obviously so in this analysis, as they do not possess a control mechanism with K as a population limiter.

The attraction of the logistic equation as a hypothetical control mechanism (Figure 8.2) is that it incorporates its purpose and goal setting in the constants r and K, with evidence for r as a goal setting already provided by the Maia hypothesis (Chapter 5). K is more obviously a goal setting for population size, as intended by Verhulst.

At the other end of the linear decline in population growth rate, we find the most highly evolved, complex and largest organisms (Figure 8.5). Here are the largest organotrophs that feed at the top end of the food pyramid of numbers. They are longer-lived, and have longer generation times (T). They avoid extinguishing their food species by self-limitation of their population density at a level that ensures prey exploitation is less than maximal. These are K-selected and colonise habitats that are stable over long periods. The higher goal of population self-limitation provided by K in the outer loop is dominant. For the reasons already given, overall control by K is necessary if the logistic control mechanism is to restrain the exponential growth that r creates.

POPULATION CONTROL BY SELF-LIMITATION

Self-limitation was implicit in the work of Alfred J. Lotka (1880–1949), who recognised the tendency for biomass, energy and its flux to be maximised, *but not so much as to exhaust organic matter and available energy*. Limitation can be of different kinds. It can be due to the density of individuals occupying one niche in one habitat, which does not involve the evolution of specific adaptations. Under these circumstances, there are negative effects on reproduction due to stress, restriction in the availability of food, limitation of space, disease or reduction in fecundity. Such limitation is seen in autotrophs. It is apparent that during evolution environmental limitation was replaced by population self-limitation, with control imposed by specific adaptations that reduce population increase, such as inhibitory exocrines in the case of freshwater hydra [4]. In this case, population self-limitation has the effect of preventing the exhaustion and possible local extinction of food species, which would jeopardise the survival of populations.

In the past there were protracted conflicts between ecologists over population regulation. During the 1950s, the issue was whether populations were limited by non-biotic or biotic factors. The debate was largely between two Australian ecologists, Charles Birch and Alexander John Nicholson, who both worked on insect populations. Birch believed that external environmental factors were responsible, while Nicholson was sure that populations were regulated by effects that depended upon the density of populations. The next step was an increase in focus. It was proposed that, not only was the density of populations important, but populations were actually self-limiting in a way that involved the evolution of homeodynamic mechanisms to regulate population density or size.

In 1962, Vero Wynne-Edwards (1906–1997) published *Animal Dispersion in Relation to Social Behaviour*, which resulted in a protracted debate that ran for over a decade [5,6]. Wynne-Edwards reasoned that populations are self-regulating:

> It must be highly advantageous to survival, and thus strongly favoured by selection, for animal species to control their own population densities, and keep them as near as possible to the optimum level for each habitat they occupy.

In this way overexploitation and local extinction of food organisms would be avoided. He argued that the maximum density of animals is limited by mechanisms that control reproduction.

A null hypothesis is provided by autotrophs, which grow and multiply until the sea is stripped of nutrients. They maximise their exploitation of nutrients by multiplying until there are not enough nutrients left to support further multiplication. Nutrients are extinguished, as it were, only to be regenerated by the demise of the great phytoplankton blooms, and the excreta or death of other organisms. We must therefore assume that organotrophs do not maximise and eliminate their food sources, implying that some form of self-constraint exists that prevents the elimination of food species.

Such control is already represented in the logistic control mechanism (Figure 8.2). A sigmoid curve is given as a product of the logistic equation (Figure 8.1b), with growth marked by a ceiling which represents a population determined by the carrying capacity of the resources provided by its habitat. For an organotroph, this could represent the amount of available living prey. The carrying capacity (K) represents the maximum population that a particular habitat can support (Figure 8.1b), which would indicate a level of constraint at which exploitation would be sustainable. In this way the logistic model brings the question into focus: K represents a ceiling to population growth that is sustainable.

The autotroph living by r-control alone knows no such constraint; the question is, what brings about such constraint for organotrophs so that they prevent themselves from exploiting their prey maximally? This would ensure that overexploitation of the prey is avoided, and that the essential biological resources are not exploited to the point of extinction. One can even see how, for the fitness of both predator and prey, some level of exploitation equivalent to the 'maximum sustainable yield' calculated by fisheries management scientists would best serve the interests of both the fishing industry and

Table 8.1. *Fecundity as indicated by the numbers of eggs or live offspring produced by individual females at each breeding event, shown for a range of species. The data also indicate the potential size of kin groups created at breeding that share at least half their genes.*

Group	Animal	Range of number of eggs/ offspring
Mollusc	Cockle	30,000–50,000
	Oyster	15,000,000–115,000,000
Insect	Mosquito	300
	Oil beetle	3,000–4,000
Fish	Pipefish	200
	Haddock	12,000–3,000,000
Amphibian	Frog	1,000–2,000
	Newt	100–400
Reptile	Lizard	5–14
	Grass snake	2–30
Bird	Swift	3
	Thrush and crow	3–5
	Mallard	8–14
Mammal	Mouse to human	1–12

the fish populations. This idea was put forward by Wynne-Edwards in support of his hypothesis. Of course, with overexploitation and falling numbers of prey, the catch per unit of effort falls, such that the cost of catching prey, as they become sparse, becomes greater than the reward. This would result in a level of exploitation that is self-limiting, simply because the prey are too sparse to justify the effort expended to catch them.

This idea is supported by the fact that many organisms are extremely fecund (Table 8.1). For species vulnerable to high levels of predation or mass mortality, much greater fertility is required to offset such losses.[1]

[1] Cockles in South Wales are exploited by large numbers of oystercatchers (as many as 15,000) in winter, which may consume as many as 400 each per day. However, their efforts are negligible compared with the severe winter of 1962–1963 which caused mortality of 90–100% over much of the cockle beds. Yet a few months later cockle larvae replenished the cockle beds at densities of up to 10,000 per square metre. The evolution of such apparent reproductive excess is apparently justified by such rare events.

Implicit in Wynne-Edwards' idea was that species evolved ways of limiting their own numbers that would be contrary to the interests of some individuals, whose reproductive potential would be delayed or denied. This idea required the acceptance of 'group selection', which implied that selection could operate on a group of individuals, such that some would be required to act 'altruistically' for the benefit of the group. Conventional thinking at that time was that natural selection acted upon individuals, so there was no case for the evolution of altruism, even though examples of altruistic behaviour were plentiful. In 1964, W.D. (Bill) Hamilton published two papers of fundamental importance on the 'genetical evolution of social behaviour' [7]. They showed how kin selection could account for the evolution of altruism. It was the geneticist and evolutionary biologist J.B.S. Haldane (1892–1964) who first put forward an argument for altruism. Haldane once said, 'I'd lay down my life for two brothers, four half-brothers or eight cousins', recognising the number of different relatives that would be required to share an equivalent complement of his own genes, and thus be of the same genetic value to him in his hypothetical sacrifice. This idea was never published by Haldane in a scientific journal, but was the seed of kin selection that Hamilton published as a formal theory in 1964. So for those who understood Hamilton's theory, the selected group could include members of an extended family, which were called kin groups.

In the 30th anniversary edition of *The Selfish Gene*, Richard Dawkins published a citation analysis of Hamilton's key 1964 two-part paper. It was uncited for 10 years, and did not reach a peak of 100 citations per year until 20 years after publication. It made the argument for kin selection, showing that the greater the relatedness of two individuals, the more genes they share, so acts of altruism between related individuals are rewarded in the improved survivorship. It has been shown since then that such arguments apply generally; selection operates on genes not individuals.

When argued, as Haldane did for his relatives, the numbers within a kin group are relatively small but, among other animal species, kin selection assumes a much greater importance because groups of related individuals within a population are much larger. One only has to consider the number of individuals produced by one female, which may be fertilised by one or more males. Although sexual systems vary, an impression of the potential size of kin groups is given by the number of eggs or live offspring produced by a female parent at each breeding event (Table 8.1). It is therefore likely that the rewards for altruism become greatest when large numbers of related individuals behave cooperatively.

Wynne-Edwards proved no match for the leading evolutionists of the time, who ridiculed his argument for group selection. In fact the case for altruism with kin groups had already been made, but took time to become accepted. Although Hamilton's paper was published only two years after Wynne-Edwards' book, he was still arguing the case for group selection in a paper published in 1993 in his 87th year. Wynne-Edwards lost his long battle for the acceptance of group selection, though the selection of altruistic behaviour within kin groups was theoretically plausible and was eventually accepted.

The focus in Wynne-Edwards' book was on birds and terrestrial vertebrates with complex social behaviour and subtle signalling systems [5,6]. For terrestrial animals using the media of sound, sight and smell, the communication of density-dependent information was difficult to demonstrate with certainty. The passage of information between one individual and another was difficult to detect, and it was difficult to establish what the information meant, let alone how it might be used to limit reproduction and population numbers. The focus on birds and mammals was because those involved were ethologists wedded to the study of more highly evolved terrestrial animals. Seemingly as an afterthought, the last chapter in Wynne-Edwards' book was devoted to self-limitation of population density in aquatic animals by the use of chemical pheromones that regulate population density by inhibiting reproduction, growth or development. It would have been a much better field in which to establish the principle of self-limitation, but during the protracted debate, the significance of aquatic pheromones was overlooked.

POPULATION CONTROL BY PHEROMONES

In some respects the most refined adaptations for self-limitation would appear to be pheromones, because chemical messengers between members of a species overcome the ambiguities of social behaviour in terrestrial animals. They are typically complex compounds that are specific to their role, and are produced in minute quantities, acting in the environment in a highly diluted form. They are often called exocrines, implying an analogous role to endocrines.[2] Sometimes just a few molecules are required for the meaning of a signal to be received and acted upon.

[2] Endocrine is another term for a hormone. The term is used here to make clear the analogous role of exocrines, which have a similar role, but act as a chemical messenger in the surrounding habitat. In this sense, pheromones are congruent with J.B.S. Haldane's important idea that exocrines gave rise to endocrines.

One problem for Wynne-Edwards' hypothesis was his contentious concept of 'epideictic displays', where the idea of the swarming of animals, such as fish or birds, as a means of establishing population strength and number was never proved conclusively. In an aquatic system, individuals in a population each contribute to the concentration of a pheromone, which by means of its concentration gives a chemical indication of population strength to all its individual members. What is more, in some cases, as population density reaches a critical level, the pheromone becomes an inhibitor and regulates population size through its effect on development or reproduction. Each pheromone is considered to be specific to its role, for its meaning to be unambiguous. Even where terrestrial animals employ pheromones, interpretation is never simple. For example, the black-tailed deer produces pheromones from seven glands, and each is thought to have a different social function. The honey bee has eleven glands producing pheromones involved in communication and social organisation.

The specificity of such agents must help in pinpointing their role, but the vanishingly low concentrations at which they are effective makes it difficult for experimental biologists and analytically challenging for chemists. The problem is less difficult with pheromones produced by aquatic organisms because water-borne pheromones can be more easily manipulated and stored for assay using some biological response, and its specificity can be established by experimenting with different species. Because experiments with aquatic pheromones on communication between organisms are technically less difficult to conduct, they are therefore more commonly reported in the literature.

What is crucial to understanding the behaviour of any feedback mechanism is the need to 'close the loop' around which information flows. Where there is debate over the meaning of behaviour, or controversy over the purpose of epideictic displays, the loop cannot be closed with certainty. It is the specificity of the chemical messengers involved in the causal cycle of feedback mechanisms that makes it possible to reduce the ambiguity of interpretation in such instances. Once the role of the pheromone is biologically identified, pheromones provide a more rigorous alternative to the interpretation of acts of social behaviour as evidence for self-limitation.

The origins of pheromonal control mechanisms may have been in the increasing accumulations of metabolites and excreta with population density. Associated signals are often implicated in studies of bacteria, protozoans, planarians, snails, crustaceans and fish. The metabolic wastes of fish, such as ammonia, do inhibit growth, but their own specific inhibitors are much more powerful in controlling growth. Excreta and metabolic byproducts may have been the precursors of pheromones, and it is not difficult to imagine how chemical agents became more effective than excreta in meeting the requirements of limiting population density.

The action of the agent and its concentration can be quantified by bioassay, although in other cases the labour of chemical extraction, isolation, analysis and synthesis to identify the molecule can be justified. An example of global importance serves to illustrate this point. Many aquatic snails produce pheromones that regulate their own growth and reproduction; in the great pond snail (*Lymnaea stagnalis*), it has been found that overcrowding leads to stunted growth and reduced reproduction. The disease bilharzia (also known as schistosomiasis) infects over 300 million people in tropical countries. It is caused by the fluke *Schistosoma*, with a freshwater snail (*Biomphalaria*) as its intermediate host. The snail releases a free-swimming phase that penetrates the skin of waders and bathers, where it develops into a fluke. The fluke spreads through the bloodstream, feeding on the lungs or liver, before reaching the intestine, where it lays eggs. The damage caused is often fatal. One way of controlling the disease that is currently being explored is to curb population growth of the snail with its own population-limiting pheromone, and a major step has been made by Chinese scientists who have identified the pheromone. The pheromone is a unique organic compound, monohydroxy-tricarboxylic acid-monoisodecyl-dimethyl ester, with a general formula of $C_{18}H_{32}O_7$. It is hoped to synthesise the pheromone and use it prevent the snail's growth and development, thus breaking the life cycle of *Schistosoma* and controlling bilharzia.

Many scattered experimental studies have shown that aquatic invertebrates produce chemicals that act as pheromones in limiting population density. Ilse Walker approached the problem from an evolutionary perspective, and studied analogous systems in a variety of species. She has identified population control systems in several distinct animal types, including the ciliated protozoan *Paramecium*, hypotrich ciliates such as *Keronopsis* (see Chapter 2), a planktonic crustacean and a parasitic insect. In the ciliate, population growth was limited by the inhibition of fission rates at high densities, while in the crustacean the production and release of ovisacs was inhibited. Walker's research on a range of invertebrates supported her contention that population control 'is a general condition of living systems', finding 'unexpectedly sophisticated mechanisms of population regulation'.

Other authors reported similar findings with flatworms and the aquatic larvae of the mosquito. Under crowded conditions, chemicals are produced which signal and slow down growth and development, producing morphological aberrations in the larvae and reducing their fitness and survival.

Simple experiments on the guppy (*Lebistes reticulatus*) provide a vertebrate example. It was found that as the number of adult guppies in an aquarium increased, the production and survival of the young decreased. This is partly because, at high densities, guppies eat their young, whatever the amount of food provided. Changing the water partially eliminates these effects, implicating a water-borne pheromone that induces cannibalism. This interpretation is corroborated by experimental evidence, which clearly shows that the equilibrium density is related to the volume of water in which the fish are kept, and therefore the concentration of pheromone. In experiments it was found that the numbers arising from one guppy would increase from one, or decrease from 50, to stabilise at an optimum population of about nine. Such work, and other more recent research with trout, provides convincing evidence of population regulation because, in aquaria of the same volume, numbers stabilised at some preferred density.

It has already been mentioned that hydra populations tend to reach a threshold density and then stop budding. Marine hydroids are typically colonial and, while the freshwater hydra produces an effective water-borne inhibitor, in the sea pheromones would be quickly diluted and ineffective. Here an inhibitor passes along the tubular stolon linking all members of a colony, and controls the growth and spacing of the uprights and polyps. After the growth of a specific stolon length without an upright, the signal fades and a new upright grows, such that the mechanism determines the spacing and density of colony members. In this way the pheromone of hydra and the inhibitory regulator of colonial hydroids have similar roles in regulating density. The adaptation of piping the pheromone as an endocrine acts as a regulatory inhibitor, and enables the regulation of hydroid density without having to counter the dilution of pheromone in turbulent tidal waters.

In nature, female frogs lay floating masses of eggs in gelatinous skeins, each female laying 1,000–2,000 eggs. Those that hatch first grow rapidly and become larger than those that hatch later. These are the first to metamorphose into froglets and leave the pond. Unlike other species, which release pheromone directly into the water, tadpoles produce cells containing pheromone in their hind-gut, which are released with faeces. The cells settle to the bottom, where they are actively sought in the sediment and eaten by other tadpoles. This refinement to the control mechanism increases its efficiency, as even in slowly moving water there is continual dilution of the pheromone. The late hatchers eventually stop eating and their growth slows; they may stop growing altogether, become stunted and die. The power of the tadpole pheromone is shown by the finding that a single large tadpole can retard the growth of other tadpoles in 75 litres of water, and such activity may last for several weeks.

The tadpole mechanism enables newly-hatched tadpoles to grow and metamorphose first, giving them time to leave the water and colonise adjacent ponds, but not all at once. As the first-hatched leave, pheromone production inevitably begins to decrease, allowing others to grow and metamorphose. This not only regulates the population in the parental pond, but also controls the rate at which metamorphosing froglets make their way to adjacent ponds. A mass breakout would attract predators.

This pheromone and regulatory inhibitor was studied by S. Meryl Rose and his wife, Florence, who recognised a negative feedback mechanism as a key factor in regulating tadpole growth. Their recognition of the importance of inhibitory regulators will feature again in Chapter 12.

LIMITATIONS TO THE INTERPRETATION OF EXPERIMENTS

Experiments on pheromone control of populations in aquaria provide experimental rigour, but there could be an oversight in failing to reproduce one crucial aspect of the natural habitat. In static laboratory aquaria containing just a few litres of water, a pheromone quickly becomes well mixed and evenly distributed. Thus all individuals are exposed to the same concentration and exhibit a similar degree of inhibition. But this is obviously not the way that water-borne pheromones behave in the environment. In a pond or small lake, much larger volumes of water are involved and there are uneven distributions of organisms, resulting in variable concentrations of any pheromone. For example, the highest concentrations will occur where the density of spawn and tadpoles is greatest. In a pond with slowly moving waters, a pheromone is slowly dispersed and, as its concentration is diluted with dispersal, a gradient of concentration is created. When a pheromone is released into the environment it carries at least three sorts of information. The first is the identity of the species, the second is a measure of population density, and the third is the direction of the source. Clearly static tanks do not allow a gradient to be set up, or the opportunity for the organism to respond to it.

The inhibitor is mildly toxic at higher concentrations, so tadpoles are unlikely to stay in noxious waters that inhibit growth, and are more likely to be driven to seek more favourable waters lower down the gradient by moving to a place where the concentrations of pheromone are lower. It seems that the main purpose of the pheromones may not be to inhibit growth and reproduction where concentrations are highest, but to disperse tadpoles towards lower concentrations, where there are fewer of their own kind. A gradient of some noxious pheromone becomes an inducement to disperse. The operation of such a mechanism acts as 'carrot and stick', encouraging dispersal; there are benefits in responding to the gradient, as well as a penalty for not doing so. Only observations of negative effects were recognised by those who experimented with static tanks, overlooking the positive ecological implications of a gradient as a dispersal mechanism.

As the young move down-gradient, the density of pheromone producers becomes lower, and it follows that pheromone concentration must also decrease, so noxious effects in nature would be avoided by the effect of the gradient. Pheromone gradients would ensure the escape of tadpoles from the hundreds of competing siblings that hatched from the same egg-mass, and provide a spur to disperse and colonise new habitats. We come therefore to the suggestion that pheromones have less to do with inhibiting growth, but instead induce dispersal by creating a noxious gradient.

A CONTROL MODEL FOR A PROTO-ORGANISM

This interpretation of how aquatic pheromones might operate finds support from a model developed by a father and son collaboration. L.R. Taylor was an established mathematical ecologist at Rothamsted Experimental Station, who worked with his son R.A.J. Taylor, a mathematical modeller, in the development of a model for population self-regulation [8,9,10]. The simulation model represented a hypothetical 'proto-organism' as a way of developing a general hypothesis for population self-regulation. The Δ-model, as it was called, satisfied the principal objectives of the Wynne-Edwards' hypothesis, but involved a different mechanism than reproductive constraint. Although the Taylors' ecological experience and data for their 'proto-organism' were drawn from large-scale insect surveys held at Rothamsted, their hypothesis had a wider application.

The Taylors thought that extrinsic regulation of numbers, such as resource limitation, that affect birth and death, cannot be expected to achieve the degree of population control that is observed in nature. In addition, they felt it could only operate at an upper limit of population density. They concluded that intrinsic factors such as behaviour must control population density and account for oscillatory fluctuations. They reasoned that the basic controlling factors for their hypothetical proto-organism are behavioural; that is intrinsic controls operating on density-dependent behaviour to induce local dispersal, followed later by congregation. They argued that such behaviour is favoured, as dispersal reduces density and therefore competition. The interpretation must incorporate the principle of feedback, and regulate population density by control of the dispersal and aggregation of organisms. In amphibians we see the primitive mechanism of water-borne pheromones driving the dispersal of tadpoles, while the terrestrial adult uses vocal signalling to call adults together in spring to breed (Figure 8.6a). The Taylors proposed that density-dependent behaviour accounted for population regulation, achieved primarily by the movement of individuals.

In developing their Δ-model, the Taylors questioned fundamental assumptions that had been prevalent. They reasoned that 'the first premise for natural selection is not maximised fecundity, as classical theory assumes, but the survival of the individual'. It is more important for an individual to stay alive, so that it can then reproduce when the opportunity arises. They had earlier recognised the paradox that 'while limiting resources are the key to classical theory, resources are hardly ever more than very locally limiting in nature'. The Taylors then attempted to show that resources are not limiting, because they are never entirely depleted in nature.

L.R. Taylor had earlier established that individuals in insect populations of all kinds are not distributed randomly, but space themselves out optimally. The Δ-model simulates populations that are continually mobile, with movement as the consequence of the pressures of density. Populations of the young disperse to develop to maturity, and then later congregate to breed. Populations are never static and so are unlikely to overpopulate their habitat; extinction, when it occurs, is due to the absence or destruction of suitable habitats. Under pressure

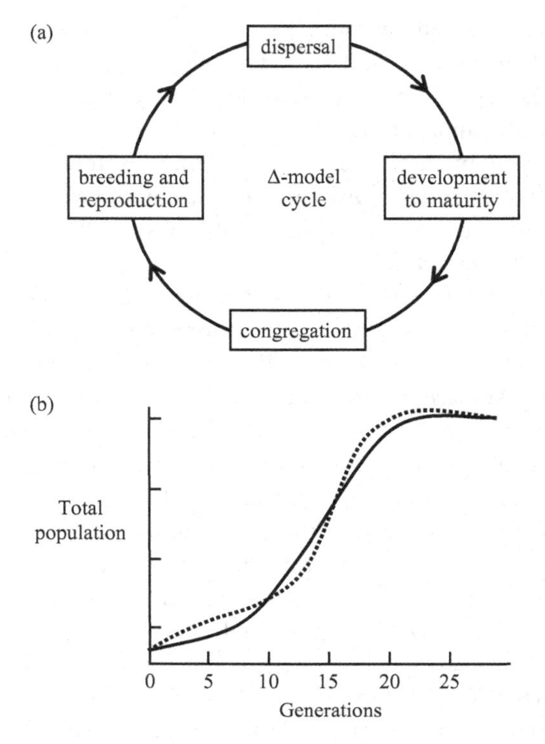

Figure 8.6. The Taylors' Δ-model of a hypothetical proto-organism, with its emphasis on movement rather than inhibition of reproduction: (a) indicates the cycle of a hypothetical population alternating between dispersal and congregation; (b) indicates the output of the model population (*dotted line*) oscillating about the logistic sigmoid curve as it converges on the sigmoid curve as its baseline. Figure redrawn by courtesy of Taylor and Taylor [8].

from density, predators or parasitism, individuals must migrate or die. The losers in social contests emigrate, which may itself lead to mortality, especially in the young.

The model itself was developed by R.A.J. Taylor; it was described as the difference between two power laws, and incorporated negative feedback. It proposed that movement is the key behaviour that relates to population density, operating through mutual attraction or mutual repulsion, leading to optimal spacing. It also incorporated a density multiplied by a distance-behaviour function, by which the model works in space and time, rather than in time alone. The output broadly follows the logistic curve as a 'baseline', but convergence to it is by a series of decaying oscillations about the baseline, as the system stabilises (Figure 8.6b).

Many runs of the Δ-model were carried out in a simulated arena, which could be adjusted to be more or less favourable, and allowed individuals to move away from, or towards, the gravitational centre of the population. Populations within the model arena were continuously mobile, and offspring often found themselves in less favourable areas than their parents. It became clear that, in a heterogeneous environment, overpopulation was unlikely but, where the habitat is rapidly changing, only species with large numbers of highly mobile offspring may be expected to survive.

This was a novel hypothesis that accepted Wynne-Edwards' principal ideas. A dynamic model provided the intrinsic density-dependent behaviour by which control of density could be regulated. The irony is that the Δ-model depends on movement and dispersion, rather than reproductive constraint, and it does not require acceptance of group selection. It is to be regretted that the Wynne-Edwards debate, as it became known, revolved around the contentious idea of group selection which, as kin selection and reciprocity theory subsequently demonstrated, should not have been obstacles to Wynne-Edwards' ideas on population self-regulation. Since those days, the Taylors' model makes clear its plausibility. They showed that individuals were not required to forego their opportunity to breed, but rather were required to move away from centres of population. In this way, the model provides control over population density of the kind which Wynne-Edwards envisaged by movement, rather than denying or delaying reproduction for some individuals. Populations stabilise at a level below that which would be imposed by limiting resources, without constraints on reproduction (Figure 8.6b).

It is regrettable that the debate revolved around terrestrial species, when aquatic species using pheromones would have provided a more satisfactory proof of concept. The Taylors' re-interpretation of population self-regulation is relevant with respect to the control of aquatic populations by water-borne pheromones. It corroborates the idea that the purpose of pheromones is not to inhibit reproduction, or to kill the young, but rather to induce movement and dispersal by means of a gradient of concentration that provides a directional cue. As the Δ-model was tested against insect data, and seems appropriate to aquatic systems, it is reasonable to suggest that it may apply more widely. The outcome creates a satisfying congruence, as experiment, hypothesis and simulation come together.

CHAOS AND ITS RELEVANCE TO CONTROL

There remains one further way in which the logistic control mechanism has a bearing on population growth. In the 1970s the logistic equation was to reveal properties of far wider significance than any before; not least because they implicated an empirical property of the equation that is relevant to its properties as a control mechanism. Robert May, now Lord May, is an Australian who trained as a physicist before becoming a leading theoretical ecologist. Later he became an effective and challenging Chief Scientific Advisor to the UK government during turbulent times for science. May has given a fascinating account of his discovery entitled *The best possible time to be alive: the logistic map*, which is drawn on here [11].

In the early 1970s, while at the University of Princeton, May carried out an intensive exploration of the properties of the logistic equation, feeding its solution back into the equation iteratively a thousand times or more to describe the logistic map. What is meant by a 'map' here is that a range of values are calculated for a term at chosen intervals, and are entered into the equation to provide a visual map of the dynamics of the equation.

If we take the logistic equation $dN/dt = rN(1 - N/K)$, divide both sides by K and define $x = N/K$, then we arrive at the differential equation $dx/dt = rx(1 - x)$. For consistency the relative growth rate is given by r, and the number in the population is x. This is the form of the equation used by May to map the logistic equation. For simplicity, the term $(1 - x)$ keeps the population size as a fraction within the limits of 0 and 1, but a multiplier can be inserted to make it more realistic for any known species population. The initial condition can be represented as $rx_{initial}(1 - x_{initial})$, the product of which is x_{next}, which is then fed back into the logistic equation for the next iteration, and so on. May used a simple deterministic form of the logistic equation as $x = rx(1 - x)$, and carried out calculations for a wide range of values for r and x. The graphic outcome is the logistic map (Figure 8.7).

The map of the logistic equation creates the so-called 'fig tree'[3] – May's beautiful plot of the cascade of period doubling that was to lead

[3] May's graphic plot of the equilibrium point (x) against population growth rate (r) was given the name 'fig tree' by Ian Stewart after the American, Mitchell Feigenbaum, who discovered a constant ratio of 4.669 between the intervals of bifurcation in the cascade of period doubling. His name is German for fig tree. This scaling determines the shape of the fig tree, and accounts for the self-similarity of the parts to the whole. The Feigenbaum number is a universal constant.

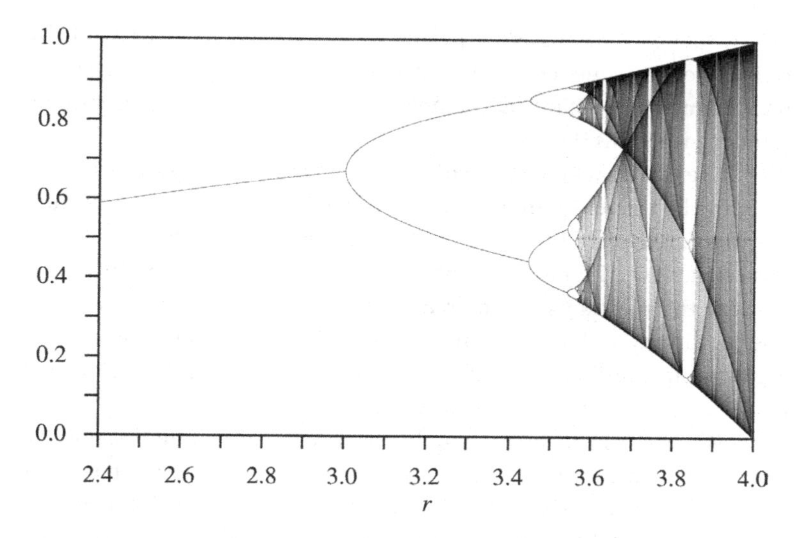

Figure 8.7. The logistic map, or 'fig tree', showing the cascade of period doubling and chaos at levels above. In islands of stability in the domain of chaos, distorted replicas of the fig tree may be seen. Courtesy of Wikipedia, http://en.wikipedia.org/wiki/Bifurcation_diagram.

to the discovery of chaos theory. May points out that his discovery could have been made at any point in the previous 200 years. As it was, he worked long and hard on the logistic map, with little more than 'paper, pencil and a lot of patience' and access to an early desktop calculator. We now consider the implications for a simulated population. With increasing growth rate, we arrive at an equilibrium population size, which is the equilibrium point setting (K) for the outer loop of the logistic control mechanism. This equilibrium point would be most apparent as an attractor if the data were plotted in phase space (as in Figure 5.7). Beyond $r = 3$, the equilibrium divides and becomes two attractors. This must have seemed surprising at first, although such instability was discovered by Eberhard Hopf in 1948 in his study of the onset of turbulence in fast-flowing water. Initially order is created by attraction of the population growth process to a single point attractor, by spiralling in on the point, in effect creating order as the population is brought to some stable equilibrium. As the attractor becomes two, the population oscillates between them. But in mapping a slightly higher rate, four attractors appear, followed quickly by eight, sixteen and so on. This is what May termed 'a cascade of period doubling', which was to lead to the threshold of chaos. As growth rates increase, populations are drawn sequentially to each attractor in turn, and into ever wilder oscillations as the number of attractors increases exponentially. The consequence was that a state of stability at one attractor is transformed into chaos by the doubling of attractors, at increasingly frequent intervals (Figure 8.7).

The reader will recall that control mechanisms typically need to be perturbed to reveal their regulatory activity, because at equilibrium the regulated state or rate simply proceeds at a constant preferred level. Likewise, the logistic control mechanism must be perturbed to activate the mechanism and create a response. Here there is no obvious perturbation, but one is provided by the initial conditions which are far from equilibrium, which itself provides the disturbance needed to elicit a control response (Figure 8.8).

We now consider the mapping as it represents different population growth rates over the critical transition from order to chaos. For rates in the range $r = 0\text{–}1$ the simulated population dies, but between 1 and 2 it quickly stabilises at an equilibrium (K). With increasing growth rates of 2–3, the population oscillates for longer before stabilising; becoming particularly slow to stabilise at $r = 3$, which is the bifurcation point at which two attractors are created. From 3.0 to 3.45 the population oscillates between the two attractors. From 3.45 to 3.54 the population oscillates between four attractors, then eight, sixteen and so on. The threshold of chaos occurs at $r_c = 3.57$, as the cascade of period doubling becomes chaotic.[4]

The discovery of chaos did not happen in a progressive way, as the account implies, but in a piecemeal fashion. In 1973, soon after taking up a permanent position at Princeton, May went to the University of Maryland to give a seminar, where he met the mathematician Jim Yorke. May had described the cascade of period doublings, explaining that he did not yet understand what happened above $r = 3.57$. At that point Yorke interjected, saying 'I know what comes next!' He even had a name for it – 'chaos'. He and his collaborator Tien-Yien Li had not looked at values below the threshold of chaos, so were unaware of the cascade of period doublings that May had found. The meeting and subsequent collaboration made it possible to assemble the two parts of the logistic map as a fig tree graph for the first time (Figure 8.7).

The discovery opened the floodgates of interest in chaos and its theory; not only did a simple deterministic equation produce such

[4] The information referred to here is drawn from the Wikipedia entry on chaos. In particular, the data are taken from a detailed analysis of the logistic map by Jeff Bryant cited in the Wikipedia entry.

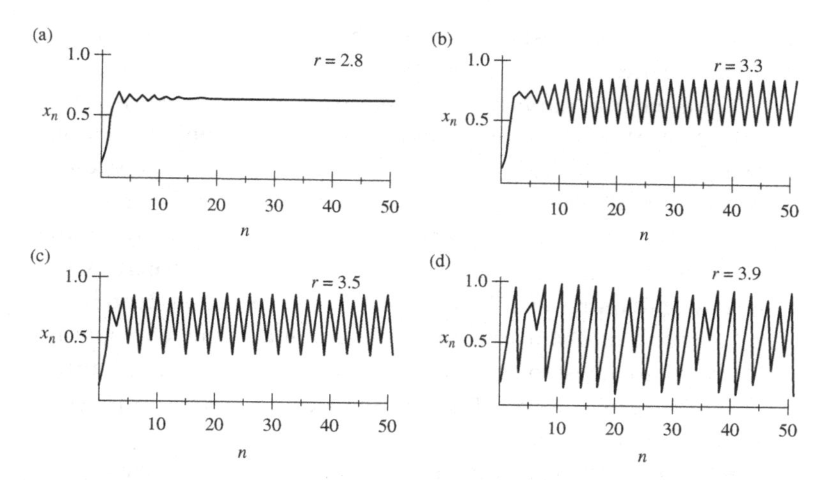

Figure 8.8. The behaviour of x (representing populations) is given with respect to t (time), using different values of r (representing the relative growth rate of the population). The onset of chaos occurs at approximately $r = 3.57$. (a) Stabilization occurs as there is one attractor equivalent to a goal setting; (b) limit cycling occurs because there are two attractors or goals and the population oscillates between them; (c) populations oscillate in a stable manner between four attractors; (d) beyond the threshold of chaos at $r = 3.57$, the cascade of period doubling ends. Any population represented by the logistic equation beyond this threshold would behave erratically and unpredictably, through a series of impossibly wild oscillations. The reader can follow the implications of changes in r by referring to Figure 8.7. Courtesy of Mathematics Illuminated, from www.learner.org/courses/mathilluminated/units/13/textbook/06.php.

complex and unpredicted control behaviour by the multiplication of attractors, but chaos too. Following the discovery, chaos itself became the focus of overwhelmingly greater appeal and importance than the control of population growth. As May explains: 'Chaotic behaviour occurs when non-linear forces are fed back into themselves. This is called non-linear feedback, and is an essential prerequisite for chaos'. The logistic map also has fractal properties. For example, a stretched version of the fig tree can be made out within areas of chaos on more detailed plots. The realisation of chaos and the beauty of fractals captured the imagination of many. The implications of the emergence of chaos from order, and the discovery of such complexity from simplicity, gave rise to fundamental theoretical questions. It was soon found that other simple non-linear equations produced complex and beautiful outputs. The list of equations that exhibit chaotic behaviour now numbers over 50.

The ecological implications of May's discovery lacked the same appeal [12,13]. What did it mean for populations and their regulation? Their behaviour at different growth rates showed the progression of increasing instability as control is lost and chaos increases (Figure 8.7). Chaos meant that its control became chaotically variable above the threshold of chaos (r_c). The interpretation of the logistic map from the perspective of control as a loss of control due to period doubling is clear. A control mechanism with two goals on a single loop appears maladaptive, so any step towards instability beyond the first bifurcation will be selected against. The control mechanism loses its purpose by having more than one goal.

A cybernetic interpretation of the logistic equation (Figure 8.2) considers the system more explicitly as a control mechanism that loses its ability to control. In an earlier chapter a presentation of growth data in phase space (Figure 5.7) shows the stabilisation of controlled growth rates at a single attractor. Following perturbation, the stability of growth rate is gradually restored as it is drawn towards the point attractor (Figure 8.8a). This phase is shown here between $r = 2.4$ and $r = 3.0$. Following the first bifurcation, two attractors are created and, as the process is drawn from one attractor to the other, oscillates with increasing amplitude as growth rate increases (Figure 8.8b). In control theory this behaviour is called 'limit cycling'. In oscillating between two attractors, the meaning of a single goal as a preferred condition defined by a single attractor or equilibrium point is lost. As attractors multiply by period doubling, control is inevitably lost because, instead of being drawn to a single point in phase space, the population is drawn to a number of attractors. Equilibration and stability become impossible due to the doubling of attractors, and the rapid shifting of the equilibrium from one point to another (Figure 8.8c, d). Any more than one attractor is obviously less than optimal, and biologically maladaptive. The control mechanism is denied a single goal that would have defined its purpose, and many goals are without meaning, leading to chaos as its capacity to regulate is destroyed.

SUMMARY

Attention is focused on the logistic equation and its historical development. A key message has been overlooked in the classical paper by Hutchinson presented in 1948 [2], the year that Wiener's seminal book

Cybernetics was published. Hutchinson argued that the logistic equation could be formally interpreted as a control mechanism. Natural populations oscillate, but the logistic equation would not mimic such behaviour, so he added time delay terms that made its behaviour more realistic. The logistic control mechanism regulates growth rate, and overcomes the positive feedback problem of exponential growth, by limiting the products of growth.

It is then asked what agents in life might bring about self-limitation of populations to impose limitation. Examples involving pheromones are considered, which are less ambiguous than findings from ethological studies, on which debate focused. But the idea of pheromones as inhibitors is replaced by one that suggests they created a gradient of toxicity that induced movement and dispersal from the natal area. The important conclusion from this chapter is that the logistic control mechanism, which combines rate control and self-limitation of the products of growth, overcomes the earlier problems of exponential increase where there is rate control alone. The logistic map of May, which exhibits chaos in a mathematical sense, is considered as it might be used to interpret the behaviour of populations at the onset of chaos, and the loss of control as the rate of population increase with period doubling.

9

Hierarchy: a controlled harmony

The actual organization of living systems into sub-cellular organelles, organs, organisms, populations, species and so forth is analogous to a hierarchy of control mechanisms.

A.A. Lyapunov

A theory of multi-level systems is indispensable for making any progress in understanding complex organizations.

M.D. Mesarovic

Reductionism means you turn a blind eye to the higher functions of hierarchy, the emergent properties of hierarchy, the goals and purposes of hierarchy.

Douglas Hofstadter

Where there is the need for a controller, a controller of the controller is also needed.

Tadeusz Kotarbinski

WHAT IS A NESTED HIERARCHY?

Animals are immensely complex and their ordering in a hierarchical manner, with a number of levels of organisation, is what helps make their understanding possible. It is almost as if life was structured as a hierarchy so as to divide up biology for the convenience of those who study it. Thus we have molecules, organelles, cells, tissues and organisms, each of which is a specialist area for different kinds of biologists, who observe the distinction between levels of biological organisation as the boundaries to their own disciplines. This is convenient, as different specialists require different technologies related to the scale and level in the hierarchy at which they work. Typically, specialists in the various fields receive different training, attend different conferences and even work in different buildings. In this way the complexities of the biological sciences have been separated by natural boundaries between them. These boundaries handicap our understanding of organisms as multi-levelled systems, and sever links between the disciplines. Such divisions have been consolidated by reductionism, which has driven biologists to look downward within the hierarchy to find mechanisms and explanation for what they observe at a higher level. This distracts the enquirer from understanding the meaning and purpose of hierarchy, which resides at higher levels. For such reasons, the published theory of hierarchies is limited (see [1]). Most scientists therefore have an approach to understanding life that looks downward within the hierarchy, rather than upward, for the meaning and purpose of what lies below.

Of course, this is a caricature of the biological sciences today. Yet all biology has been divided by levels of organisation as natural boundaries of investigation, because the consequence has been our failure to understand how hierarchical levels relate to one another. So what is the significance of hierarchy from the perspective of the organism itself?

Surprisingly, nearly all dictionaries of biology omit an entry on 'hierarchy'. This is a remarkable omission, given the pervasiveness of hierarchies in the organisation of living organisms, but it reflects the extent to which biologists have failed to adopt hierarchy as an organising principle. We must remind ourselves that hierarchy evolved through natural selection and not for the convenience of reductionist scientists in order to partition biology. Each level represents a major evolutionary advance that contributes to fitness. The author and a colleague posed the question, some years ago, as to whether growth is controlled by the hierarchical system [1]. Whether this is true or not, we must expect hierarchy to be adaptive.

COMMONPLACE HIERARCHIES

The concept of 'hierarchy' is so deeply embedded in our daily lives that it is difficult to imagine life without it. Business, the armed forces, the church, the civil service and so on are all organised hierarchically. In conversation, the implications of hierarchy pass almost unnoticed: high and mighty, high-ups, High Court, High Priest, high-ranking, Royal Highness. Conversely there are low-born, low-class, the Lower Chamber and the lower deck, all implying some underlying level of a hierarchy. Any attempt to overthrow a government, and its hierarchy, quickly leads to chaos, until some new hierarchy is established and order returns. In a broader sense, the association of hierarchy with order is important, since a hierarchy in any context organises, and so creates order.

The usual metaphors for a hierarchy, such as library classification systems and the organisation of the military, only go so far in helping us to understand what is meant by hierarchy in biology. Biological hierarchy incorporates a particular twist, in that it is physically 'nested'. To depict what a nested hierarchy is and does is difficult, because no single analogy fits it precisely. A Russian 'matryoshka' doll is a good first step, as it gives emphasis to nesting one level within another. But we have to imagine that the largest doll contains many dolls at the next lower level, and each one of these contains many smaller dolls at the next level down, and so on. The number of dolls of each size rises sharply with each reduction in size, reaching astronomical numbers as the dolls become microscopic. If we were to take the dolls apart and lay them out in an orderly way on a large table, we would then have a pyramid of numbers with the largest doll at one end of the table, and the rest arranged by size in rows across the table. Now we would have a military type of hierarchy, except that all the officers and all the men would need to fit inside the commander-in-chief!

We need to use the concept of modularity in describing hierarchy, since life exists in units of different sizes, which are the building blocks of hierarchy. Modules of different sizes constitute the levels of organisation, yet do not constrain the variety of novelty. Components can be assembled for different purposes, just as Lego blocks or house bricks can be used to make toys or buildings of any kind. A limited number of shapes of building block (say ten) is capable of being assembled in a large number of ways to meet a greater variety of purposes. Clearly there are many advantages to modularity.

Any level in a nested hierarchy must be large enough to encompass a number of modules of the level below it (Figure 9.1). At the same time, modules must be small enough for a number to be enclosed by the level above. So the modules that make up a hierarchy have a wide range of sizes (Figure 9.2). The highest level has one module, and from the highest to the lowest the number of modules that make up each level increases sharply. Better-known cells like the yeast cell

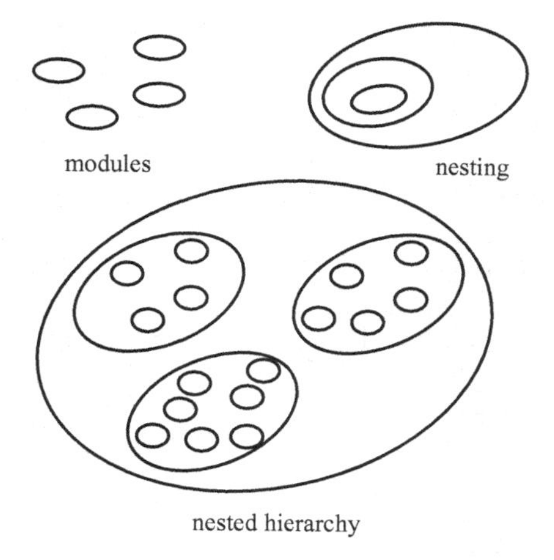

Figure 9.1. Diagram showing modules, nesting of modules and a hierarchy of nested modules.

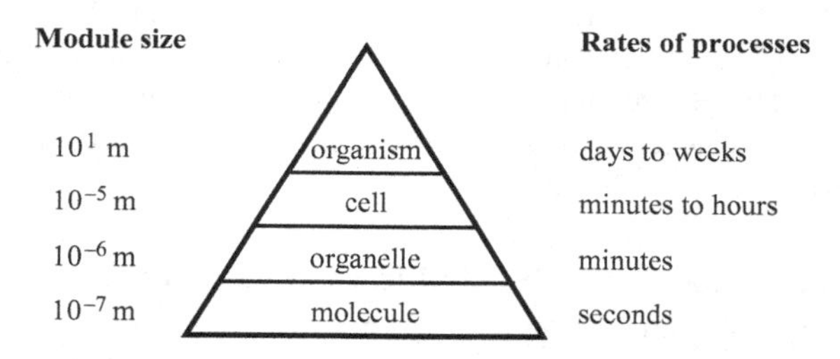

Figure 9.2. A nested hierarchy is shown as a pyramid, with the area of the segments indicating a greater number of smaller modules at lower levels, decreasing to unity at the highest level. The numbers on the left-hand side are rough approximations to indicate that the size of the modules increases significantly at each level to contain all those beneath. The diagram also shows that the rates of processes slow down towards the apex (right side) because each level includes all the processes that occur in the levels below.

may contain several thousand mitochondria and 200,000 ribosomes, for which ~16 million protein molecules are synthesised with each cell cycle.

It is useful to think of a hierarchy as a pyramid, reflecting the number of modules at each level (Figure 9.2). The area of each stratum reflects the number of modules at each level, so at the vertex individuality is represented by a single module. A hierarchy gives ownership of the many modules below to the highest level in the

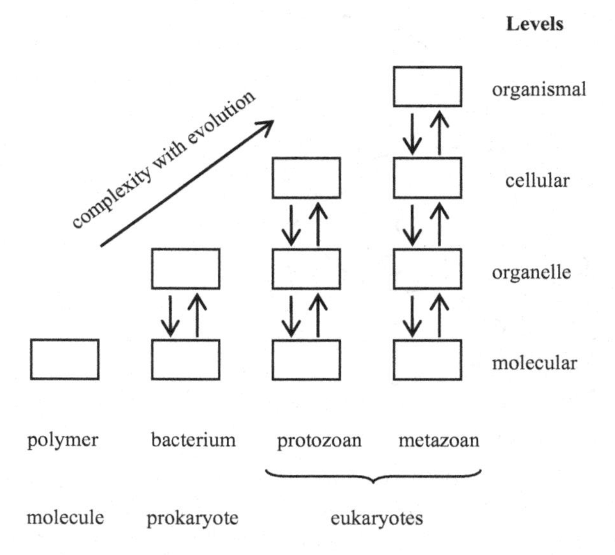

Figure 9.3. Diagram indicating the levels of organisation in different major groups of organisms (*vertical*), in relation to levels of organisation (*horizontal*). The 45° line shows the way in which, in evolving, organisms have acquired new levels of organisation, and so have increased in complexity.

hierarchy. The pinnacle of a hierarchical pyramid represents the level of greatest biological complexity, because it encloses all the levels of organisation beneath it. Hierarchy cannot be separated from the fact that life evolved from something small and simple to become something large and complex (Figure 9.3). During the course of evolution, the addition of new levels to the hierarchy has been through a series of major steps and increasing size, each of which has opened up new evolutionary possibilities. At each step, a new level of organisation is added by superimposing itself upon pre-existing organisms or, more realistically, the functionality of simpler organisms is adopted by enclosing them within a larger organism, which imposes its own purpose upon them. In the evolution of cells this process is called endosymbiosis. The result is that the seat of individuality moves upwards with each new level. It is at this level that overall direction and control originates, and purpose is to be found.

If you are finding it difficult to comprehend what is meant by a nested hierarchy, you will begin to see why so few have sought to try to understand the meaning of hierarchy in biology. There is simply no accessible model to help understand nested hierarchies. We begin to see why there is so little published information about how they work and what they do. The furthest we can go is to take two conceptual models: the first is the Russian doll, where each doll fits inside another; the second is to superimpose the idea of a pyramid of numbers, where in opening up each size of doll, one actually finds many smaller dolls, such that at each downward step the numbers increase.

We need to add one more property to the organisational pyramid, which is that each module is a replicator. Each module is the product of replication, and is able to make copies of itself. Modules are unitary components of life, with different modules at each level of organisation. Self-replicating modules may be molecules like DNA, organelles like mitochondria or the cells of all metazoans. But the stability of the modular hierarchy depends upon the control of replication at each level. So we begin to see that modularity is also a growth problem, as unlimited replication would threaten the organisation and coherence of the whole.

SOME THEORY OF HIERARCHY

Before we become drawn further into the question of why and how biological hierarchies are so important, we should consider briefly what is already known about them in terms of hierarchy theory. What are the innate properties of a nested hierarchy? At different times over the last 50 years the concept of a 'hierarchy' has been of interest to philosophers, systems theorists and mathematicians, as well as biologists.

It was the philosopher and biologist Joseph H. Woodger (1894–1981) who proposed a method based on axioms from which theories and their implications could be deduced. In the 1930s, using this method, he explored hierarchical order in biology, identifying various kinds of hierarchies [2]. The 'spatial hierarchy' is nested such that any level consists of modules of the level below, and constitutes the modules of the level above. At its simplest, one can visualise four replicators that constitute the main levels of organisation: molecules, cell organelles, the cell itself and the multicellular organism (Figure 9.2). So molecules make up organelles, which make up cells, and cells make up metazoan organisms. The modules that create each level originated from free-living organisms, and retain the ability to replicate themselves. It is their ability to multiply and make identical copies of themselves that concerns us, as there must be limitation to any potential exponential growth.

Ludwig von Bertalanffy (1901–1972) was an Austrian theoretical biologist who emigrated to Canada. He was the co-founder of general systems theory,[1] and soon adopted Woodger's logical analysis of hierarchies into his systems view of biology [3]. Woodger had defined hierarchies of descent by cell division, of spatial hierarchy and a genetic hierarchy. Von Bertalanffy recognised that any hierarchy of parts also needed to be supplemented by a hierarchy of processes. This created an image of interlocking subsystems: an idea that we shall return to. He also later folded cybernetics into his general systems theory, which is another reason for Wiener's original concept of cybernetics falling into disuse.

We turn next to James K. Feibleman (1904–1987), a self-taught philosopher, psychiatrist, author and poet, who became head of a university philosophy department. He specialised in symbolic logic and epistemology, and wrote a paper in 1954 that summarised the essential features of hierarchies [4]. For the time, he probably took the understanding of hierarchies as far as it would go by thought and reason alone.

The number of modules in a hierarchy is most numerous at the lowest level, decreasing to one at the highest level of all. Each level is organised by the one above and controls the level below. The higher level provides direction, while the lower one has the mechanism that responds to authority from above. Each pair of adjacent levels is in this way interdependent. What emerges from these simple rules is a hierarchy controlled by a single module at the highest level, which is the seat of autonomy and 'ownership'. It is to this level that the whole hierarchy belongs, and at which the purpose of the hierarchy is located and can be explained. Causality in a nested hierarchy is from the bottom up, such that anything which impinges on a hierarchy has an impact on molecules at the lowest level, which affect organelles, which affect cells, and so on. Thus a toxic effect at the lowest level cascades upwards through the hierarchy.

[1] General systems theory (GST): this is an approach, rather than a theory, to an interdisciplinary field of science devoted to the study of complex systems in nature, society and science. It adopts a holistic rather than a reductionist approach, as the properties of systems are destroyed by reducing them to their component parts.

A hierarchy increases in complexity upwards, not simply because of the added complexity of a new level, but due to nesting, as each level encompasses the complexity of those beneath it. It seems that no level of organisation has been lost during the course of evolution. In this way the complexity of organisms increases cumulatively with each additional level of organisation, such that ownership, and with it purpose, changes during evolution.

It is possible to 'reduce', in an analytical sense, a higher level to a lower one, by saying, for example, that genes consist of nothing but DNA molecules, but this does not help us to understand hierarchies. Feibleman referred, 50 years ago, to the 'fallacy of reduction'. A reductionist view of a hierarchical organism would be to consider one level, then to seek understanding in the level below. There one would find the mechanism for the level in question, but it is the level above that provides its purpose. Without detracting from the understanding of living systems achieved using a reductionist approach, reductionism has imposed a blinkered effect on the study of hierarchies, and life in general, overlooking those emergent properties of organisation and complexity that are impossible to discern from the components alone. It becomes clear why reductionism is little help, and why hierarchies have not been better understood, while reductionist science has provided the predominant approach to acquiring new understanding. As the limits of reductionism have become better understood, investigators have become more confident in moving upwards or downwards within hierarchies, or in both directions.

Feibleman recognised that 'the higher level processes in a hierarchy will necessarily extend over longer time intervals, since they require the preliminary completion of lower level subprocesses'. This implies that the more levels that constitute an organism, the slower the response time at the highest level (Figure 9.2); the lag at the lowest level is added to the lag in the level above, and so on. So delay in the system is due to the cumulative delays of each of the control systems in the hierarchy, and the time required for some response from the hierarchy increases as it progresses upwards through the levels. Consequently, the more levels, the more delays, and the longer it takes to complete higher-level processes. It also follows that within a metazoan, more rapid responses are found by accessing lower levels of organisation.

More recently the contribution of Herbert A. Simon's (1916–2001) insights to complexity and hierarchy should not be overlooked. He was an interdisciplinarian who won the Nobel Prize in economics for his work on decision-making and uncertainty when information is less than perfect. His interest in hierarchy is well known from his paper on *The architecture of complexity* [5]. He considered different kinds of hierarchies, and concluded that complexity could not have evolved without each level being composed of elements of a lower level. This is tantamount to saying that a nested hierarchy is the only kind of complexity, as there appear to have been no others. But such hierarchies can be created in more than one way. In nature the opportunity for hierarchy has been ever-present, as replicators in replicating do not part immediately and so create something new and bigger. If it confers advantage in that form, it must survive and evolve.

An early step in evolution was the evolution of cells whose living descendents are called Archaebacteria. These are the simplest units capable of metabolising and replicating. Some were adopted by a process called 'endosymbiosis', an idea made credible by the American biologist Lynn Margulis, and which has occurred several times during evolution [6]. The precise mechanism can never be known, but it has to begin with a large cell engulfing a smaller cell. The larger host organism had to pre-exist, so size alone had survival value, without the added value of the smaller organisms that it adopted. Perhaps the smaller cell was mistaken for food and was therefore engulfed, although it was really a parasite looking for a host. Either way, the relationship evolved to become a permanent one of mutual benefit to both. Forever captive and subordinate, the adopted modules are protected and live a more stable life, perhaps with better prospects for survival inside a larger organism. So it was that bacteria became modules within eukaryote cells, which in time themselves became the building blocks of all multicellular life. So the addition of new levels of organisation was achieved by superimposing new higher levels upon the older smaller units (Figure 9.3). In this way cells brought together various functions of what we now know as the eukaryote cell, which is so adaptive and multifunctional that it is shared by all multicellular organisms.

When the number of all possible relationships between entities or modules is considered, a hierarchical relationship proves to be the simplest. It may be that hierarchy provided the most favoured organisational structure because a hierarchy reduces the complexity of decision-making to a proportion, rather than a power function, of the number of tasks. Although life has become exceedingly complex, natural selection opts for parsimony at each small step. Yet as Rupert Riedl (1925–2005) exclaimed, the consequences of hierarchical order 'can be so complex as to reach the limit of conceivability'. So while the organising principle may be simple, the complexity that can be embedded within it has evolved to become massive. The ubiquity of hierarchy as the organisational structure for living things leads one to the idea that, if natural selection did not create anything better than a hierarchy, perhaps nothing better existed. H.H. Pattee, the eminent writer on hierarchies, posed a question that is not as trivial as it might seem [7]. He said: 'If asked what the fundamental reason is for hierarchical organisation, I suspect most people would simply say, "How else would you do it?"' It is a challenge to any thinker to imagine a simpler, a more efficient or a more effective system, but none is apparent.

IMPLICATIONS OF HIERARCHIES

We now consider more closely the rules by which modules are used to create organisms. Modules are of different sizes, and a simple model can be constructed where the modules that constitute each level of organisation increase approximately tenfold in size over the one below (see Figure 9.2). For instance, the DNA double helix is 0.002 microns wide, organelles such as mitochondria are up to 5 microns long, cells are 10 microns across and multicellular organisms, like humans for instance, are 2 metres tall, or 2 million microns. So it is that modules at each level are of sufficiently different sizes to fit inside one another and become 'nested'.

The more levels within a hierarchy, the slower the whole system operates, due to delays in commands being passed down through the hierarchy, and delays in the upward feedback. Therefore, in growth processes, the lag in control mechanisms increases, and the frequency of oscillations decreases with the addition of more levels (see Figure 9.6). During evolution the frequency of growth rate oscillations decreases with the number of levels, and the size of organisms, with the slowing in the rates of metabolic and growth processes. So it is that biochemical systems may cycle every 0.5 seconds (8×10^{-3} min), the frequency growth rate oscillations for yeast are 3–4 hours (21×10^{1} min), for hydroids 12.5 hours (7.5×10^{2} min) and for humans 118 days (17×10^{4} min).

Multicells are thought to have evolved from free-living unicells at least ten times during the course of evolution. The control of organelle replication within the cell must therefore be regulated within the new domain bounded by the cell wall. In addition, the cell has to superimpose overall control for organelle density to best fit its own purposes. It was an essential evolutionary step for the host module to impose limitation and control over replication of the adopted organelles. Each kind of module must also have its own population control mechanism, for without it, uncontrolled replication of the adopted module would have disastrous results for the host. Any module that assumes its own autonomy threatens the survival of the whole organism (Chapter 12).

Connectivity between modules of equal status is essential; without connections, all modules would act in their own self-interest, and competition would result. Each level of modules is subordinate to the level above. Modules also contain smaller modules at a lower level than themselves, over which they impose control. At its simplest an organismal function is likely to require the involvement of the modules of which it is made, its constituent modules and their constituent modules, and so on. Subordination, and coordination, must be built into nesting. It follows that for a hierarchy to work as one, under the command of the highest levels of organisation, communication and the flow of information is vital. These must include commands flowing downward through the hierarchy, and coordinating links between each level, as well as links between modules at each level. For an organism to respond to some stimulus or change, contributions are required at the molecular, organelle, cell and organismal levels, each of which depends on the part played by the levels beneath. The result is a multi-level response that is, of necessity, sequential.

A hierarchy is not a centralised system; it is a devolved system within which the business of each level is controlled at its own level, but with commands from a higher level that can intervene to meet higher purposes. 'Housekeeping' jobs, like the replacement and repair of damaged or worn-out modules, are dealt with locally. In our own bodies there is little awareness, at the highest level, of liver function, the replacement of cells or the healing of wounds, yet they proceed continually without conscious intervention. Other processes like breathing, and to a lesser extent heart rate, can be brought within the ambit of conscious control, but usually the control of such processes is delegated to lower levels. Pain receptors and awareness are restricted to those processes over which consciousness and higher centres in the nervous systems have some control.

ORGANISATION OF CONTROL

From this consideration of the properties of hierarchies, there emerges the realisation that there must be an organisational structure of

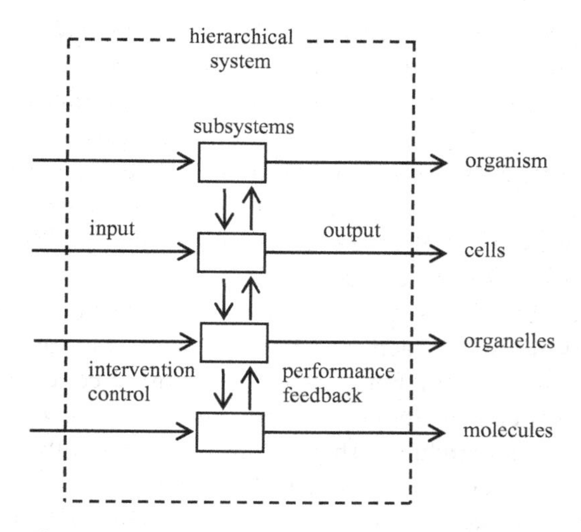

Figure 9.4. A diagram, modified from the model of Mesarovic *et al.* [9], indicating inputs and outputs at each level (*horizontal*), and the minimal links for information flow between levels (*vertical*).

command and control. Independent organisms control their own multiplication, but as endosymbiotic modules their replication is directed to meet the host's purposes. Subordination of lower levels by higher levels means that there is control of some kind running throughout the hierarchy. Endosymbiosis provides a new purpose for the adoptive module, and hints at other properties of control. As we begin to consider a 'control hierarchy', we turn to another important figure in the understanding of hierarchies. This is the system theorist and modeller, Mihajlo Mesarovic. With the first Club of Rome report, *Limits to Growth*, in 1972, he was involved in understanding global change, but realised that instability was being created at a regional level. He gathered a team together at Case Western Reserve University in Cleveland, Ohio, to study global problems at a lower level, showing how national policies were contributing to the instability of the world.

His earlier work had addressed the question of control processes within hierarchies [8,9]. He developed a mathematical theory of hierarchical systems that was 'structural, formal and independent of the field of application'. He also recognised that the principle of nesting could not work unless there was just one module at the highest level, with overall control. Mesarovic and his colleagues set out a generalised organisational structure for a hierarchical system (Figure 9.4). They proposed that the management of interdependencies between levels requires at least two channels of communication, one for downward commands and one for upward feedback of information on the performance of lower levels. The constant exchange of information between levels provides the mechanism for overall coordination.

While the vertical fluxes within Mesarovic's hierarchy consist of a flow of information, the horizontal fluxes of the subsystems, acting as input–output systems, deal in energy and matter. These fluxes reflect Waterman's view that energy and information are the two aspects of life that epitomise what organisms do. Clearly hierarchies are not merely systems of structural organisation, in the way that Woodger thought about them, but are also the means of controlling processes within systems. A single-celled organism has its own purpose in survival but, as a module within a metazoan, these cells serve the purpose of the higher levels as well as their own.

The vertical links between levels indicate the routes by which information flows within the hierarchy, and give the hierarchy its overall behaviour and properties. The arrows pointing downwards implement 'intervention controls' and can overrule the local control system at the level below. They have priority of action over the local processes that run through the hierarchy, imposing control by command from above. 'Performance feedback' is provided by the arrows in the upward direction, which channel information back to higher levels about the performance of each lower level.

At this level of abstraction, one could gain a false impression of regularity and equivalence at each level. But Mesarovic emphasised that each level has its own laws and principles, and its own variables and factors, so it is necessary to treat the functioning of any one level as different from all the others. Vertical links carrying information between levels may not necessarily pass through each intervening level, but may skip levels to increase the speed of signalling. A good example would be a reflex response, which is not the result of thought by the brain, but is an automatic response that goes no further that the spinal cord for its command for action. Many processes at any one level are self-controlled, and may involve a thousand feedback loops in each cellular module. The boxes representing each level may contain any number of subsystems responsible for the autonomous control of processes and functions at that level.

Each control mechanism is likely to have the function of optimising some process, or maintaining variables at set levels, like the feedback mechanisms considered in earlier chapters. There is an asymmetry in the vertical fluxes of information in each direction, which ensures the overall command of the highest modules. What Mesarovic calls 'intervention control' is made possible by the imperative of signals from above, which ensures overall command. The flow of information from below, as 'performance feedback', simply informs higher levels about the progress of processes below. The downward flow of information is active, while the upward flow is passive. Overall command and control originates at the highest level, and control information flows downwards to each level from above, while performance information on output is fed upwards from beneath to inform of the consequences of commands.

One can obtain a more detailed understanding of mechanisms by moving down through the strata (reductionism), but a deeper understanding of their overall significance and purpose by moving upwards (holism). The understanding of hierarchies requires one to cross the levels of organisation, rather than to observe the boundaries between levels. One can go further and say that it is not the boxes that matter most in the Mesarovic model (Figure 9.4), but the links between them.

In Chapter 6, the use of a control mechanism for internal growth processes was introduced, based on the logistic equation. The origins of this control mechanism for population growth are given in detail in Chapter 8. There is reason to believe that such a control mechanism, used here more as mathematical metaphor than a model, may have evolved more than once, and might have applications in different contexts. This is not difficult to accept, because the requirements of any replicator to control its own growth and multiplication are essentially the same, involving control over the underlying rate process and limitation to the cumulative products of growth. We know, from numerous examples of convergent evolution in unrelated species, that if the selection pressures are the same, similar outcomes will result. It is just that, typically, most known examples of evolutionary convergence relate to form rather than process.

It can be appreciated that a free-living and independent organism may have evolved a control mechanism resembling the logistic control mechanism. The evidence suggests that control mechanisms with the properties of the logistic control mechanism are to be found in ecology, and as the regulatory mechanism for the control of modules internally. We are already familiar with the properties of the logistic equation that enable it to be widely used in modelling in population ecology, in microbiology, in cancer biology and in cell culture. We now suggest that the logistic control mechanism has other applications that are equally general.

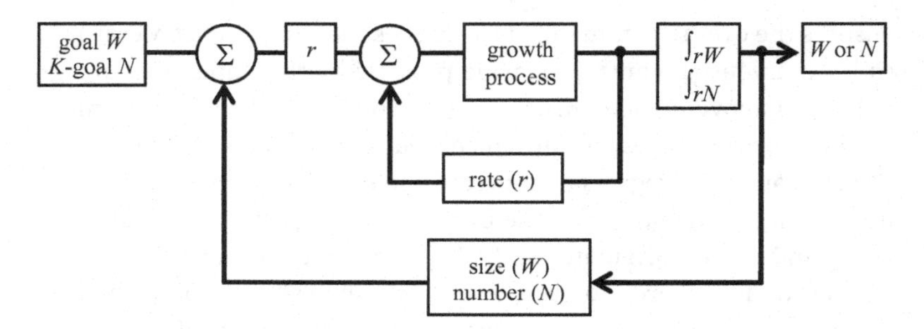

Figure 9.5. In this form, the logistic control mechanism is capable of representing the control of cells or populations as specific growth rate (r), and the outer loop limits the tissue size (W), or population number (N) in relation to the carrying capacity (K) of the habitat. The generality of this form of growth control module indicates that it is capable of representing control in replicators and organisms of different kinds.

It may never be known how endosymbiosis came about, but an organism could have survived ingestion and became an organelle of a larger organism, as Margulis has proposed [6]. Its control mechanism would then have been internalised too. The inner loop providing rate control of biosynthesis would remain the same, but the purpose of the outer loop providing limitation of numbers would change. The idea that exocrine glands secreting pheromones to the exterior became endocrine glands secreting hormones to the interior was first proposed by J.B.S. Haldane, which he put forward – like some of his other more provocative ideas – with little further development. If such speculation were to be true, any inhibitory exocrine limiting population numbers could become an endocrine or hormone responsible for limiting the numbers of modules within the organism. More will be made of inhibitory regulators in Chapter 12, as well as the link between such agents acting as exocrines in populations and as endocrines within mammals. Inhibitory regulators are now known to limit the cell population size in mammalian tissues. The point is that such a lineage provides further corroboration of the idea that a mechanism with the properties of the logistic control mechanism may well have evolved for the control of populations in ecology and ended up performing a homologous role within higher organisms.

We therefore adopt the logistic control mechanism (Figure 9.5) as the minimal unit of control within the hierarchy, from populations to subcellular biosynthesis (Figure 9.6). In this way, a linked series

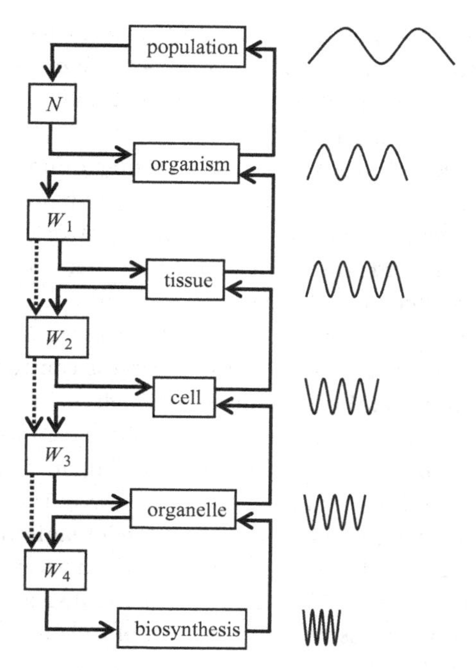

Figure 9.6. A hierarchy of growth control mechanisms for the different levels of biological organisation using the logistic control module given in Figure 9.5. While each level of organisation is semi-autonomous in controlling its own growth, overall control of size (W) or number (N) is determined for the highest level by intervention control (*dotted line*). An indication of the frequency of control oscillations shows that the rate of processes, and the frequency of oscillations, decreases with each additional level.

of logistic control mechanisms would operate at each level of organisation. The dual loop module divides its functions between those at each level of organisation and those that relate to the organism as a whole. Thus the inner loop providing rate control of biosynthesis regulates the growth process at its own level (within each box). Feedback on performance passes upwards through the hierarchy from the lowest to the highest level. Such loops relate to the size of the population of modules, which is of necessity limited by the size of the population in the level above. The growth of new modules is limited locally, but allows for intervention control from above: this applies from the lowest to the highest level of organisation (dotted line from W_1 to W_4 in Figure 9.6).

From the lowest to the highest levels, growth rates become slower, as does their frequency of oscillation. This is to be expected because the operation of each mechanism is slowed down by the delays in its constituent modules in the level below. It follows that delays accumulate. This hierarchical system allows semi-autonomous control, such that there is local control at each level of growth process, to allow for repair and regeneration. But the overall control of the products of each level is limited from above, ensuring that no level outgrows the level within which it is nested.

It should be noted in each case, for example the control of biosynthesis or cell proliferation and so on, that the process is controlled by regulating its rate by means of the inner loop. Although the process may be controlled at a constant specific rate, the outcome results in exponential growth unless size limitation is superimposed upon it, thus preventing continuous acceleration and unsustainable growth. The output, or product, of each process feeds into the next level of organisation, providing the right number of cells or size of tissue, while preventing unlimited and autonomous growth. This allows us to visualise a hierarchy of minimal control mechanisms for each level of organisation, linked to operate sequentially. This also suggests how growth control across the levels of organisation might be coupled and coordinated, feeding into the overall growth process of a metazoan. While growth at any level may be related to local requirements, overall growth contributes to the same process of metazoan growth, as it contributes to cells, tissues and organisms.

Within this hierarchy, it seems likely that intervention control, as proposed by Mesarovic, is effected by an overall goal setting provided by the highest level, which is linked to each of the subordinate goals at lower levels. We know, for example, that overall control is affected by growth hormone from the pituitary, which reaches down through the hierarchy to play a subcellular regulatory role (see Figure 6.5). Performance feedback is provided by the products of replication and growth at each level providing the modules for the level of organisation above (Figure 9.5).

Whereas the logistic control module is associated with independent life forms, organotrophs have an external loop. This works by secreting an exocrine into the organism's habitat, which is contributed to by all the individuals of its species. Each organism also senses the exocrine concentration surrounding it, which provides a collective index of the population density. As exocrine concentration increases with density, it has the effect of limiting reproduction at a level that is within the carrying capacity of the system.

In his later years, Jacob Bronowski (1908–1974) held a fellowship at the Salk Institute in California, and was interested in the evolution of complexity. He believed that advances as important as the addition of a new level of organisation to an organism could not be reversed because no single mutation, or cluster of mutations, could possibly undo such a major evolutionary step [10]. This, he argued, would ensure that such advances would be conserved. As he believed such steps were too large to be reversed, they provided 'a barb to the arrow of time': a ratchet to evolution. This is an important idea because it implies that evolution must be progressive, increasing in complexity by the cumulative addition of new levels and greater functionality in creating a hierarchy.

The coordination of an integrated system such as an organism is easily disturbed, since the nesting that ensures integration also makes it vulnerable. Bronowski thought that the addition of levels of organisation during evolution might increase stability so that, as an organism became more complex, its 'stratified stability' increased, with each layer adding extra resilience to the effects of environmental challenge. However, he did not expand on what he might have meant by this idea. The extent to which the disturbance is transferred from one level to another depends on the level of integration, the strength of the disturbance and the neutralising effect of adaptive responses. Any disturbance introduced at one level cascades upwards through the hierarchy, simply because the modules at one level constitute the level above in the hierarchy.

These four approaches summarised from Woodger, Feibleman, Mesarovic and Bronowski, spread over 70 years, provide some understanding the function of biological hierarchies, indicating advances from a descriptive classification of the kinds of hierarchies to a more rigorous formal description. More important, they lead us to think about an integrated system for the control of processes that involves all levels of organisation. The hierarchy is at the heart of command and control processes throughout the organism. However important hierarchy may be as a nested organisational structure, what emerges is that hierarchy is a framework upon which hangs the control of processes within the organism, although here the emphasis is upon the control of growth. Hierarchy provides the means of coupling and coordinating all levels, with the highest level responsible for overall command.

COMPLEX ADAPTIVE SYSTEMS (*CAS*)

Significant advances in understanding hierarchical organisms have been provided by the study of simulated adaptive complex systems. In 1984 the Santa Fe Institute was set up to give a broad interdisciplinary focus to the study of complexity and its meaning. The Nobel Laureate, Murray Gell-Mann, was a leading instigator, who had recognised the way in which computer-based learning systems, or neural nets, had proved so successful in detecting patterns and trends in large data sets [11]. Gell-Mann brought together a small group of investigators to capture the essence of natural selection from biological evolution, so that it could be used to help businesses, economies and societies. First, it was necessary to abstract the principles within a computer simulation model, so that it could be used to simulate non-biological processes. Their aim was to study man-made complex systems, whether institutions, ventures or cities, using the example of Darwinian selection as a model for adaptation and the creation of novelty. They called them 'complex adaptive systems', or *cas* for short. One outcome was to invite John Holland to develop a computer game to convince players of the creative power of evolution by natural selection, 'to produce a highly visual model that would illustrate the creation of complex structures by natural selection'. The result was the ECHO model described in Holland's book *Hidden Order: How Adaptation Builds Complexity* [12]. He captured the essence of evolution by natural selection, and put it to work in a computer. He thought of his model as a caricature of reality, and laid the foundations of the field of 'genetic programming'.

Holland's achievement is important here because he also answered the question as to how hierarchies might have come about, since hierarchy is a necessary feature of the organisation of *cas*. He first had to distil the principles of Darwinian natural selection down to a level of abstraction that could be applied to non-biological systems. Holland proposed a system of rules based on a minimal set of general principles.

In nested hierarchies, the modules at each level constitute the level above. Holland reduced the interactions between hierarchical levels to a set of laws relating the relationship to the levels above and below. Each level was assigned the same laws, providing an interlocking cohesion and the strength of shared algorithms. He saw hierarchy as the organising principle of *cas*.

He identified three mechanisms and properties that he believed to be central to the task of creating a model to mimic natural selection.

The first of these is *modularity*, and has already been considered. Nature is modular at all levels. The second property is *aggregation*. Populations of modules must be connected to one another, if each is to be part of an integrated whole. Every module cannot connect directly to all other modules, since connectivity in large populations of modules would soon become impossible. Nevertheless, modules are connected to adjacent modules, and thereby to other modules in a population. Thus signals are passed from module to module.

There are also the benefits of number and of size that come from the aggregation of modules. There is literally 'safety in numbers', and the whole is resilient to damage and some losses; because the module can also replicate itself, regeneration is at hand. Identical building blocks are created by the iteration of the same genetic sequences, so modularity is genetically economical and efficient. Modularity is an essential principle, as it would be impossible to have a genome capable of coding each element separately. As it is, any number of identical units can be replicated by the same genetic algorithms. So modularity in biology, or technology, represents a huge economy in the use of information. Identical modules may be assembled in different ways to perform different functions, and modified for different purposes.

The third principle is that of *tagging*, which facilitates the interaction of like with like, and defines the identity of the individual. This makes possible the adoption of 'self', and the rejection of 'non-self'. Tagging is epitomised in flags, icons, trademarks and so on. Identity distinguishes those modules that are different, and with whom there is competition, from those that are the same, and with whom there is cooperation. Identity provides recognisable individuality, which indicates variation, and differences that are the basis of adoption or rejection, or the selection of friends and foes, allies and enemies. Without tagging, individuals would be indistinguishable from one another, and all life would share a single identity. Without individuality, and variation between individuals, there can be no basis for natural selection.

Modules replicate and may aggregate, or simply stay attached to the individual that gave rise to them, consolidated by the advantages of aggregation. Aggregation allows modules to recognise that they have the same identity, aggregates come together to create 'agents'. This is a term that Holland borrowed from economics to avoid preconceptions related to biology. 'Agent' means an active element in the sense that an agent is extant and responds to change, or reacts to a stimulus with a response.

Holland reasoned that such evolutionary steps must have survival value to be conserved. For example, the new level of organisation must contribute to the fitness of the organism by increased efficiency, or some other competitive advantage, which is of benefit to the whole. This process is repeated, such that agents as modules come together to form meta-agents. Similarly meta-agents come together to form meta-meta-agents. Shops may 'aggregate' with other shops to form a chain, and chains may aggregate to form a cooperative or group, which ultimately becomes so large as to form a conglomerate. We can see that the basic mechanism required to create a hierarchy requires modules, aggregation and tagging, and Holland concluded that these three properties are sufficient to make possible the development of hierarchies.

Processes within a *cas* include flows and fluxes. Agents can be seen as processors within a network, connected such that interactions are possible between them. Through such networks, materials, energy and information flow. Among the important properties of flows is the multiplier effect, which amplifies some resource such as money or signal strength, as it passes from node to node. Another is the recycling effect, whereby raw materials are re-used, creating savings and economies which contribute to the success of hierarchical systems.

As Holland explains, internal models that are capable of prediction are a 'pervasive feature' of *cas*. In the simplest case, imagine a pattern over time that is followed by an event of significance. With memory, just a few repetitions are required until the event can be anticipated by the pattern that precedes it. Here then is a simple internal model by which occurrences can be anticipated (see Chapter 7). Holland's model is of necessity scale-free, as it must be relevant to *cas* of all sizes. However, one disadvantage of size is the tendency for a system to slow up because of the limitations of the rate at which modules within an aggregation can communicate with one another (Figure 9.2).

The final and most exciting property of *cas* is that they create variety and diversity. These are the features that the instigators of ECHO most wished to capture. This is because, as variety and randomness are introduced into the process, it is as if there is a roll of the dice. This variety is then exposed to selection, such that adaptations that add to fitness are rewarded, and the maladaptive are not. Novelty is continuously created when variety is exposed to selection, which Holland described as the hallmark of *cas*. Holland concluded that in ECHO 'we now have a way that complex multi-agents can evolve'. He felt that his was as rigorous a model to simulate natural selection as von Neumann's theoretical demonstration of a self-reproducing machine. He has drawn together general principles and mechanisms common to all complex adaptive systems. For us, Holland's rules and mechanisms provide a hypothesis for the way in which hierarchies might have evolved.

Holland's model is important because in distilling Darwinian natural selection from first principles into an algorithm for a model, he could see the process in the raw. He also found hierarchical organisation to be fundamental, as one might expect given how common it is in non-biological *cas*, let alone in all forms of life. The process of examining hierarchy in this critical way enabled hierarchy to be recognised for its intrinsic worth, rather than simply describing hierarchy as a feature of pre-existing systems.

AULIN'S LAW OF REQUISITE HIERARCHY

We return to the question of extracting hierarchical function from hierarchical structure. We have already discussed briefly the cohesiveness and resilience to disturbance provided by hierarchies. In Chapter 7 it was shown how Ross Ashby's Law of Requisite Variety indicated that homeodynamic mechanisms have a capacity to resist the effects of disturbance. Any external perturbation to the controlled process is countered by a neutralising response, with a capacity to negate the effects of those fluctuating variables in its habitat that were responsible. It is in this way that such systems provide the adaptive template by which each organism fits its niche. The question here is, how might a hierarchy of coupled control mechanisms operate together to better provide the resilience required?

Ashby's law was presented in terms of a graphical model (Chapter 7). Here a hydraulic model has been adopted which can first be used to demonstrate inhibition of one control mechanism at one level (Figure 9.7a). We can think of the toxic load (I) upon an organism as water flowing into a vessel. The load is countered by a response (R) which is represented by the flow of water out of the vessel from the lower outflow pipe, such that the outflow is equal to, or greater than, the inflow ($R \geq I$). It is significant that the effect of toxic load is considered as a rate. The water, or load, runs away at a rate that does not accumulate, so the level does not reach the overflow pipe (E). If inflow is greater than the rate at which it can run away, the water soon overflows from the higher outlet pipe and induces an effect (E). This

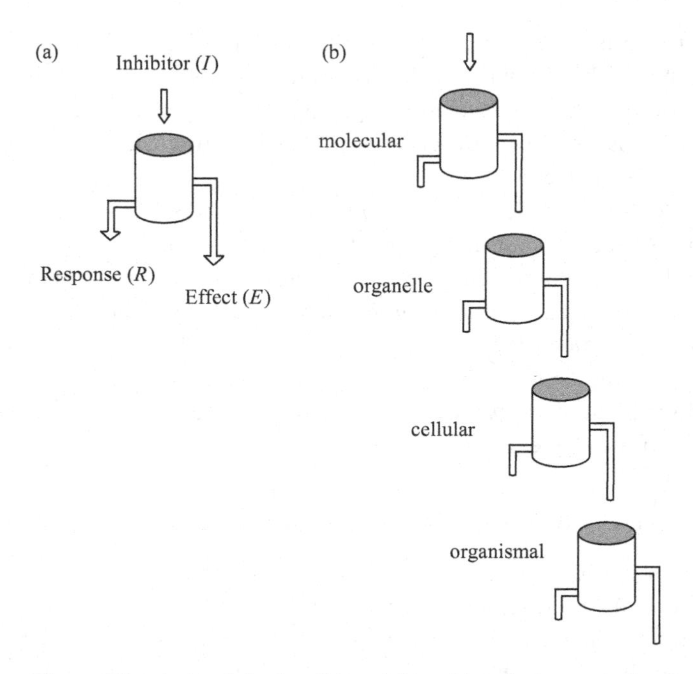

Figure 9.7. A simple hydraulic model used to mimic a cascade of perturbation through a biological hierarchy. Within a single level (a), the input and outputs are used to mimic toxic effects (I), the response (R) and the overall effect (E), as given by Ashby's Law of Requisite Variety; (b) shows multiple units of the same model to simulate Aulin's Law of Requisite Hierarchy. (In the diagram, the hierarchy is reversed to allow the abstraction of water to flow downward through the levels to the highest hierarchical level at the bottom.) Redrawn from Holdgate (1979).

represents overload and toxicity because the flow representing load (I) exceeds the outflow representing counter-response (R), so that the vessel overflows, and the overall effect (E) exceeds the toxicological threshold such that poisoning occurs. In this way the hydraulic model resembles Ashby's Law, which can be expressed so that it states that the overall effect (E) of an appropriate response would be $E = I - R$, as described in Chapter 7.

The hydraulic model becomes more instructive when it is used to mimic the behaviour of a hierarchy of control mechanisms. The Finnish cyberneticist, A.Y. Aulin-Ahmavaara extended Ashby's thinking with her complementary Law of Requisite Hierarchy [13]. She stated that 'the lack of regulatory capability can be compensated for ... by greater hierarchy in the organisation'. For simplicity, Aulin assumed that a number of mechanisms at different levels within a hierarchy each have an equal capacity to respond to perturbation. She provided a formal expression of her interpretation (given here in the same terms used in the earlier discussion of Ashby's Law). The capacity of a hierarchy to resist disturbance is assumed by Aulin to be the sum of the responses of the levels in the hierarchy of the control mechanism:

$$E = I - (R_1 + R_2 + R_3 + \dots R_n).$$

But we are already aware that all the levels in this control hierarchy do not act simultaneously, as Aulin's model implies. We realise from what is known about hierarchies that the impact of some disturbance passes through the hierarchy as a cascade. We can therefore think of Aulin's model within the hydraulic model as a series of vessels through which the water passes sequentially, as a better representation of reality (Figure 9.7b). In this way each level of organisation is represented by a separate vessel, with the residual outcome at each level passing onto the next level and vessel in series. We can now see this system more realistically as a sequence of steps, where the excess inhibition at each level is passed onto the next, which may in turn do the same thing. The hydraulic model shows how responses at each level might neutralise part of the load, and pass on what remains to the next level. In this way total inhibition is eroded sequentially, at each level of the hierarchy in turn, with a possibility of neutralising it completely at the lower levels, so preventing toxicity from reaching higher levels of organisation.

There is a significant advantage in the sequential response to inhibition. While toxic perturbation might begin as a cascade that could overwhelm the organism, some lesser perturbation might limit the effect. A few cells might be destroyed, the tissues might become inflamed, but the effect does not become systemic, as the progress of toxicity can be brought to a halt at an early stage. A sequential mechanism has the effect of localising toxicity to a few thousand cells, or a tissue, without involving the whole organism. How much better, and probably more realistic, to localise some toxic perturbation in this way, rather than allowing it to become systemic?

Aulin also recognised the importance of the membranes that bound each level of organisation as, with each level, another membrane is added, limiting the penetration of foreign substances. Aulin, in her formal expression, included terms for responses at each level of organisation, and for each membrane, to provide a realistic model for the overall resistance to disturbance or toxic stress.

A HIERARCHY OF RESPONSES AND EFFECTS

But what might such responses be in actuality? What are the kinds of responses that could be embedded within Aulin's simple model? Some 20 years ago, as part of a volume written with colleagues (Bayne et al., 1985), I proposed a hypothetical scheme in the concluding chapter, using the hierarchical concept to pull together our disparate techniques into a common framework (Figure 9.8). It was known from our work on the sub-lethal effects of toxic agents that organisms possess different mechanisms to counter the effects of toxic agents at different levels of organisation. A hierarchy of responses and effects was presented in a way that suggested how they might work together within an organism to resist stress and toxicity. The diagram indicates how the responses at each level might ameliorate harmful effects by resisting the upward cascade of perturbation from the lowest level upwards. In this way chronic toxicity would be dissipated by responses at each level in such a way that the effects might peter out before threatening higher levels of organisation within the organism. At the time there was little evidence to support such an interpretation, but now there is more justification for the explanation that was put forward.

We can consider how responses at different levels might react to toxic substances. Metals are ubiquitous elements that occur in low concentrations on land, in the air, rivers and the sea. Some are micronutrients that are necessary for life, while others are toxic. From the time that life originated, organisms have learned to live with metals, using those that are needed and rejecting those that are not. Sometimes membranes fail to separate the essential metals from the toxic; and for those that can be both, their concentrations have to be regulated. If we consider an influx of a toxic metal, it will enter the organism and hierarchy through semi-permeable membranes to have an effect first at the molecular level, then organelles and cells, before cascading upwards through the hierarchy (Figure 9.8).

Some organisms are adapted to regulate the concentrations of essential metals, but tend to be less able to regulate those that are toxic. On the other hand, there exists a system that detoxifies metals by binding them to proteins that have this purpose. Metals may then be sequestered in vesicles within the cell, or may ultimately be deposited in jaws, shell and other hard parts. Elemental metals cannot be degraded, and metal compounds degraded to their ionic form may become more toxic as a consequence.

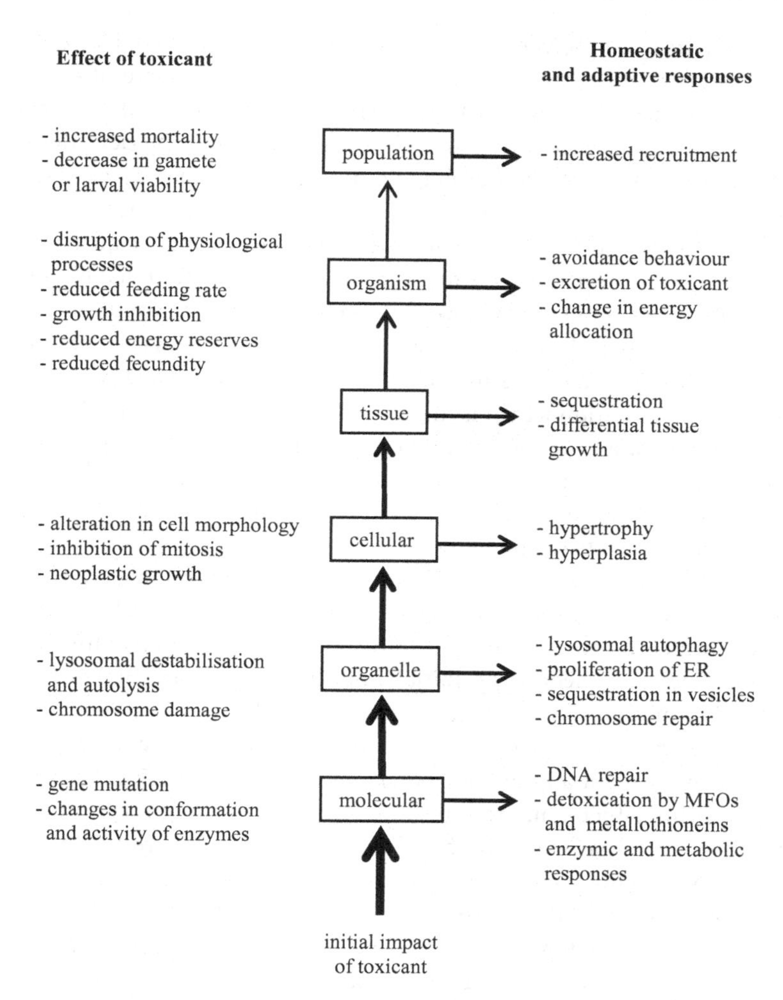

Figure 9.8. A schematic chart showing the effect of toxicity cascading through a biological hierarchy, its effect being ameliorated by neutralising responses at each level of organisation. Redrawn from Bayne et al. (1985).

In response to mutagenic agents, damage to genetic coding is corrected by DNA repair mechanisms. Cell damage is repaired by breaking down cells by a process of apoptosis, autolysis and regeneration. The liver may be the principal organ of toxicant degradation, as here detoxification of organic molecules may be achieved by reducing them to constituents that are water soluble, which can then be excreted in urine.

Apart from the various mechanisms that deal with toxic agents within cells and tissues, there is also the need to regulate affected processes. This is the role of homeodynamic mechanisms which control physiological processes. Control relates to neutralising deviations of the process, rather than dealing with causes, so such responses are generalised and respond in just the same way whatever the cause. Homeodynamic responses work to keep key processes under control, by countering any deviations due to external causes. So we find that inhibitory effects are countered and neutralised by stimulatory responses.

The highest level of response is avoidance behaviour of the whole organism, and the use of memory and intelligence to move away from a location which is unfavourable due to high levels of toxic contaminants, and to find another that is more nearly optimal. An 'internal model', linked to sensors and memory of previous experience, provides what is needed for the organism to head in the right direction along a toxic gradient.

There are numerous mechanisms that counter toxicity. They are organised within the biological hierarchy to have an orchestrated effect in neutralising the impact of toxicity on processes (Figure 9.8). From hierarchy theory, the initial impact is at the lowest level of organisation before cascading upwards through the hierarchy, with adaptive responses deployed at each level, sequestering and degrading toxins or maintaining homeodynamic control. Hierarchical organisms, within which numerous responses are arrayed, might - for these reasons - be expected to be fitter, and to have greater resilience to meet challenging conditions. Such fitness translates into organisms that can disperse and colonise more demanding habitats.

The idea of how a hierarchy of adaptive responses might be more effective in responding to toxic stress was intuitively an attractive idea, and attracted some interest when first published by Bayne et al. in 1985 (Figure 9.8). Augmented by contributions from various sources, it now has the underpinning it badly needed.

MAIA AS A HIERARCHY OF GROWTH CONTROLLERS

Maia emerged from toxicological work with hydroids and corroborative modelling suggesting that the output of the control mechanism closely resembled a single loop control mechanism (Chapter 5). It later became apparent that the rate-controlled system had to be part of a dual loop control mechanism that was responsible for creating tissues and organs that grew, and then are maintained at a constant size throughout life (Chapter 6). Adaptations to the mechanism required two feedforward loops to account for facets of the adaptive behaviour of a rate-sensitive loop. They account for the minimisation of delays in responding to perturbation, and adaptations to sustained perturbation that give acquired resistance to perturbation (Chapter 7). It was later shown that a nested dual loop control mechanism could be derived from the logistic equation (Figure 9.5).

Here the logistic control mechanism is taken further. The control mechanism, though considered first in the context of populations (Chapter 8), now seems equally appropriate for the regulation of any replicator. Whatever kind of modules they represent, replicators have certain fundamental properties in common. Each has the ability to make identical copies of itself. In each case it seems likely that the underlying process is controlled by its specific rate (r), which is likely to be constant, although it has the consequence of replicating modules at an exponential rate. We have seen such rate control in the growth of hydroid colonies (Chapter 5), in populations of hydra (Chapter 8) and in yeast cells (Chapter 12).

In terms of the integrated products of replication, in effect this becomes a positive feedback system, which to be contained requires a negative feedback loop to be superimposed upon the inner rate-control loop. The inner loop provides rate control of replication, while the outer loop provides limitation of the modular products of replication. Each control module has two goals, one for size (W or N) and for rate (r), yet both exert control on the same underlying process. Dual loop control of this kind is a necessary alliance that must control the growth and multiplication of any replicator. However important the exponential increase provided by the inner loop may be, it must ultimately be limited by the outer loop. Any positive feedback mechanism requires a negative feedback to impose a ceiling on its products, once exponential increase in numbers becomes maladaptive. Within a hierarchy, rate-control is necessary for the process at one level, while state-control limits the numbers of modules required to constitute the level above; yet both loops feed back to control the same process.

The logistic control module can be used as the minimal basic module for control at each level of organisation (Figure 9.5). It can be seen that this proposal fulfils the requirements of the Mesarovic scheme (Figure 9.4) in that higher levels control the modular products of the mechanism by providing top-down control, while local rate-control is specific to the process of replication at each level within the hierarchy. This allows local control of the processes of cell turnover, growth during development, adaptation to load, as well as repair and regeneration. Such control relates to ephemeral environmental control requirements, but overall control of size (W) is subject to control from the level above, and all the levels are ultimately subject to control from the highest level. Each dual control module spans two levels; the process of replication is at one level, while the provision of the requisite number of modules creates the level above. This, in effect, is the 'performance feedback' of Mesarovic. Although Holland did not make this suggestion, his description of a hierarchy as providing 'an interlocking cohesion and strength of shared algorithms' is appropriate.

The logistic control mechanism has two goals, which define its purposes. The first is the goal for specific growth rate (r), which determines the baseline growth rate for biosynthesis. This is the goal setting for the inner loop, against which the effects of perturbation are measured ($r\%$) to reveal the oscillatory response of the control mechanism following perturbation (Chapter 5). The second goal, for populations of organisms, is the carrying capacity (K), which is the maximum number of individuals of the species that the habitat can sustain indefinitely. Where the control mechanism is used in relation to internal replicators, K could refer to the number of replicators that are genetically prescribed for some purpose at that level. For example, the number of cells is specified for a specific tissue block (see Chapter 12), although local control can be overruled by intervention from a higher level.

It is also notable that the frequency of oscillations in control output decreases upwards through the hierarchy, decreasing with each level and growth rate, as the operation of each level must incorporate the delays in the operation of each level below. Thus the creation of processes takes longer as one ascends the hierarchy (Figure 9.6). We shall see later that frequencies in the data presented here range from 0.4 cycles per hour in yeast (Chapter 12) to one cycle per 27 years in a population of sheep (Chapter 13). It can be assumed that some microbial processes will oscillate faster, and some animal processes will oscillate more slowly.

Starting from a simple rate-controlled mechanism, the hypothesis was given the name Maia (Chapter 5). We now have a hypothetical system that crudely represents control at all levels of organisation, each level with its own replicator and requiring a specific growth control mechanism. The hypothesis now encompasses the control of growth from molecules to populations, as a coupled and coordinated hierarchy of control mechanisms. We can assume that all growth processes that have the potential to grow exponentially must also have controls to limit such growth because, without control, the ability to grow exponentially has the potential to destabilise and destroy any organism (Chapter 12). Doubtless control mechanisms at each level incorporate their own complexity, but it seems likely that they should, as a minimum, incorporate the features of the system described here.

Maia, by way of yeast and hydroid output providing evidence of complex control behaviour, has led us to the adoption of the logistic control mechanism because it incorporates rate control coupled with size limitation. The two elements of the dual mechanism are linked, because it has been established that any constant specific growth rate will always produce exponential growth, which at some stage must be limited, otherwise it ultimately becomes maladaptive. The two elements can therefore not be separated, but together constitute a unit of control at each level of organisation within the biological hierarchy.

Such assumptions can be made because there is a sound rationale for believing that, from the control of biosynthesis to the limitation of population increase, the basic unit of control has properties shared with the logistic control mechanism. Here it is used as a metaphor and first approximation for a unit of control for any biological replicator. It is rooted in the knowledge that the specific growth rate has been shown experimentally to provide realistic control output (Figure 5.5), and that the carrying capacity as a maximum population size equates to known maxima for experimental populations of organisms, and for populations of cells that constitute a tissue.

SUMMARY

The logistic control mechanism is adopted as a unit of control appropriate to each biological replicator, as it satisfies the principal control requirements of each. It is therefore adopted as a module of control for each level of biological organisation. As Lyapunov said, an organisational hierarchy is therefore also a control hierarchy. The properties of the hierarchy can be considered in this way, as it is likely to be its functional properties that it was selected for, rather than its structural properties alone. Homeodynamic control of growth processes implies resistance to external factors that perturb those processes. Toxic stress passes as a cascade through the hierarchy from the lowest level upwards. Stress is counteracted a little at each level, neutralising toxicity and repairing damage. The impact of such challenges may peter out before higher systems are affected, and the organism survives. Hierarchy has survival value as the context within which adaptive responses to the impacts of challenge and change are neutralised, providing stability by layering adaptive responses to stress.

10

History of hormesis and links to homeopathy

All things are poison and nothing is without poison, only the dose permits something to be not poisonous.

Paracelsus

Everything is poisonous, nothing is poisonous, it is all a matter of dose.

Claude Bernard

THE ORIGINS OF HORMESIS

Between 1421 and 1423 great exploratory fleets sailed from China; hundreds of junks crewed by many thousands of men set out from Nanching. Gavin Menzies, in his book *1421*, vividly describes how the captains and their fleets literally explored the world, riding the great ocean currents, making discoveries that have long been attributed to later explorers. They circumnavigated the globe in great voyages that took two years and more to complete. By that time, Chinese culture was scientifically advanced and capable of making silk, brass, gunpowder and porcelain. They were also very knowledgeable about medicine, minerals and metals.

The crews enjoyed a rich and varied diet from the cultivated plants and animals kept on board the huge junks. Menzies records that they killed bugs and insects with arsenic but, remarkably, used the same poison to promote the growth of plants and silkworms. They had discovered that arsenic, while toxic at higher concentrations, also had stimulatory and beneficial effects on growth at lower concentrations. They might have assumed this to be a specific property of arsenic, but we now know that stimulation due to low concentrations of toxic substances is a general phenomenon. This discovery has been made repeatedly over the centuries for many toxic agents and a wide range of animals and plants. Yet it remains a conundrum to science and medicine, so in this and the next chapter, what is known about the stimulatory effects of low concentrations of toxic agents will be summarised. We will later find that hormesis has a cybernetic explanation, but this chapter will deal with the history of hormesis, and the difficulties in gaining acceptance for an idea that many scientists now accept as commonplace. The next chapter will then review the progress that has been made in recent times to establish hormesis as a general phenomenon, and to provide a homeodynamic mechanism that provides an explanation for hormesis.

The beneficial effects of toxic substances at low doses were well known and widely used in the eighteenth century. In 1785, Dr Thomas Fowler of Stafford concocted a general tonic containing arsenic, which he called 'Fowler's Solution'. It was prescribed by doctors for malaria, cholera, tuberculosis, syphilis and other diseases, until it was found to cause bladder cancer and cirrhosis of the liver. Charles Dickens is said to have taken it, and Charles Darwin used it often. It was a 1% solution of potassium arsenite and must have been responsible for many deaths through overdose, when taken by those who wrongly assumed that more is better.

In 1932 the Australian racehorse Phar Lap died unexpectedly when in America. Post-mortem analysis of tissues suggested that the horse was given a large dose of arsenic 35 hours before death, so poisoning theories spread rapidly. Sadly it seems that the trainer had given Phar Lap a double dose of Fowler's Solution, which he had used routinely at lower doses to stimulate the horse's appetite and performance.

It was in 1943 that low-dose stimulation was given the name 'hormesis', which we use today. It originated from the work of two botanists, C.M. Southam and J. Ehrlich, studying the resistance of red cedar heartwood to fungal decay. They rediscovered that at extreme dilutions, instead of inhibiting fungal decay, the heartwood extract actually stimulated growth (Figure 10.1a) [1]. 'Hormesis' comes from the Greek *hormaein* meaning to excite, describing the stimulatory effect of any toxic substance at concentrations less than those that inhibit. Such effects are as old as the use of medicinal preparations, and certainly as old as the first alcoholic drinks. In medicine it was called 'pharmacological inversion', meaning that the beneficial effects of a drug at low levels may be reversed at higher levels. But this was no explanation. Inversion might be considered a common side-effect of many drugs, as beneficial effects are expected of prescribed doses but they are often toxic at higher doses. Opium was widely used for

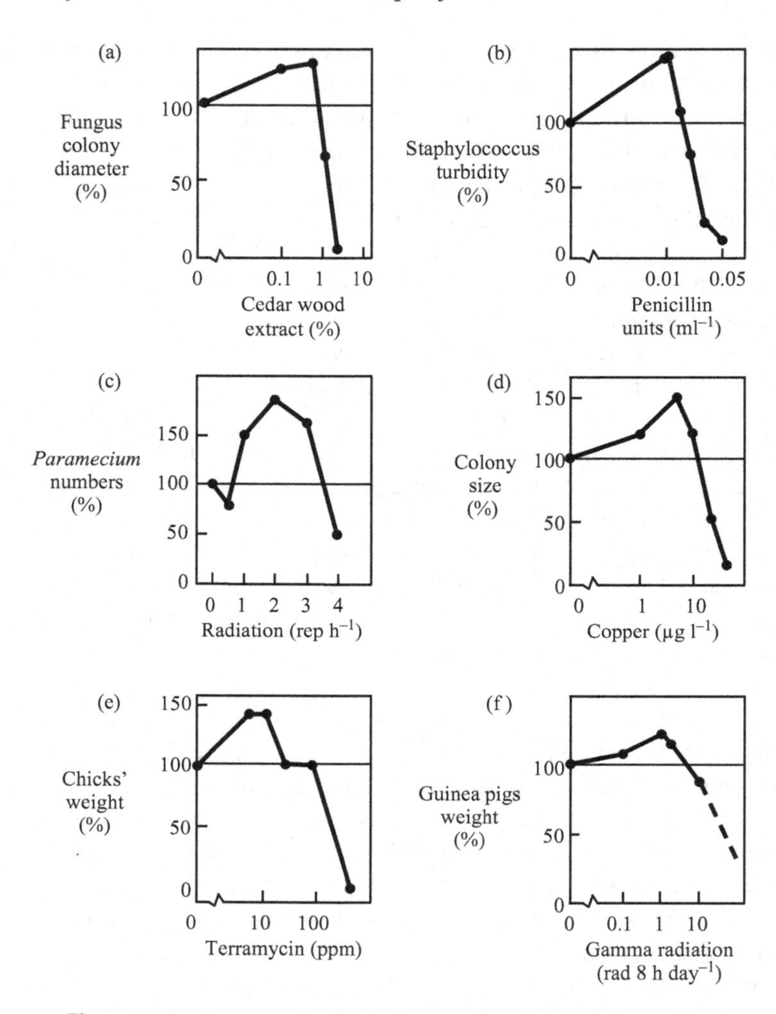

Figure 10.1. A range of published examples of growth hormesis illustrating the variety of organisms and toxic agents that exhibit similar effects. Redrawn from Stebbing [1].

medical purposes in Victorian times; it stimulates at small doses, soothes at moderate doses and kills at large doses. This is not such an inaccessible idea, and has been discovered and rediscovered many times. Today we accept that everyday drugs like caffeine, nicotine and alcohol are used to stimulate at low doses, but become toxic at higher doses. This ambiguous property has been known through the ages, but as the quotations at the beginning of the chapter indicate: what matters most is the dose.

THE CONFUSION OF HORMESIS WITH HOMEOPATHY

Homeopaths have claimed that hormesis provides a scientific basis for believing that the potency of their drugs increases with dilution. But we need to critically consider the questions posed by homeopathy, which claims that its remedies become increasingly more potent with extreme dilution. The dependence of homeopathy on hormesis has done much to handicap the acceptance of hormesis as a scientific phenomenon, despite there being so much evidence in support of hormesis and so little for homeopathy. It is time to divorce the two for, as we shall see, those who accept and practice homeopathy have misrepresented the meaning of hormesis in serving their own interests.

Where does the idea of low-dose stimulation in homeopathy originate? There is an important and early distinction between two kinds of medicine: conventional medicine is termed 'allopathic', from the Greek, literally meaning that treatments should induce the opposite condition. This approach involves trying to counter the symptoms, as it is the symptoms that are assumed to be undesirable. Homeopathy assumes the opposite interpretation: that the symptoms are the body's own responses to illness, and so should be encouraged.

The Greek physician Hippocrates (460–370 BC) created a school of medicine on the Aegean island of Chios. More than 50 books (called the Hippocratic collection) have been attributed to him, but it is uncertain which were actually written by him. In them we see the emergence of a rational medicine from magic and religion. They advise that the physician should interfere as little as possible with the natural healing process. Where necessary, pains or ailments may be relieved by agents that have an opposite effect to the symptoms, while fevers are cured by agents that create a similar effect to the disease. This can be recognised as the origin of homeopathy. Hippocrates believed that treatments with 'similarity' or 'opposition' were equally valid options, with nature as the principal healer. The key decision for a doctor was whether the symptoms were the harmful effects of disease, and should be opposed, or the body's response to the disease, which should be supplemented. Nothing better was written for nearly 2,000 years, and the works won widespread support. Doctors today, on completing their training, may still opt to take the Hippocratic Oath. Then, and for a long time afterwards, what we now call allopathy and homeopathy were considered equally valid approaches to healing. The term 'homeopathy' was taken from the Greek by its founder Samuel

Hahnemann (1755–1843), to mean 'like the disease', and uses remedies that on their own would have effects similar to the disease.

ENLIGHTENMENT AND EXPERIMENT

The Age of Enlightenment of the sixteenth and seventeenth centuries marked the beginning of emancipation from superstition and ignorance, but crucially this period also saw a rejection of magic, mysticism and divinity. The use of experiment was a way of providing new understanding, such as William Harvey's demonstration of the circulation of the blood driven by the heart as a pump, or Isaac Newton's use of prisms to show that white light consists of all the colours of the spectrum. Not only did experiment become an important way of answering questions, but new and more demanding criteria were adopted for the explanation and acceptance of the truth of findings.

Before asking what influence this approach to knowledge had on medicine, we need to consider the health of populations across Europe in the eighteenth century. Disease was ever-present and always ready to take advantage of natural defences depressed by squalor and hardship. In France it was said that there was 'chronic morbidity', from which the whole population was suffering. Living conditions in towns and cities were filthy and sordid; disease sprang from contaminated water and open sewers, and biological vectors of disease were always present. As we know from the tale of the Pied Piper of Hamlin, rats often reached plague proportions, and no-one escaped lice and fleas. Typhoid was common due to the contamination of drinking water; stagnant water and swamps harboured infestations of mosquitoes carrying malaria. Syphilis was widespread. Adult life expectancy was less than 40 years, so the predominance of contagious diseases rendered other ailments of less importance; the average person did not become old enough for the diseases of aging to become a factor.

The Black Death was caused by a bacillus transmitted by rat fleas, which did not discriminate between rats and humans, and infected fleas spread the disease from person to person. The Black Death swept across Europe killing as many as three out of four, and 25 million in all. Infection was a death sentence and few survived. The Great Plague of London of 1664 and 1665 took 70,000 from a population of 460,000. Descriptions of the disease were vague and 'cures' were ineffective. Although there were advances in medical understanding, the treatment of illness remained largely ineffective. The Scottish physician William Cullen (1710–1790) wrote: 'We know nothing of the nature of contagion that can lead us to any measures for removing or correcting it'.

In principle it is impossible to provide a reasoned treatment for disease without a cause, and the microbial causes of contagion remained invisible until the late nineteenth century. Yet Taoist alchemists, living in the mountains of south Szechuan in China, had apparently known how to inoculate against smallpox from the tenth century. It was the dying wish of Prime Minister Wang Tan in 1017 to prevent the disease spreading to other members of his family. The call for help brought a holy physician, a Taoist hermit from the mountain O-mei Shan, who knew how to inoculate against smallpox. In the tenth century, the Chinese used the term *chung tou*, meaning 'to inoculate'. The choice of person from whom the inoculum was taken was crucial, preferably someone who had suffered a mild form of the disease (*Variola minor*), and even better if it had been passed through many individuals without the disease developing into more than a few scabs. It became known that the contagion could be attenuated still further by carrying pustules in a bottle or a small box around the neck for a month or more, allowing the scabs to dry slowly at body temperature. The powder would then be blown into the nostrils of the person to be inoculated, or inserted into the nose on a plug of cotton.

By 1720, inoculation against smallpox had spread from China to Turkey by way of the Silk Road and was adopted by many in Europe. Lady Mary Wortley Montagu (1689–1762), the wife of the British Ambassador to Turkey, was one of the people responsible for introducing the technique to England. She was a well-known beauty who had been badly marked by smallpox and hoped that others might somehow avoid the disease. She had her own children inoculated, and wrote to friends in England in April 1716, explaining how a mild form of the disease might be induced, and so give immunity against later infection. Every year thousands of Turks underwent the procedure without mishap. Earlier, E. Timoni had published an account of this practice of 'variolation' in the *Philosophical Transactions of the Royal Society*. It became widely used in Europe, providing immunity against smallpox.

In 1796 Edward Jenner's independent discovery using cowpox to inoculate against smallpox was remarkable; yet his attempt to publish his findings in the *Philosophical Transactions* was rejected; 'he should not publish it lest he should injure his reputation', was the advice! It is puzzling that it was Jenner's findings that Louis Pasteur (1822–1895) knew about, and to which he gave the name 'vaccination' in Jenner's honour. And it was Jenner's finding which led Napoleon to make inoculation compulsory in the French army, although 'variolation' had been widely practiced in Europe more than 50 years earlier. The British Royal Family were inoculated and Jenner even sent instructions in a letter, together with some of the dried vaccine, to the third President of the United States, Thomas Jefferson, to inoculate his family and friends.

HOMEOPATHY CONVERGES WITH ORTHODOX MEDICINE

Samuel Hahnemann was the founder of modern 'homeopathy', and in 1796 put forward his 'Law of Similars'. He set down that a homeopathic remedy should be capable of causing a similar state in a healthy subject, but at a lower dose causes a counter-reaction in the patient that is stronger than the pathological stimulus of the disease itself. Furthermore, the remedy should use a dose that is the minimum required to be effective in stimulating the body's natural defences. The reader will notice that this sounds like a rationale for the orthodox inoculation against infectious disease.

If we look at the history of medicine, one principle recurs. The Hippocratic School tells us that a fever will be caused and cured with the same agent. Many anecdotal cures incorporated the same idea. In fact the principle of treating 'like with like' has been used to treat contagious disease throughout the history of medicine, unwittingly stimulating the immune system with an inoculum of the causal pathogen. The origin of the homeopathic rationale for extremely low concentrations of inoculum becomes clear. If we assume that its application was primarily for the treatment of infectious disease, extreme dilution was essential to avoid infecting the patient. This makes sense of Hahnemann's requirement that a 'minimal effective dose' be used. Repeated dilution and 'succussion' or 'dynamisation' in an unfavourable diluent is likely to have helped to attenuate the pathogen. Both the 'succussion' and the medium are likely to weaken the pathogen. While he knew nothing of immunology, there was an expectation that such procedures would stimulate the body's natural defences.

In the eighteenth century, physicians had a limited range of medicines, including camphor, mineral medicines, calomel and other salts of mercury. Blood-letting was routine – at best harmful and at worst lethal. Homeopathy was superior to conventional medicine in treating infectious disease, partly because the treatment itself was not harmful, but also because of the fact that within its methods lay an empirical understanding of the principle of immunisation. Treatment using very small doses of some preparation involving the disease pathogen was to some degree effective in treating epidemics of yellow fever, typhoid fever and cholera which raged across Europe at that time.

In the nineteenth century, homeopathy converged with orthodox medicine. Wilhelm Lux, a veterinary surgeon who practised in the 1830s, promoted isopathic homeopathy and used the contagious products of disease (secretions containing the pathogens) to prepare therapeutic products to treat animals (anthrax and distemper), assuming that every contagious disease had within it the means of a cure. In 1865, Father Denys Collet, a doctor and Dominican friar, rediscovered isotherapy and published *Méthode Pasteur par voie interne* in 1898. Here we see a convergence of homeopathy with the work of Louis Pasteur, which was soon to become accepted as medical orthodoxy.

The rationale for homeopathy is most similar to orthodox methods for treating infectious diseases, which nowadays use some form of the causal agent, attenuated to prevent infection, and administered at great dilutions. Historically homeopathy was a method that had proved effective in treating contagious diseases. Hahnemann's Law of Similars tells us that the remedy is capable of causing a similar disease state in the healthy. This is exactly what conventional medicine would tell us for the prevention of infectious disease. The disease pathogen is the means of inducing immunity and, were it not attenuated and diluted, would cause the disease that it is intended to prevent. A minimal effective dose is essential to avoid the risk of causing the disease. How much better to inoculate a patient with an extreme dilution than to infect them! It seems that there is no more plausible explanation for extreme dilutions than that minute doses are required to stimulate the immune system without causing infection. This is perhaps the origin of the paradoxical assumption that the potency of homeopathic remedies for complaints that do not involve disease pathogens must also increase with their dilution. It also helps us to understand why today there are apparently no dose–response relationships for homeopathic remedies that span the full range of concentrations. All that was required was to use a low enough dose to avoid killing the patient, but sufficient to stimulate a natural response. It must be remembered that at this time disease pathogens were unknown, and the difference between infections and other diseases was much less apparent. In the absence of such a distinction, one would expect that an approach to infectious diseases that worked would also be effective when applied to other diseases.

This would explain why homeopathic medicines were used to treat non-contagious diseases but, without the same rationale, they were unlikely to be effective. Their supposed efficacy seems to depend solely on using remedies at minute doses sufficient to stimulate the immune reaction; therefore, for diseases that involve no immune reaction, homeopathic remedies are unlikely to be effective. In recent times, scepticism about homeopathic medicine has led to experiments conducted by researchers to be exposed to stringent scientific examination. It is pertinent that the majority of experiments to test the efficacy of homeopathy in recent times have tended to relate to infectious disease and diseases of the immune system, which are stimulated by low levels of the agent responsible.

THE ORIGINS OF HORMESIS AS SCIENCE

The Arndt–Schulz Law provided evidence that the potency of remedies might increase with their dilution. It states that concentrations of toxic substances below those that inhibit biological processes have the capacity to stimulate them. Scientists now know the Arndt–Schulz Law as hormesis, but to understand the connection of hormesis to homeopathy, we must go back to the origin of the Arndt–Schulz Law in the late nineteenth century.

Traditionally Hugo Schulz (1835–1932) has been credited with discovering hormesis, as he provided the first experimental evidence that demonstrated toxic stimulation. At university he had developed an early interest in pharmacology. He was an assistant at the Bonn Institute of Pharmacology and in 1878 began to search for a non-toxic antiseptic for wound treatment. It was not until he was 47 that Schulz became a full professor at the University of Greifswald, and was able to determine the direction of his own research, only to discover on taking up his appointment that his Institute of Pathology was a single room, with 'one window, two gas burners, but no water'. Although his circumstances were enough to drive him to despair, and his appeals to the university were initially unsuccessful, he resolved to do what he could. He decided that in an institute for pathology, a study of fermentation would be of interest, looking for agents that might inhibit fermentation. He worked with yeast, using the production of carbon dioxide as a measure of growth and metabolism. He considered the effects of a range of concentrations of formic acid and other bactericidal chemicals upon the growth of yeast. Regardless of the agent he chose, he found that high concentrations inhibited growth, while low concentrations stimulated growth. Stimulation occurred whatever agent he tested, and Schulz found that iodine, bromine, chromic acid, arsenic acid and mercuric chloride all stimulated yeast growth at low concentrations.

It was not long before he began to work on an extract of the herbaceous hellebore plant, which contains a poison called veratrine. It was once used to poison the tips of arrows, and was later found to be an effective insecticide. It had been known that gastrointestinal infections could be treated with veratrine. Schulz began to investigate the effects of low doses on cultures of the bacterium *Salmonella*, working with a colleague who had access to better facilities. They were surprised by the outcome. Despite what they thought would be a lethal amount of veratrine in the culture medium (0.1%), the bacterium actually grew. Though such evidence was slim, Schulz argued that if a concentration of a toxin estimated to kill the pathogen allowed it to thrive, veratrine must stimulate some mechanism that counters the effect of the toxin. As we shall see in the next chapter, this is a similar interpretation to the modern homeodynamic explanation for hormesis. Schulz's experiments with yeast and *Salmonella* would determine the direction of his career.

Rudolf Arndt (1835–1900) was the other contributor to the Arndt–Schulz Law. He was trained as a medical doctor and went into private practice, but was soon drawn into military service. He later took up an academic career, and in 1873 became a professor at the University of Greifswald. Schulz and Arndt became friends and it was in 1885, as they walked together, that Arndt first explained to Schulz his 'laws of biology', which became known as Arndt's Rule. It is more accurately referred to as a set of linked propositions. They state that weak stimuli excite physiological activity, moderate ones favour it, strong ones retard it and stronger ones arrest it.

It is improbable that Arndt, as a professor of psychiatry at this time, was not familiar with W.M. Wundt's book on physiological psychology, published in 1874. In it Wundt describes the relationship between the degree of arousal and pleasure (Figure 10.2a) [see 2], which indicates that arousal produces feelings of pleasure at low levels, a moderate level of arousal has an intermediate effect, while higher levels cause increasing displeasure. Arndt may well have been influenced by the analogy of Wundt's curve when he put forward what became known as Arndt's Rule, as it provided a relationship that similarly took a biological response from the positive to the negative as a continuum. Arndt proposed that the first phase of a response to a toxic agent is stimulatory, which then passes to an inhibitory phase, finally arresting the response.

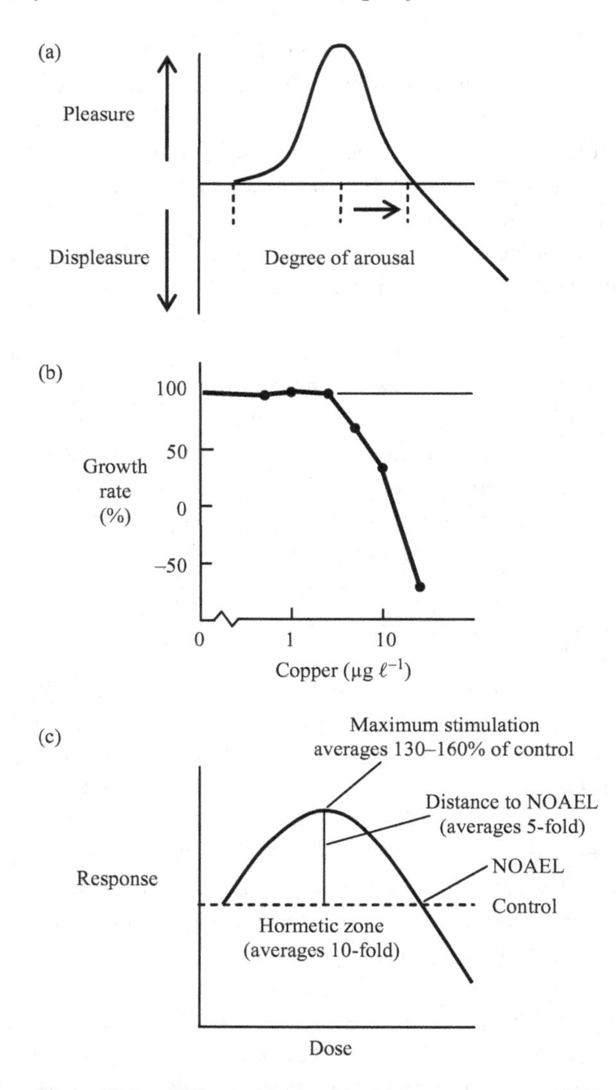

Figure 10.2. Different kinds of dose–response curves: (a) Wundt's curve; redrawn with permission from Young [2]; (b) a typical α-curve showing the inhibition of growth rate by copper in an experiment with populations of *Hydra*; (c) diagram indicating an idealised β-curve. NOAEL = no observed adverse effect level. Redrawn from Calabrese and Baldwin [5].

However, Arndt's Rule became obsolete because the exceptions to it were too numerous. Nevertheless, the form of Wundt's curve followed closely what we now refer to as the β-curve, with a form and origins that will be given later. At the time, their conversation was a revelation to Schulz, as he could see that his experiments supported Arndt's ideas. Schulz explained this in an autobiographical essay:

> If this first theorem, that weak stimuli promote the vital activity of organs and organisms, is correct, a toxin in sufficiently reduced quantity ... should have a beneficial effect on the substrate.

He concluded that his results with yeast

> elegantly proved the correctness of the first Arndt theorem. Later on it was also repeatedly confirmed by others, even with the use of other starting materials.

It was Arndt's first proposition, and Schulz's experimental evidence, that came together as the Arndt–Schulz Law, which is the stimulatory effect of low concentrations of toxic substances. What brought notoriety to Schulz was his claim that his findings would prove common to all organisms and toxins. This was too bold a claim; he had worked with a number of chemicals, and even a biological toxin, but much more data were required in order to make any claim of universality. Besides which there was no known mechanism that could have brought about a positive response.

Conditions in Schulz's 'institute' were so unfavourable for experimental work that Schulz spent much of his time reading the old literature, including the works of the founder of homeopathy, Samuel Hahnemann. In pondering his own results, Schulz realised that 'veratrine intoxication in humans can have an eerie similarity to an attack of salmonellosis'. Perhaps with Hahnemann's Law of Similars at the back of his mind, he believed he had found something of therapeutic importance, and he began to work on human subjects: often himself and his trusting students.

The similarity of his results to homeopathy threatened to damage his scientific reputation. Schulz wrote:

> It [stimulation] smelled very much like homeopathy! However, after I sent him my veratrine work, my friend Ernst Weber wrote me: 'Now you will have the hounds after you!' He was right. Since the publishing of this work I have not only had the homeopathic lay organizations on my heels. Even some of my specialist colleagues have used this and some of my subsequent work to assist me in acquiring the honorary title of the 'Greifswald homeopath'.

Even then, homeopathy was not accepted as science, and the reputation of any scientists whose work extended towards homeopathy was liable to be tarnished. Idealistically, Schulz believed that his 'purpose was to obtain the truth, and let the result come out as it may'. However, his high principles did not prevent him from being sidelined; his link with homeopathy is probably why Schulz's career never advanced beyond Greifswald. The Arndt–Schulz Law became a 'scientific pillar of homeopathy' and is used in its support to this day. At the same time, Schulz's work provided the earliest experimental evidence for hormesis. The association would contribute to the marginalisation of hormesis, while lending some credibility to homeopathy. The use of the same experimental evidence to support contradictory systems weakened both.

SUPPORT FOR AND CRITICISM OF HORMESIS

The Arndt–Schulz Law attracted some support from the eminent German microbiologist Ferdinand Hueppe (1852–1938). In 1896 he published a textbook which was translated into English as the *Principles of Bacteriology* [3]. In it he gave an account of his independent discovery of the same principle of stimulation by low doses from his work with bacteria. Before publication, he became aware of Schulz's work and acknowledged his priority, and helped to defend him against criticism. Hueppe expressed the view that Schulz's work on the stimulatory effects of low doses should stand, and not be rejected simply because its central finding had been adopted to support the doctrines of homeopathy.

Later criticism came from the eminent Scottish pharmacologist Alfred J. Clarke (1885–1941), best known for his influential *General Pharmacology*, which was published in Germany in 1937 [4]: 'It is true that such (stimulatory) effects are often observed, but there is no necessity to postulate any mysterious property of living tissues'. A phenomenon without a mechanism to account for it is always vulnerable to criticism. Clarke gave examples of enzyme systems that exhibited concentration–effect curves with a stimulatory phase followed by an inhibitory phase, claiming that hormetic curves could easily be explained in this way. In citing an example that illustrates the Arndt–Schulz Law, he showed that it could be explained on physico-chemical grounds alone. He wrote that

> although our present knowledge is adequate to explain most of the diphasic actions (i.e. stimulatory at low levels and inhibitory at higher levels) met with in more complex systems, yet there seems no reason to consider them peculiarly mysterious.

He also wrote, 'This law is in accordance with homeopathic doctrines and hence has maintained a certain popularity', adding that 'many of the effects that appear to support this law have found simple explanations'. Clarke assumed that each example of hormesis could be explained by its specific causes, and that there was no justification to invoke some general and unknown mechanism to account for all the examples of toxic stimulation.

Edward Calabrese, who is now a leading expert on hormesis, writes that - almost single-handedly - Clarke marginalised hormesis. His text long remained influential and had the effect of turning away from hormesis many influential figures in toxicology who might have carried the subject forward. But he was right in pointing out the absence of a general mechanism to back up the idea of hormesis as a general phenomenon.

At that time, reductionism was the dominant paradigm for research. Specific effects were most likely to have specific causes, and one could not be expected to accept some phenomenon as being general without some equally ubiquitous mechanism to account for it. So the very idea of the generality of a response was viewed with lofty scepticism. A.J. Clarke's criticisms of the Arndt–Schulz Law, and thereby hormesis, remained influential into the 1970s, due to his reputation and the use of *General Pharmacology* by generations of students, who would not have been able to remain ignorant of his repeated and telling criticisms of the Arndt–Schulz Law. In some senses, Clarke was right. The consistent shape of the curve need not imply a single mechanism, even if different examples of the same mechanism would be expected to share a common form. Now, due largely to the efforts of Calabrese and his colleagues, there exists a large database of published examples exhibiting what is referred to as the β-curve (see Figure 10.2). The experimental observations cited are statistically significant, and the phenomenon is seemingly universal.

What process, that occurs as widely as hormesis, might explain it? Homeostasis is ubiquitous, so is an eligible candidate mechanism to account for the β-curve. A homeodynamic hypothesis is put forward in the next chapter to account for the Arndt–Schulz Law, hormesis and the β-curve. As we shall see, such an idea is plausible because of the generality of homeodynamic systems in biology, and at all levels of organisation. What is more, they respond in a non-specific way to any agent that disturbs the process they control, which suggests that a single interpretation might account for the same effect due to a variety of agents.

Reductionism was the traditional approach to research even as late as the 1980s, when I attempted to publish ideas on hormesis as a general phenomenon. The utility of holism only gained ground with the acceptance of systems theory, which had remained peripheral to orthodox biological thinking for many years. It required recognition that complex systems had important properties not discernible from a scrutiny of their component parts ('emergence') before holism could take its place alongside reductionism as an equally valid method of investigation. In Clarke's era, reductionism was the dominant approach, to the exclusion of any other. In recent times, as Nobel prize winner, Sydney Brenner is reported to have robustly remarked, the best approach to research within the biological hierarchy of organisation is neither the reductionist's 'top-down' approach, nor the holist's 'bottom-up', but 'middle-out'!

IS HOMEOPATHY SCIENCE?

So how might the debate between the mutually exclusive paradigms of hormesis and homeopathy be reconciled? Arndt was a passionate supporter of homeopathy, and Schulz was a scientific researcher in pharmacology.

While homeopathy was the oldest and most respected of the complementary medicines, and cited the Arndt–Schulz Law as providing scientific evidence, it was also the birthplace of hormesis. So there has long been a need to tease these two fields apart. As in Schulz's day, homeopathy is not accepted as science, for reasons that will be explained.

Both Arndt and Schulz were well aware of the homeopathic principles, yet viewed Schulz's data from a scientific perspective. Schulz had read deeply in the homeopathic literature, yet was carrying out research to the scientific standards of his day. His methods were sound, and experiments were carried out 'blind'. He was prepared to accept the truth of his experiments, wherever they led him. Schulz's results were accepted as scientifically valid; he had found that low concentrations of many chemical agents are capable of stimulating biological processes. From a homeopathic perspective, this was taken to mean that their potency to benefit biological systems increases with dilution. But there was one important way in which Schulz's results did not support homeopathy. To reach that point, we must look first at 'posology', or the dilution of homeopathic drugs.

The relationship between the dose or concentration of some toxic agent and its biological effect is a central concept of toxicology. It is expressed graphically as the so-called 'dose–response' curve (Figure 10.2b). Such curves are used to present the biological effects of toxicity in relation to the concentration of some toxic agent. Stimulation at concentrations lower than those that inhibit is now recognised as a common, if not a general, feature of dose–response curves. John F. Townsend, then a student of T.D. (Don) Luckey, published a paper in 1960 reviewing the various dose–response curves found in the literature. The basic dose–response curve was called the α-curve (Figure 10.2b), while the β-curve indicates hormesis in a biphasic curve (Figure 10.2c). The typical form of the β-curve has been defined by Edward Calabrese and Linda Baldwin; it has a range of 30–60%. Stimulation declines into inhibition as a continuum, the threshold is exceeded and the substance becomes toxic [5]. This much is known and has been confirmed by thousands of examples of the β-curve.

Data that adhere to the β-curve show that the effect is stimulatory, in the same way as homeopathic drugs are expected to become more potent as they are diluted. This is the hormetic range of dilutions that behave as homeopathic drugs are expected to, and lends some support to the expectations of homeopaths. But it is also known from numerous examples that hormesis typically extends over a tenfold range of concentrations (see Figure 10.1) and does not extend into the wide range of dilutions that homeopaths consider therapeutic. This indicates that what hormesis demonstrates, and what homeopathists claim, is not the same thing.

Homeopathic drugs are diluted on decimal or centismal scales. For example, drugs are diluted by 1 in 10 at each step, denoted by 'x', the Roman numeral for ten. The centismal scale involves dilutions of 1 in a 100 and are denoted by 'c'. So a solution of 11x, as a dilution of the original drug, would be 0.000000000001 or 10^{-12}. This is assumed to be a homeopathic 'safe point' from which to create therapeutic dilutions. Therapeutic dilutions extend to much lower concentrations, and are assumed to become more potent with dilution. The choice of dilution depends on the nature of the ailment. Over this scale, a linear relationship of concentration to effect is implied, although there appear to be no published dose–response, or concentration–effect, relationships to confirm that the efficacy of homeopathic remedies increases over such a range of dilutions. As the entire hormetic peak typically falls within just one order of magnitude (see inset in Figure 10.3), it is clear that

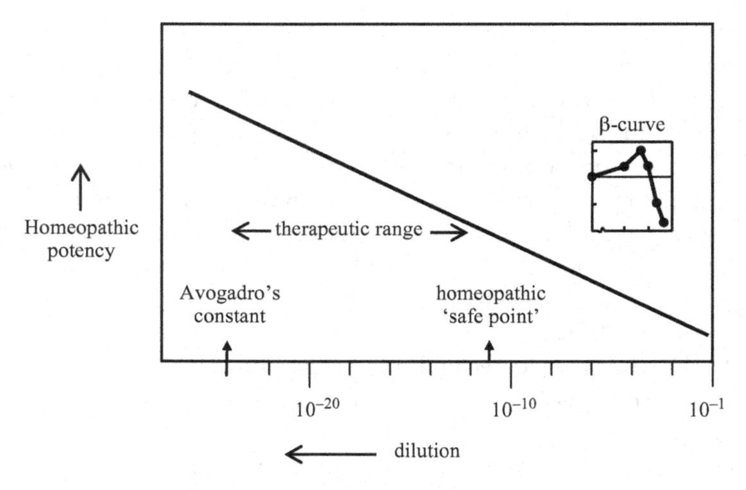

Figure 10.3. Diagram indicating the full range of homeopathic dilutions and the therapeutic range on a logarithmic scale as multiples of ten. Inset is a typical β-curve, positioned above the range of dilutions likely to apply, and to approximately the same scale.

the Arndt–Schulz Law or hormesis cannot be used to support what is claimed for homeopathy.

One critic of homeopathy calculated that a 12c (24x) dilution would be equivalent to a 'pinch of salt in both the North and South Atlantic Oceans'. But it was Amedeo Avogadro (1776–1856), the Italian physicist and chemist, who established the dilution of a substance that contains none of the original atoms or molecules, and can be calculated using the Avogadro constant ($6.02213567 \times 10^{23}$). A straightforward calculation demonstrates that dilution of any substance beyond 10^{24} means that it is improbable that any atoms or molecules of the original compound remain. Once it became clear that the most 'potent' homeopathic concentrations contained none of the original medicine, some 'memory' of it in the diluent was claimed, but no hypothesis of this kind has ever withstood scientific scrutiny. Homeopathic dilutions cover a range of dilutions millions of times lower than those that cause hormesis. Obviously, if the effect of homeopathic medicines and hormesis were to have the same interpretation, then they would fall within the same range of concentrations. Clearly they do not. The homeopathic scale of dilutions in concentration extends over 24 orders of magnitude, while hormesis typically spans just one order, or a tenfold range of concentrations (Figure 10.3). The relationship is presumed to be linear, and increases in potency with the presumed benefits of increasing dilution. The therapeutic range of a homeopathic drug begins at the 'safe point' (11x) and extends to dilutions of 24x, in which there is a negligible residue of the original drug. Clearly these dilutions extend far beyond the point at which biological systems on the β-curve show any stimulatory effect (Figure 10.3). Homeopathy and hormesis cannot share the same explanation because they occur in ranges of the dilution curve that are many orders of magnitude apart. There is no overlap of the sort that would be expected if the same mechanism was responsible.

The 'dose–response' curve is a key concept in pharmacology and toxicology, yet homeopathists do not use it because they are apparently unable to demonstrate a concentration-related effect over the many orders of magnitude that they assume to be therapeutic. There are apparently no experimental data for any homeopathic drug that relate its potency to a concentration over 20 orders of magnitude. It is a routine expectation of pharmacologists and toxicologists to express their data with a graph of concentration or dose against time, yet homeopathy does not recognise that in failing to fulfil this expectation, they invalidate any claim to be scientific.

From their published reviews, and knowledge of thousands of examples, Calabrese and Baldwin leave us in no doubt that hormesis is a genuine, consistent and reproducible phenomenon [5]. Homeopathic practitioners, on the other hand, have no basis for invoking the Arndt–Schulz Law (or hormesis) as a scientific principle to provide grounds for homeopathic dilutions and their potency. Their extrapolation over a huge range of dilutions has never been demonstrated by providing concentration–effect curves for each remedy, which is a routine requirement in toxicology and the study of hormesis.

The battle between science and homeopathy has run for more than a century. But, using this analysis, it must be concluded that homeopathy does not conform to the normal expectations of science. Furthermore, it is an abuse of the scientific basis of hormesis to misinterpret its meaning in order to lend homeopathy the scientific credibility it lacks. To do so casts doubt upon a large body of sound evidence in support of hormesis of a kind that does not exist for homeopathy. For those researchers involved in the study of hormesis, it hinders publication, impedes the funding of research and obstructs its acceptance by scientist and lay person alike. Unfortunately, the historical adoption of the Arndt–Schulz Law in support of homeopathy is still being perpetuated, with homeopaths citing current papers on hormesis, including my own, as though the foregoing arguments were not plain for all to see. The Chief Scientific Advisor to the UK Government, Sir David King, is justified in stating that 'there is no scientific evidence whatsoever supporting the use of homeopathy'. In the UK, the groundswell of informed opinion has led to the National Health Service withdrawing provision of homeopathy to its patients.

EXAMPLES OF GROWTH HORMESIS

We tend to think of antibiotics as medicines, but they evolved in fungi as chemical weapons which they used in their fight against other microbes in competition for space and nutrients. The problem was to find antibiotics that are toxic to microbial pathogens, but not to the humans they infect.

Penicillin was discovered by Alexander Fleming (1881–1955) as a contaminant mould spotted on a discarded petri dish, around which the growth of bacteria was arrested. It was his trained mind that was necessary to realise the significance of antibiosis, as he had been looking for some time for an agent that would kill pathogenic bacteria. Penicillin is relatively innocuous to man because it arrests processes that are specific to bacteria. In 1929 Fleming published a paper on penicillin, describing the bactericidal agent. He had earlier hoped that an enzyme found in tears, called lysozyme, might prove to be what he was looking for. He had tested it thoroughly and found it wanting. So when the opportunity came, he immediately recognised that penicillin had the required properties, and he already had techniques ready to assess it. For some years after the initial discovery, a small team continued to work on extracting penicillin in a form that could be used experimentally, but with the death of the team's mycologist the work petered out.

It was not until the outbreak of war in 1939 that Howard Florey (1898–1968) put forward a proposal to look into Fleming's discovery of the bactericidal properties of penicillin. His collaborator was the German biochemist Ernst Chain (1906–1979), whose task was the extraction and purification of penicillin. Its action was not by killing bacteria, so much as blocking cell division, which proved to be effective at concentrations as low as one in 200,000. When the amazing powers of penicillin first became better known in 1941, it was obvious that production on an adequate scale, even for use by military forces, could not be carried out in Britain. Florey approached visiting representatives of the Rockefeller Foundation, and it was not long before he flew to the USA armed with records, data and cultures, accompanied by the redoubtable Norman Heatley to assist him. All of this was offered freely to any pharmaceutical company that might be interested in scaling up production. Production methods were rapidly developed in the USA, and large amounts of penicillin were available by D-Day. Fleming, Florey and Chain received the Nobel Prize for medicine in 1945 for the discovery and development of penicillin.

In the same year it was discovered that low doses of penicillin could also *increase* the growth of staphylococci. Surgeon Lieutenant Commander W. Sloan Miller, and his collaborators at the Royal Naval Medical School, demonstrated that concentrations of penicillin immediately below those that inhibit bacterial growth actually stimulate growth. At its peak this amounted to an increase in density to a level nearly 40% above those cultures that remained untreated (Figure 10.1b). Experiments with mice showed that low doses of penicillin given to infected mice resulted in *lower* survival rates than in the untreated mice, because penicillin at low doses stimulated the growth of *Staphylococcus*. During the early post-war years, when penicillin was in short supply, doctors wished to make the new drug available to as many patients as possible; but by reducing the dose, the growth of the pathogen was stimulated, with consequences that were sometimes fatal.

The hormetic effects of penicillin generated great interest beyond its medical applications. Such stimulatory effects may have prompted the investigation in America of the possible industrial use of dietary antibiotics as growth promoters. This work began in T.D. Luckey's laboratory in 1943, in which he achieved growth stimulation by feeding *Penicillium* mycelium directly to chicks. This soon led to experiments with germ-free chicks kept under aseptic conditions to establish that growth enhancement was due to some dietary effect, rather than some indirect effect on the intestinal microflora. Over the following 20 years, dietary antibiotics were found to increase growth rates of turkeys, pigs, calves and chicks (Figure 10.1e). The hormetic effect was greatest in the fast-growing young, and was due to improved efficiency in the conversion of food into flesh. Typically the addition of ~10 grams of antibiotic was required per ton of feed. Increases in growth varied between species from increments of 5–10% in calves to 30% in turkeys, while a 20% increase in the growth of chickens could be achieved with a small addition of penicillin to their drinking water.

It later became clear that hormesis was not limited to antibiotics, but growth could be promoted with other agents that were toxic at higher concentrations. For some years, metals were used as growth promoters in proprietary animal feeds. The metalloid arsenic was investigated and compounds were used on the same range of animals for which antibiotics had been so effective. Copper sulphate was found to stimulate the growth rate of pigs to the same extent as those fed antibiotics. A range of metals have been shown to stimulate the growth of hydroid colonies, the increment increasing during exposure to over 50% (Figure 10.1d). Hormetic effects of antibiotics on the growth of plants were found by 1947, although much earlier findings in the botanical literature are now recognised as hormesis. The most marked effects related primarily to germination and seedling growth, and early experiments showed a doubling in seedling weight.

No mechanism was found that might account for hormesis, although ideas were put forward that it was due to overcompensation in response to inhibitory stress (see Chapter 11). The variety of agents that could be used to stimulate growth, in the form of the β-curve, strongly suggested that stimulation was not due to the chemical nature of the toxic agents, as they were chemically unrelated. How could this be? The hypothesis that will be explored in the next chapter is that all such toxic inhibitors elicit a stimulatory response at low concentrations which exceeds that necessary to neutralise inhibition. Before we can consider this idea in detail, we need to look at yet another totally different kind of toxic agent. It was the range of chemical and physical agents capable of causing hormesis that first suggested the idea that hormesis was due to a biological response to some effect that each of the agents caused, rather than to the properties of the agent itself.

RADIATION AND GROWTH HORMESIS

We now consider some examples going back over the last century showing the stimulation of the growth of organisms by toxic substances. The French physiologist, pacifist and author Charles Richet (1850–1935) was awarded the Nobel Prize in 1913 for his work on anaphylaxis. In 1906 he had published a paper describing the effects on lactic fermentation of radioactivity, just seven years after Marie Curie had given it that name. In fact it was Marie and Pierre Curie who gave him radium for his experiments. The effects of radium were observed through a wide range of low levels and times of exposure. These experiments may have been the first to study the biological effects of radioactivity, resulting in the stimulation of fermentation at low levels, and inhibition at higher levels. Such effects of radiation were reproduced in many other biological systems in later studies, to the extent that radiation hormesis became a field of study in its own right.

Radiation has been a factor in the evolution of life, with mutation and random genetic change creating biological variability upon which natural selection operates. It would seem that ionising radiation[1] would be a most improbable agent to be capable of stimulating biological growth. X-radiation is actually a stream of disintegrating atomic particles, which pass through the tissue at velocities approaching the speed of light, but are absorbed by bones and dense tissues to create the negative of the X-ray image. Another consequence of radiation is the creation of free radicals in the body. Free radicals are harmful – they are created when unpaired electrons steal electrons from another molecule, making them unstable and so causing a chain reaction. This quickly results in molecular damage and disruption of biochemical pathways, and can cause cell death. When life began, cosmic radiation from outer space, and radiation originating from soil and rocks, were factors that organisms have had to adapt to, but at that time levels of radiation were 3–5 times higher than the present. The same sources of radiation exist to this day, so it is not surprising that DNA repair mechanisms have evolved that can make good nearly all radiation damage to the genome.

The sensitivity of organisms to radioactivity depends upon their size and complexity. Bacteria can tolerate massive amounts of radiation, while mammals are sensitive to levels a thousand times lower. Rapidly dividing cells of the young are the most sensitive, so that the aged show much greater tolerance. Homeodynamic systems of various kinds respond adaptively to radiation and its effects, repairing chemical damage and countering the inhibition of growth rates. Luckey found that radiation hormesis occurs at roughly one hundred times above ambient levels, or at about one-hundredth of the dose that causes significant harm. With the dropping of the first atomic bombs to end World War II, interest in radiobiology grew rapidly.

The testing of nuclear weapons in the Nevada Desert in the 1950s revealed that a variety of biological processes were apparently stimulated by exposure to low levels of radiation. These included stimulation

[1] In this brief consideration of the biological effects of radiation, the units by which dose has been measured have been changed several times. The correct SI units today are becquerel, gray and sievert, in place of curie, roentgen, rad and rem used previously. To avoid such complications, I have used qualitative terms rather than unfamiliar units of radiation.

of the growth of desert plants and vegetation around nuclear weapon test sites, and even the rate at which small mammals reproduced.

Experimental work followed, and the same stimulatory effects were found in protozoa and other micro-organisms. A study in 1953 involved the culture of a protozoan *Paramecium* on cavity microscope slides, so that the living cells could be observed under the microscope. It was found that *Paramecium* cultured on the slides manufactured in Czechoslovakia had higher growth rates, cell division and reproduction, and it was discovered that the Czech microscope slides were slightly radioactive. Later experiments showed that calibrated levels of radiation could stimulate growth and reproduction over a range of radiation levels, giving rise to the characteristic β-curve (Figure 10.1c).

Large-scale apparatus was required at that time to expose even the smallest organisms to controlled doses of radiation. At the Marine Biological Laboratory at Woods Hole in Massachusetts, the apparatus was substantial, standing 3 metres high and weighing 4 tons; its bulk was necessary in part to shield the researchers. Even such large apparatus afforded little space for experimental organisms, so much of the early work during the 1960s on the biological effects of radiation was carried out on micro-organisms. Experiments with the duckweed *Lemna*, which was a relatively small and convenient experimental plant, showed that low doses of radiation stimulate reproduction. Other studies showed hormesis in the regenerative growth of crab limbs, and later the growth of the brine shrimp.

Perhaps the most important early work was published in 1954 by Egon Lorenz and his colleagues at the National Cancer Institute at Bethesda in Maryland, USA. Their classical series of long-term experiments was conducted as part of the Manhattan Project to develop the first atomic bomb. They exposed small mammals to low levels of gamma radiation over a range of doses for periods as long as 2.5 years [6]. Low doses significantly increased the growth rates of guinea pigs to attain weights 24% greater than the untreated control animals, yielding a β-curve of typical proportions (Figure 10.1f). Small mammals often become experimental surrogates for humans, and one finding of these classical experiments was that mice exposed to low levels of X-radiation had significantly longer lives than those that were untreated. Nearly 2.8 times more mice survived to the 3-year maximum lifespan for mice. Recent work on longevity shows that a variety of stress-inducing agents are capable of increasing the longevity of animals in laboratory experiments, in a way that is now attributed to hormesis. Such experiments, it is hoped, may provide insights into the process of aging at the subcellular level.

It is not surprising that early in the history of hormesis, the stimulatory effects of low doses of radiation found commercial applications. In 1946 the Russian scientist P. Breslevets, working at the Institute of Physics in the USSR, proposed the term 'radio-stimulation' for the positive effects of radiation. The effects on seeds included quicker germination, rapid early growth, greater vigour and an earlier harvest. Radio-stimulation became widely used in Eastern Europe. By 1984, 100,000 acres of crops had been planted in Hungary with radiation doubling the ears on wheat in standing crops, and doubling the yield of peas and potatoes. The seeds or tubers were passed through a mobile irradiation unit taken to the field at planting time. The best results were achieved with hatching eggs, potato tubers, winter wheat and sunflower seeds. In 1985 an initiative was taken to adopt radio-stimulation of farm crops in Canada, but public health and other concerns became an obstacle. Interest has since turned to the hormetic effects of the relatively innocuous ultraviolet light at particular wavelengths (UV/C). In recent times it has become a developing technology in horticulture, with the aim of increasing yields using hormesis.

Here emphasis is focused on those examples that relate to biological growth processes. The β-curve is the norm, so preliminary experiments covering a wide range of levels or concentrations are required to identify the peak of the β-curve (Figure 10.1). It is often not appreciated that the greatest stimulation of growth occurs with low doses over long periods (see Chapter 11), as in Lorenz's experiments with small mammals. When these requirements are met, hormesis is found to be a robust and reproducible phenomenon.

THE LINEAR NO-THRESHOLD (LNT) HYPOTHESIS

Controversy surrounds radiation hormesis because it is incompatible with the central concept of radiobiological protection. The 1950s was a period of intense nuclear weapons testing. Radioactive fallout from the tests was a factor used by the peace movement to create pressure for nuclear disarmament. Apart from the possibility of war, there were widespread fears of people becoming contaminated by nuclear fallout, and increasing frequencies of cancer. With hindsight it can be seen that the route to a test ban treaty was fuelled by a simplistic understanding of the risks of nuclear radiation that is scientifically unsound. It was widely thought that a single ionising particle could cause damage to the DNA, and a mutation could initiate a tumour. Into this context came a simple, robust and precautionary concept that offered protection from radioactivity.

In 1959 the International Commission on Radiation Protection (ICRP) accepted the 'linear no-threshold' (LNT) hypothesis[2] as the relationship between radiation dose and its biological consequences. The LNT relationship assumes that the effects of low doses of ionising radiation can be estimated by linear extrapolation from effects observed at higher doses. The LNT concept implies that there is no safe dose, and that even the lowest doses of ionising radiation pose a harmful biological threat. The LNT hypothesis offers the greatest protection from radiation, as it predicts harm at any level, giving people protection from the release of any level of radiation. In the prevailing political climate the concept served its purpose, providing more than adequate protection, but the LNT is now known to be a flawed model of the scientific relationship between radiation dose and biological effect.

Supporters of the LNT hypothesis refuted any data that were non-linear, as indicated by a threshold, and therefore rejected any evidence in support of radiation hormesis, as such data typically exhibited a threshold and hormesis. As the effects of radiation are assumed to be negative and harmful to life, the LNT concept does not admit any positive or stimulatory responses by exposed organisms. Linearity also implies that no dose of radiation, however small, is without its proportionate effect. The LNT relationship implies that biological systems are inert and unresponsive to radiation exposure. This is now known to be untrue, and there are a variety of adaptive responses that respond to radiation by repairing radiation damage and countering its effects. Many adaptive responses to radiation damage are now known. Without the evolution of adaptations that counteract, neutralise and repair radiation damage, life as we know it would not have survived. Such responses create non-linearities in dose–response relationships, apparent as thresholds of effects and stimulation causing hormesis. As the LNT implies linearity in all such relationships, data that are non-linear are not accepted by those who still adhere to the LNT. While it might have been reasonable to believe, when the LNT hypothesis was first conceived, that biological responses to radiation played no part in ameliorating its effects, this it is no longer so, and a large body of evidence to the contrary has been accumulated over the last 50 years.

[2] It is appropriate to term the linear no-threshold (LNT) concept a hypothesis, as those who derived it overlooked much evidence for non-linearity.

The status of radiation hormesis was reviewed by Luckey in his book *Hormesis with Ionizing Radiation*, published in 1980, which included references to about 3,000 reports on stimulation by low-dose irradiation. The first international conference on radiation hormesis was held in 1985, and the variety of evidence added much weight to the case for rejecting the LNT hypothesis. In 1994 the United Nations Scientific Committee on the Effects of Atomic Radiation (UNSCEAR) recognised and endorsed the existence of radiation hormesis, a move that shook 'radiology's ethical and technical foundations'. The US Radiation Science and Health Foundation used the evidence for radiation hormesis as sufficient reason to call for a re-evaluation of the LNT hypothesis. Nevertheless, in 2000 the US authorities re-affirmed their acceptance of the LNT model and its implications, despite admitting exceptions. The main flaw in the LNT hypothesis is that it makes a false assumption, stating that the effects of low doses can 'be estimated by linear extrapolation'. Linearity was assumed because initially there were few data considered reliable at low doses. But as a matter of scientific principle, extrapolation beyond the data must not be assumed to create new information; it is a false assumption and bad science.

It is recognised now that dose–response relationships at low levels are almost invariably non-linear (see Chapter 7). There are many adaptive responses of organisms that effectively counteract radiation effects, including homeodynamic responses that counter the effect of noxious agents at low levels of radiation. Any such response that counteracts the harmful effects of radiation, and creates thresholds in dose–response curves, displaces the linear relationship, deferring harmful effects to a higher concentration, so producing an α-curve (Figure 10.2b). The threshold in any dose–response curve is the critical point that makes it possible to calculate the capacity of the biological response to resist the effects of radiation. The capacity of the adaptive response is given by the interval in the dose–effect curve between the lowest level of radiation that elicits an adaptive response and the threshold of inhibition which marks the onset of overload (Chapter 11), producing the β-curve (Figure 10.2c). Typically, over time, the α-curve develops into the β-curve; these two curves are the typical forms of dose–response relationships.

The LNT hypothesis denies the possibility of any adaptive response to low levels of radiation, which creates these curves. For this reason, the LNT model is an obstacle to the acceptance of radiation hormesis, or any rationale that accepts that adaptive responses play a part in creating it. The LNT hypothesis rejects any positive observations such as the stimulation of growth, development, fertility, longevity and such like, which are typical biological indices of sublethal effects. The acceptance of exceptions to linearity, which are now accepted by those for whom the LNT is reality, misses the point. In 'dose–response' relations, Calabrese and co-authors now accept that non-linear hormesis and the β-curve is the norm. But even if this were not true, 'dose–response' relationships typically follow the α-curve, which with its threshold is also non-linear. Non-linearity is a feature of all α- and β-curves which define the forms of concentration–effect curves in toxicology, so the LNT hypothesis is at odds with all such toxicological data.

RADIATION HORMESIS AND HUMANS

We have considered the biological effects of radiation from protozoans to guinea pigs, which all exhibit the β-curve, but what evidence is there for radiation hormesis in humans? One cannot ignore the much earlier anecdotal evidence for the beneficial effects of low doses of radiation on humans. There is a long history of spa resorts, where subterranean thermal or mineral waters have been used for drinking and bathing. The practice of 'taking the waters', as beneficial for the health, was popular during the nineteenth century. Hydrotherapy was widely adopted by the Victorians, who paid visits to fashionable spas such as Malvern, Ilkley, Leamington and Bath. People took in large quantities of water, exercise, fresh air and wholesome food. Hydropathic doctors also believed that much illness was medically induced, so all tonics and medicaments were denied. The panoply of Victorian medicines and tonics often included opiates, mercury, arsenic and alkaloids such as colchicines, quinine, veratrine and aconite. The patient was also denied alcohol, snuff and tobacco.

There was one important factor common to all waters of subterranean origin of which the Victorians were unaware. While radioactivity had been discovered in the laboratory much earlier, it was not until Hans Geiger (1882–1945) developed and refined the Geiger counter in 1928 that measurement of levels of radiation in the environment could be made. One beneficial constituent of subterranean water is radon, a gas generated by the radioactive decay of radium. The history of 'radon spas' goes back to warm mineral springs arising on the volcanic island of Ischia off Naples. They were used 2,500 years ago for therapeutic purposes. According to archaeological evidence, the thermal springs at Bad Gastein in Austria may have been used 5,000 years ago, though it was not until the Middle Ages that water was redirected to wooden tubs for people to take the waters. Such spas exist in the UK, Finland, Germany, France, Austria, Brazil, Russia, Japan and, more recently, in the USA and New Zealand. People went to spas to drink radioactive water and stay in caves to be irradiated for hours. For those who could not afford the luxury of staying at a spa, bottled water was exported. During the 1920s and 1930s, drinking the radon dissolved in drinking water was popular and widely endorsed. One spa, in five years, sold over 400,000 bottles of spa water containing radon.

One important reason for the benefits of 'taking the waters' is now considered to be radiation in water rising from within the Earth, which is about 6–10 times higher than background levels. The treatment was effective in relieving inflammatory diseases (arthritis), and long-term benefits were found for asthma, gout, high blood pressure and many other ailments, although more rigorous evidence is required as to the specific benefits. The explanation is that low doses of radiation stimulate the rates of physiological processes, resulting in a range of non-specific benefits to health.

Epidemiological evidence for the beneficial effects of radiation is different but just as compelling, and relies on selected groups of people among nuclear industry workers. Large numbers allow statistical tests to be carried out to determine the probability of differences that are not due to chance. Careful selection of the statistical groups for comparison gives such techniques the rigour of controlled experiments of a kind not permissible on human beings. T.D. Luckey, in 1994, assembled – in a meta-study – data from various earlier epidemiological studies relating cancer frequencies to exposure to ionising radiation. The study, with data from the UK and Canada as well as a number of American studies, included over 7 million person years [7]. One group provided a baseline, and was not exposed to a level of radiation any higher than those in non-nuclear work. The other group included those exposed to low levels of radiation due to their work in the nuclear industry. Low levels of radiation reduced the frequency of cancer in those exposed in the workplace to just under half the frequency (48%) of that found in those individuals who had not been exposed to radiation in the workplace.

Bernard Cohen and his many collaborators studied the relationship between lung cancer and concentrations of radon in houses, representing the dwellings of 90% of the people in the USA [8]. The work involved the collaboration of 450 physics professors to make measurements of radon across 1,700 counties. Again the results showed

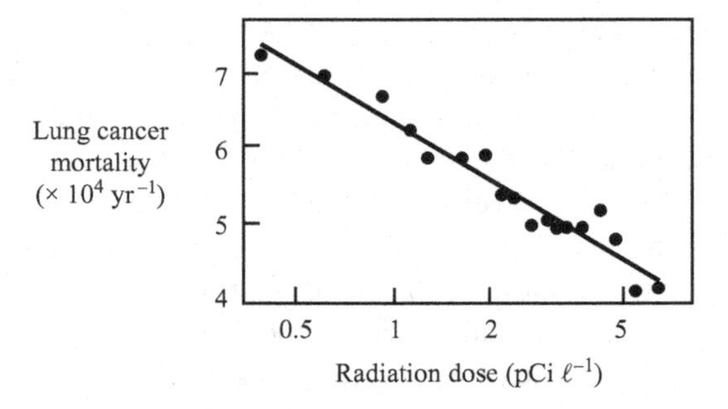

Figure 10.4. An epidemiological study of lung cancer mortality rates in the USA. Each data point represents a US county. The mean line through the data is orthogonal to the line representing the prediction of the LNT hypothesis, indicating that the results are the opposite of what would be expected. Redrawn from Cohen [8].

that lung cancer mortality is *reduced* by low levels of radon. Cohen has shown that for low levels there is a significant *inverse* relationship between radon inhalation and cancer deaths (Figure 10.4). Cohen's study also uncovered evidence that the average lifespan of people exposed to low levels of radon gas may exceed that of the general population. Comparable results were obtained in the UK and elsewhere. Both these examples, the meta-study of dockyard workers assembled by Luckey and the radon study by Cohen, exhibit radiation hormesis in the form of the enhancement of longevity due to low levels of radiation. In particular, they demonstrate that low doses have a protective effect against cancer.

It seems likely that the reason for the apparently protective effect of low levels of radiation against cancer is that the DNA repair mechanism is inducible. Elevated levels of radiation, in causing a greater frequency of mutations, initiate an adaptive increase in the capacity to repair DNA damage. As long ago as 1984, D. Billen showed that each cell suffers 70 million spontaneous DNA damage events per year [9]. Recent data for UV light show as many as one million molecular lesions per cell per day, yet this amounts to less than 0.0002% of the 6 billion bases of the human genome. Such damage to the genome is continuously corrected by DNA repair mechanisms.

The system has been shown to be inducible. White blood cells can be exposed to a moderate experimental dose of radiation such that chromosomal breakages amount to 30–40% of the whole genome. But if the cells are first pre-exposed to a smaller dose of radiation, which is one hundred times less, before the higher dose, then chromosome

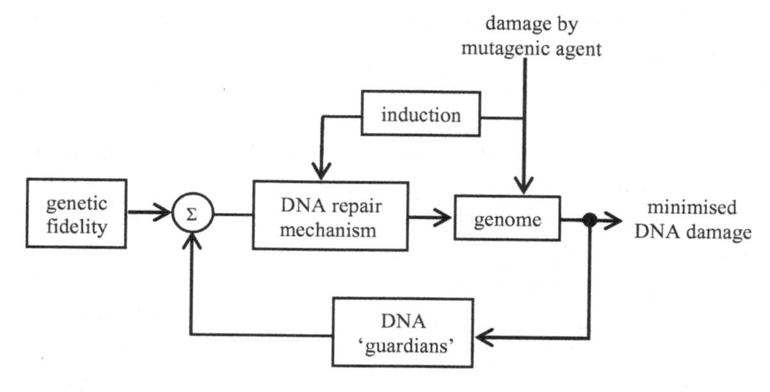

Figure 10.5. A diagram of DNA repair as a hypothetical feedback mechanism. Induction of DNA repair is indicated by a feedforward loop. The implication is that such principles apply to DNA repair mechanisms.

breaks are reduced by 15–20%. The DNA repair mechanism is induced by the preliminary low level of radiation, giving it a greater capacity to repair DNA damage due to the second and more potent dose [10].

A variety of repair mechanisms are involved to make good the damage. What is more, low levels of radiation have the effect of increasing the output of the system in response to an increased work load. If we accept that one form of aging is due to the accumulation of genetic mutations, the induction of DNA repair mechanisms by low levels of radiation is likely to increase the rate of repair and so reduce aging. Such an interpretation would account for Lorenz's classical findings of increased longevity in mice exposed to low levels of radiation for long periods. The same would apply to the epidemiological findings of others on human populations, such as those studied by Cohen and the meta-study of Luckey.

As DNA repair mechanisms are adaptive and inducible, they can be considered as a cybernetic system, or a control mechanism regulating genome fidelity (Figure 10.5). Fidelity is maintained by using the DNA strand that is undamaged to provide the template to repair the damaged strand. In effect, the undamaged strand provides the goal setting as well as the means of error correction, by providing the template for recreating genome fidelity. A key element in any control mechanism is the sensor of system performance that is fed back to relate to the goal setting. So-called 'DNA guardians' are large proteins that monitor the genome signalling damage, summoning the correct repair system and preventing the cell from dividing while damage persists. If repair is not possible, as with double-stranded breaks, the guardian proteins initiate apoptosis.

When the capacity of the mechanism to repair genetic damage is exceeded, it is preferable for the organism to kill the cell rather than to attempt to repair damage to its DNA. Overload of the repair mechanism would leave damage to DNA, so apoptosis and cell death is a safer option for the organism. Cells and organisms that retain DNA damage, such as those due to double strand breaks, age rapidly. For this reason the induction of DNA repair mechanisms can be expected to reduce aging.

One obvious implication is that the DNA repair mechanism has a capacity to repair DNA damage equivalent to its workload. That is to say, if damage occurs at a rate of a million molecular lesions per day, the repair mechanism must be capable of exceeding the damage rate for the organism to maintain the integrity of its genome. Using Ashby's Law, the equilibrium the system is represented by $G_{\text{damage}} - G_{\text{repair}} = G_{\text{fidelity}}$, so while the rate of repair exceeds the rate of DNA damage, the genome (G) will remain accurate. As the rate of damage must be at least matched by the rate of repair, the biological response to radiation is a crucial factor.

There is an additional factor. When radiation reaches a level at which the repair mechanism become overloaded, high output failure will inevitably result in a threshold. As this threshold due to overload causes cell death, it is a critical point in determining the biological effect of radiation.

The linear no-threshold (LNT) hypothesis is incompatible with positive responses to toxic agents, as they inevitably create thresholds in concentration–effect curves. The LNT is flawed in that such a hypothesis cannot admit any adaptive biological responses that oppose or neutralise the effects of radiation.

At Chernobyl the LNT model failed to provide an accurate forecast of human casualties, with substantial overestimates of casualties. The health and fitness of animals around the site of the nuclear power station after the event was a surprise to ecologists working in the area. It appears that a missing factor in the model calculations was the role of DNA repair mechanisms. Not only was their capacity unknown or overlooked, but so was their increased capacity induced by long-term low levels of radiation.

In the next chapter we look at more recent research on the growing weight of evidence for hormesis, before going on to explore the possibility that a homeodynamic mechanism is responsible for hormesis: an acceptable mechanism to account for hormesis and the β-curve. So we now move on to consider the implications of the Maia hypothesis which, it is hoped, will provide a satisfactory explanation for hormesis.

SUMMARY

Hormesis is the stimulatory effect of low levels of toxic agents. Despite claims to the contrary, it is shown that homeopathy does not rest on the principles of hormesis. The implications and applications of hormesis are explored through the use of antibiotics and radiation, as both have been used commercially as hormetic stimulants in agriculture for enhancing the growth of farm animals and crop plants at planting. The beneficial effects of low levels of radiation extend to workers in the nuclear industry, and to populations exposed to low levels of radon; these benefits include increased longevity and reductions in the incidence of cancer. Such benefits are at odds with the outdated concept used in nuclear industry radiobiological protection – the linear no-threshold (LNT) hypothesis used to predict the effects of radiation. The LNT assumes no adaptive responses, but the example of DNA repair is sufficient to highlight the error in this assumption. It is reported that there are one million lesions per cell per day due to UV light alone and so, for life to continue, repair must occur at a matching rate. The fact that such systems are inducible may help to explain why low doses are beneficial and increase longevity while reducing the frequency of cancer. In the next chapter, an interpretation will be sought showing that hormesis is a byproduct of an adaptive response to toxic agents.

11

Maian mechanisms for hormesis and catch-up growth

What is hormesis and why haven't we heard about it before?

Leonard Sagan (1985)

The hormetic model is not an exception to the rule - it is the rule.

Calabrese and Baldwin (2003)

HORMESIS IN RECENT TIMES

The first meeting on radiation hormesis was held in California in August 1985. Although I had not worked on radiation hormesis, I had published a review a few years earlier, which included examples of radiation hormesis. This was an attempt to put over the idea that hormesis was a general phenomenon that I believed would be shown to have a general explanation. So I had deliberately included in the review examples of hormesis that were as diverse as could be found. They encompassed a wide range, including different toxic agents and organisms from many phyla, some showing the hormetic effects of radiation. The examples chosen were all of growth hormesis, because that was my interest, although the reader should appreciate that hormesis occurs in a number of other biological processes as well, from the fields of biochemistry, physiology, behaviour and others. The examples of growth hormesis from the literature had been published in many different forms, so in my review they were plotted in the same way, showing that they matched the β-curve. Even the examples exhibiting radiation hormesis ranged from bacteria and protozoans to insects and guinea pigs, although the majority were due to low levels of other toxic agents of the form characterised by the β-curve (Figure 10.1).

The meeting was organised by the Electric Power Research Institute, to examine critically the challenge that radiation hormesis represented to the dominant paradigm of the LNT hypothesis (Chapter 10). Even then, the power industry had been committed to the concept of LNT for nearly 30 years, and needed to learn more about radiation hormesis and the phenomenon in general. Leonard Sagan opened the meeting by posing the question 'What is hormesis and why haven't we heard about it before?' Sagan had assembled contributors involved in hormesis from an unusually wide range of scientists from cancer research, cell biology, plant physiology, agriculture, marine and freshwater biology and beyond [1]. His choice of quotation from Thomas Kuhn was revealing: 'Normal science ... often suppresses fundamental novelties because they are necessarily subversive of its basic commitments', accepting that new ideas which do not fit the predominant paradigm were likely to be repressed. It seemed that Sagan suspected, even then, that there was something important about hormesis, as well as the growing embarrassment that hormesis presented to his industry.

Even at that time the LNT model was a concept of international importance, adhered to by the USA and her allies in regulating exposure of personnel and the public to radiation. Governments are not used to responding, as science does, to flawed hypotheses by discarding them. It seems that the LNT model was not seen merely as a hypothesis when first adopted, but it became obviously so with hindsight. For Einstein, 'a hypothesis is a statement whose truth is temporarily assumed'; it is a tool to be discarded when it is no longer supported by the facts, to be replaced by new hypotheses that fit the current state of knowledge. The problem was that the LNT hypothesis was the basis of precautionary measures that ensured radiobiological protection for an industry, and the population at large, both in peace and in war. But it was also a scientific principle, upon which models to forecast biological and human risk were built. In its favour, it offered a degree of protection that was more than adequate, as LNT implied that radiation had some deleterious effect, however low the dose; but, on the other hand, it was fundamentally flawed, as exposure to low levels often had no harmful (and sometimes had beneficial) biological effects. It has already been mentioned that calls for the LNT hypothesis to be replaced by a concept that reflected scientific knowledge more accurately had repeatedly been rejected.

Chernobyl is the most recent example to show that the LNT hypothesis is a defective concept. As we saw in the last chapter, epidemiological evidence from large studies has demonstrated that there is something wrong with the LNT hypothesis (Figure 10.4). Even though biological 'dose–response' curves are typically non-linear, the clarion call for a more realistic model was opposed. There was too much invested by governments and their institutions in the LNT hypothesis to accept change. To be excessively cautious might not be a bad thing in itself, but the cost implications of the LNT have been, and continue to be, colossal. Remember, the LNT predicts that *any* level of radiation is considered harmful, despite the fact that we live constantly subjected to easily measured background levels, and that life evolved in a world with background radiation 3–5 times higher than it is today.

A TOXICOLOGICAL PHENOMENON SHEDS LIGHT ON BIOLOGICAL RESPONSES

As we have seen, in its early years hormesis was linked with homeopathy and was criticised for the association, because homeopathy does not adhere to scientific principles and makes claims that cannot be proved. For many years hormesis was sidelined. Without general acceptance in journals and textbooks, hormesis has been 'rediscovered' many times. To come across hormesis anew is not just surprising, it also presents a paradox as to how toxic agents at low concentrations can cause biological stimulation. It is an observation that does not stand to reason: how could toxic inhibitors *stimulate* biological processes? Such a paradox is always likely to be doubted, particularly when the common-sense expectation is that toxicity should increase with concentration. In the 1930s and 1940s, the threshold model (α-curve) was the dominant expectation of toxicological data, but even then there were many exceptions that exhibited hormesis, which were ignored because they were at odds with the existing tenets of toxicology.

In the 1960s, toxicology was a young science; one would have not expected hormesis to be discussed at, let alone be presented to, scientific meetings at the time. Toxicology is also an applied science and most toxicologists, it seemed, were employed in the chemical industry or in government laboratories concerned primarily with monitoring and legislation. The observation that low levels of toxic substances might have beneficial effects on organisms was contrary to accepted thinking and ran counter to the expectations of industry and the legislature. Nevertheless, hormesis was a recurrent observation, as it had been for a century. It was doubtless noted by opponents of hormesis that it would be difficult to establish the concept of hormesis without a mechanism to account for it. But the scientific climate of the time remained reductionist, with individual toxic agents being treated as if they each had their own specific effects. The idea of a general explanation would not have been well received.

The primary aim of toxicologists has been to establish the efficacy of poisons in eradicating pests and diseases, while protecting mankind from their effects. Pharmacologists developed drugs that owed their efficacy to discoveries of differential toxicity between human beings and pathogens, parasites or cancer cells. Environmental toxicology introduced the new objective of protecting the natural flora and fauna from the effects of poisons that are used in farming (pesticides and herbicides) or that find their way into effluents and drainage water and into the aquatic environment. The shift in emphasis was to identify the effects of much lower concentrations of chemicals that, as environmental contaminants, might have adverse biological effects. Complex man-made chemicals were found to cause subtle and sometimes cumulative effects at low concentrations due to their action as mutagens, carcinogens or endocrine disruptors that mimicked hormones.

The sub-lethal effects of chemicals at lower concentrations began to provide examples of hormesis with growing frequency. Such examples complicated the assumption in the minds of many that poisons simply become more toxic with concentration. Hormesis showed this to be only partly true, as poisons stimulate biological processes and, confusingly, cause 'beneficial' effects at low concentrations, as well as causing 'harmful' effects at high concentrations. At low concentrations, their effects as poisons are gradual enough to observe that organisms have adaptive responses to toxicity. Physiologists have long used low levels of toxic substances to help understand the normal effects and responses of organisms to perturbation. One learns little about an equilibrium without perturbing it.

While reductionism was the predominant approach to research, funding agencies could be expected to avoid supporting efforts to establish general hypotheses such as hormesis. Nevertheless, toxicologists investigating concentrations of contaminants at environmental concentrations found examples of hormesis with growing frequency. Similarly in pharmacology, studies of low doses became necessary to establish the minimal doses that were effective, in order to reduce side-effects, and in the process revealed more instances of hormesis. For an increasing number of chemicals (especially carcinogens), the LNT hypothesis had been applied, implying that no concentration, however low, was considered to be harmless. The actuality was that 'dose–response' relationships were frequently non-linear, and hormesis revealed that often the reverse was true, and low concentrations were actually beneficial. It became logically impossible to accept both the LNT hypothesis and hormesis.

Hormesis was now well enough known as a concept to attach the name to discoveries of stimulation that were unexpected, and often a distraction, falling beyond the usual ambit of toxicologists to investigate. Sometimes hormesis was ignored or discarded because it was unwanted, or the reverse of what had been expected. But hormesis was real enough, and during the 1980s was being found with greater frequency as the number of toxicologists increased, and as their interests were drawn to the effects of lower concentrations of toxic agents. From this time onwards, interest in hormesis began to grow rapidly.

By considering lower concentrations of toxic agents, it became common to find those levels of toxic agents to which organisms had evolved adaptive responses able to counter, neutralise, ameliorate or repair the consequences of toxicity. Increased interest in low levels of toxic agents, and the way that organisms responded to them, led to a better understanding of the metabolism, degradation and excretion of toxins. Other responses to organic agents provided mechanisms of detoxication by chemical binding and sequestration. Cellular responses included hypertrophy and hyperplasia or, at higher toxicities, autolysis and autophagy. Damage to the genome initiates DNA and chromosomal repair. But the ubiquity of homeodynamic systems in regulating physiological processes meant that each worked to counter the inhibitory effects of all toxic agents. It is to the homeodynamic control of growth that we shall later return in seeking a general explanation of hormesis.

CALABRESE'S CONTRIBUTION TO HORMESIS

An American scientist presented a paper on hormesis at the first international meeting at Oakland, California in 1985. Edward Calabrese, who was to play an important part in the resurgence of hormesis, had rediscovered hormesis for himself as a junior, nearly 20 years earlier, at Bridgewater State College. He was in a plant physiology class, and his laboratory group had been told to dose some peppermint plants with the herbicide Phosfon. The task was to establish what dose was required to stunt plant growth. Calabrese's experiment apparently failed, and instead of shrivelling up, the plants grew green and luxuriant. 'Either you treated the plants with the wrong chemical, or you mislabelled them', the professor said. 'God forbid you discovered something new'. When the young Calabrese tried to repeat the experiment, the peppermint shrivelled as expected, but when the lower dilutions first used in error were tested again, the plants thrived. By every measure – height, weight, root length – the plants grew about 40% better than the controls that were not sprayed with Phosfon. A poison that is lethal at high doses may have the opposite effect at low doses. The young Calabrese *had* discovered something.

It was not until 1976 that Calabrese published his paper with a colleague under the title 'Stimulation of growth of peppermint (*Mentha piperita*) by Phosfon, a growth retardant'. A decade after serendipity had smiled on the student with his personal discovery of hormesis, he gave a review paper at the Oakland meeting, drawing together a number of examples of chemical hormesis that supported the generality of his curious finding [1]. He had already experienced the doubts of others about his discovery and decided at an early stage that an inductive approach which depended on the weight of evidence would be more persuasive. He knew that the only way to be believed was to provide overwhelming evidence for hormesis.

Calabrese's approach attracted the funding to create a comprehensive database for hormesis. From the start, he realised that it was necessary to redefine what was meant by hormesis, providing the statistical criteria that needed to be met by instances that could be accepted as hormesis. Stimulation was expected to be 30–60% higher than the controls and to span a tenfold range of concentrations (Figure 10.2.). Most examples drawn from the scientific literature met these broad criteria in conforming to the β-curve and focused their reviews upon results that could withstand scientific scrutiny. Calabrese, with Linda Baldwin as his co-author, set about establishing hormesis as a scientific phenomenon by means of a comprehensive bibliographic search for instances from every field. Through the sheer weight of evidence, they were able to show that hormesis is reported with surprising frequency in the scientific literature. As part of Calabrese's strategy to make hormesis credible to the scientific community, his *BELLE Newsletter* became important. It dealt only with the 'Biological Effects of Low Levels of Exposure' and was distributed to those in the field who were interested, but for Calabrese it also created a network which ensured that he discovered who was working on hormesis, making his coverage of the current literature on hormesis more complete.

From the 1990s onwards, Calabrese published a series of papers and reviews that have re-established hormesis and brought it into the mainstream of toxicological thinking. From the early stages, he recognised the importance of setting hormesis in its historical context. Later he was able to write of the re-emergence of hormesis [2], and to describe the β-curve as a dose–response model of fundamental importance. Given the generality of hormesis among both animals and plants, he argued for a biological interpretation of hormesis, favouring the idea of hormesis as the result of overcompensation of homeostatic mechanisms.

In his hands, the hormesis database and his review articles became the principal weapons in the fight to get hormesis accepted by the scientific community. By 1997 there were 350 examples of hormesis in the database. This number grew rapidly, and by the year 2000 there were 6,000 examples, including 1,000 relating to metals and their compounds alone. The reviews took a broad ambit, reviewing hormesis according to the classes of chemicals that induced it (inorganic chemicals, peptides, ethanol, chemotherapeutics). Due to the level of interest, radiation hormesis had been an earlier priority, to which he returned. This time the reviews took on a biological emphasis rather than focusing on chemical causes, covering the fields of cancer and immunology, and other examples of hormesis in medicine. Having established that the β-curve is the more common form of concentration–effect model, the principle of hormesis is now becoming accepted and is being applied in research.

The scale of the bibliographic exercise to assemble the database was little short of heroic. As a result of this major effort by Edward Calabrese and Linda Baldwin, no-one could have been left in any doubt that hormesis is a general phenomenon. Any effect of toxicity that occurred so widely had to be of fundamental importance, not only to toxicology, but to biology and medical science in general [3]. The indices that exhibit hormesis include various kinds of biological effects, including molecular, metabolic, physiological, reproductive and behavioural. There appeared to be no taxonomic limitation, such that hormesis within the β-curve may be found in any group of organisms, from bacteria, protozoans, invertebrates and vertebrates through to mammals.

It also seemed that any kind of toxic agent is capable of causing hormesis. Such agents include both heavy metals and organic chemicals, both natural and man-made, including antibiotics and pesticides, and different kinds of ionising radiation. In making this generalisation, other kinds of stress also appeared to be capable of inducing hormesis, including osmotic stress due to salinity extremes, turbidity for aquatic organisms and the pruning of plants. There is enough evidence to indicate that hormesis, as the predominant feature of the β-curve, is a general phenomenon induced by stress of every kind.

For those who might wish to seek their own examples of hormesis, there are some pitfalls to avoid. Hormesis due to toxic agents may be a secure generalisation, but certain conditions must be met for those who wish to observe hormesis in the laboratory. Even though the organism may be capable of exhibiting hormesis, it might not necessarily be observed. It is essential that the indices of biological effect adopted are capable of exhibiting a positive effect. It is also essential that the range of concentrations chosen is wide enough, with concentrations spanning at least two orders of magnitude below the threshold of toxicity, at intervals of a half order of magnitude or less, to guarantee that enough concentrations are used to define the β-curve. The concentrations need to be statistically significant, if the peak of stimulation is to be accepted as hormesis. The emergence of hormesis and the β-curve also takes time to develop. In hydroids, 3 weeks is required for the maximum stimulation to be observed. Failure to meet these requirements can result in hormesis being overlooked; but the absence of evidence for hormesis can obviously not be taken as proof that it does not occur.

It now seems that for indices that provide a sub-lethal measure of biological processes, any toxic agent is likely to elicit a β-curve. Calabrese drew the conclusion from the wealth of data now available to him that hormesis and the β-curve are the rule. As the concept of hormesis has seen a resurgence, a schism has opened up in the toxicological community. However, those who continue to reject the generality of hormesis must now examine the evidence that Calabrese and his colleagues have assembled, and find some alternative explanation for the wealth of data that exhibit hormesis and the β-curve. The generality that Hugo Schulz claimed prematurely a century earlier can now be accepted, in the knowledge that a mass of data have been brought together that meet rigorous criteria and support the original claim.

With the evidence for hormesis that has been gathered and published, the profile of hormesis has been raised, with accounts of hormesis being published in leading journals in the USA and Europe. In bringing hormesis back from the wilderness, the conventional wisdom was challenged. Those involved in environmental protection and agencies that are responsible for legislation must begin to address the implications of the acceptance of hormesis as a general toxicological phenomenon, and the β-curve as the typical concentration–response curve. Nevertheless, complete acceptance is unlikely to occur without a satisfactory hypothesis and mechanism to account for hormesis.

A MECHANISM FOR HORMESIS

Some would believe that hormesis has many explanations, and there have always been agents known as 'stimulants' due to their own specific properties. But if hormesis is the rule, as the evidence now suggests, there must also be a non-specific interpretation. Without a satisfactory mechanism, the accumulation of examples of hormesis is phenomenology, and the generalisation implied by the term 'hormesis' has no explanatory power.

A generalisable hypothesis[1] based on work with hydroids is summarised here. Colonial hydroids proved, at an early stage, to be sensitive and precise bioassay organisms. They could be cultured indefinitely in the laboratory as a clone. They proved to be good models with which to study growth and its control, because colony biomass could be accurately determined in a non-sacrificial way. Experiments to establish the sensitivity of hydroids to environmental concentrations of metals required a series of experiments on their effects on colonial growth rates. It was not long before copper was found to stimulate the growth of hydroids at low concentrations, giving a concentration–effect curve which later became recognised as the β-curve. Stimulation was unexpected, but it appeared that copper might be deficient in the uncontaminated offshore waters, so its addition might meet some requirement as a micronutrient and thus stimulate growth. Experiments with other non-essential metals (including cadmium and mercury) showed that this interpretation was wrong, as they too caused hormesis. This gave rise to the idea that metals generally might

[1] The obvious problem for experimentalists attempting to establish a biological generality is that they must start with a single organism, yet hope to learn whether a particular phenomenon applies to other or to all organisms. The key step is publication, which allows other scientists the opportunity to verify or disprove the hypothesis and to challenge the question of generality by using other species. But there is another factor that also makes a difference. If one is considering fundamental processes that have their basis in the lower levels of biological organisation, there is such consistency in the structure and function that little is thought of the problems of extrapolation. Such reasoning gave Monod the conviction to say that 'what is true for *E. coli* is also true for the elephant'.

be the causal agents of hormesis, but this was refuted by experiments with toxic organic compounds that likewise induced hormesis.

It soon became clear that the discovery of stimulation by toxicity was not new: the Arndt–Schulz Law revealed that such stimulation was known from work carried out a century earlier. Papers on the Arndt–Schulz Law and hormesis were gathered from the literature in the hope of finding an explanation. There was some speculation, but nothing detailed or definitive emerged. So how might hormesis be caused?

Experiments with salinity led to a completely different interpretation. Bioassays were required for estuarine waters with reduced salinity, as it was necessary to know the point at which reductions in salinity alone might become critical for the hydroid. Much pollution occurs in estuaries, but how far upstream could one go before the hydroid would be more influenced by reductions in salinity than by the toxicity of the water? The long-term strategy was that, once the water became less saline a brackish-water hydroid would be needed, and then a freshwater hydra where the estuary became fresh water, in order to provide continuity from the sea to the river.

But, remarkably, reduced salinity also resulted in the stimulation of growth rate, to create a β-curve like that found for the metals. This experiment, more than any other, showed that growth hormesis was not necessarily due to agents that caused toxicity, but by anything that could induce stress and inhibit growth. In this instance the stress involved *removing* something essential for life – salt from the water. Hormesis was induced not by exposing hydroids to low levels of a metal, but by the dilution of seawater. As reduced salinity could also cause hormesis, perhaps the stimulation of growth had more to do with a reaction of the organism to inhibition, rather than some chemical property of the agents tested. It began to look as though hormesis was the result of a *response* to inhibition, rather than some stimulatory *effect* shared by each of the toxic agents. Each agent tested was a growth inhibitor at higher levels, suggesting that the cause of hormesis was the consequence of an adaptive response to growth inhibition.

Hormesis seemed to imply something fundamental about the growth process, and perhaps of wider interest. With this in mind, I became immersed in the study of biological growth, and the Maia hypothesis emerged and developed over time. Logically, a stimulatory response is required to negate inhibition. Perhaps a stimulatory response was related to the control of growth, and the response was due to some homeodynamic regulator? Interest in what might be the cause of hormesis was the starting point for a study of growth control. It led to the development of a method of perturbing growth and filtering out the consequences from raw cumulative data (see Chapter 5). This gave the output of the rate-sensitive control mechanism, which provided an explanation for hormesis [4,5]. The result was dynamic data for growth rate that revealed the characteristic oscillatory behaviour of feedback mechanisms (Figure 5.5). The aim here is to interpret a particular kind of concentration–effect curve, the β-curve, for which hormesis was the key feature. Given the wealth and diversity of the data exhibiting hormesis, a single interpretation held a fascination because of its explanatory potential.

ASHBY'S LAW AND HORMESIS

We now need to reconsider the idea of control, but not from the position of accessing the dynamic output that bears the signature of cybernetic behaviour, which has been covered in an earlier chapter (Chapter 5). Instead we consider the outcome of control from the perspective of what homeodynamic control means and does for the organism over time. This was most succinctly stated by Ross Ashby in his misunderstood Law of Requisite Variety (Chapter 7). This law is more appropriate to the capacity of control systems rather than the variety of responses, as it seems that this is not quite what Ashby actually meant. When the dynamic output of a growth control mechanism is integrated over time, what is the final outcome? It is this which determines the contribution of control to organism fitness. The law states that the overall effect is the sum of inhibition minus the counter-response. This succinctly represents what homeodynamic systems do. Their purpose is not so much to maintain a process at some preferred rate, but rather to oppose the effects of extrinsic disturbance on the controlled process. It is the capacity to resist perturbation that bestows fitness upon organisms.

What becomes clear is that the nature of any *response*[2] to inhibition is of necessity stimulatory, such that a stimulatory response is required in order to neutralise disturbance or inhibition by toxic agents. Within normal toxicological relations of concentration plotted against *effect*, the result is an α-curve, which has already been

[2] The distinction between the term 'response' and its use as 'dose–response' was made in Chapter 7, but needs repeating here. 'Response' refers to the reaction of the mechanism, rather than the final outcome, for which 'effect' is used. Italics are used in the text to emphasise the difference.

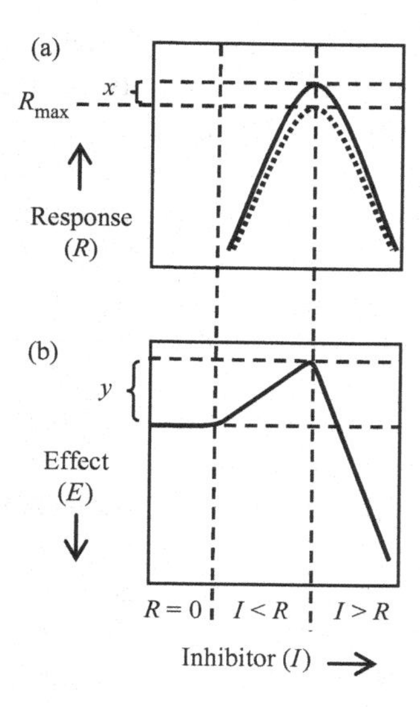

Figure 11.1. A hypothetical diagram representing Ashby's Law of Requisite Variety. An earlier diagram provided an interpretation of the α-curve (Figure 7.2). This provides an interpretation of hormesis due to an adaptive response to inhibition (a), with hormesis as a byproduct in the β-curve (b). The dotted line in (a) indicates the pre-adapted response (*R*) curve to the inhibitor; the continuous line shows the effect of adaptation which results in an elevation of growth rate with a peak indicated by *x*. In (b) the effect (*E*) of this adaptation, with a peak height given by *y*, results in a typical hormetic stimulation of growth. From Stebbing (2009), with permission.

explained using Ashby's Law (Chapter 7). Over the sub-threshold range of the curve, where inhibition is neutralised, the rising concentration of inhibitor is matched by a counter-*response*, such that the outcome is as if – within limits – the organism is unaffected. Clearly any mismatch between inhibitor and the organism's *response* will result in over- or undercorrection. Undercorrection ($I > R$) will result in inhibition, while overcorrection ($I < R$) will result in stimulation. Here we can see how the paradoxical stimulation comes about.

We use a graphical summary of Ashby's Law that separates the three components of inhibition, response and effect, to provide a hypothetical interpretation of the β-curve (Figure 11.1). It is represented by two linked curves, showing the *response* (a) and the consequential *effect* (b) in relation to the level of inhibition, or disturbance in Ashby's terms. We have already discussed in Chapter 7, with respect to the α-curve (Figure 7.2), how a graphical presentation of Ashby's Law requires not only an *effect* curve but also a *response* curve, as the *effect* is the inhibition minus the *response*. Here these two curves are used to illustrate the origins of the β-curve. As before, the *response* curve (Figure 11.1a) rises, with increasing concentration, to neutralise inhibition due to the toxin. Once the maximal *response* (R_{max}) is exceeded ($I > R$), as a consequence of what is termed 'high output failure', the response curve falls with inhibition, as the organism is exposed only to inhibition ($E = I$). The form of the *response* curve follows from the assumption that growth is controlled by a homeodynamic mechanism (Figure 11.1a) that first rises to respond to inhibition, and then falls with concentration as the mechanism becomes overloaded.

As before, it must be emphasised that the *response* is not observable directly, because it only occurs in reaction to inhibition, and is dissipated in doing so. The only direct observation of the *response* is a relaxation stimulation upon the removal of load (see below). Comparison with the interpretation of the α-curve (Figure 7.2) shows that, over time, this system becomes adaptive (Figure 11.1a). What differs is that over a range of concentrations to which the system can adapt, growth rate increases (x). The response to sustained inhibition is to adjust the growth rate goal setting, conferring increased resistance to inhibition and a greater capacity for neutralisation (Figure 7.4). The adaptation is, in effect, also anticipatory because the adjustment is sustained, and is not merely to help neutralise the inhibition that initiated it. So any future toxic inhibition is resisted, and the adjustment is conserved and growth rate maintained at a higher setting. However, in the absence of any further inhibition, growth rate continues at the elevated level, with the inevitable consequence that growth is stimulated (y). In this way colonies become much larger than those that had not adapted their growth rate in this way (Figure 11.1b). It is the adjustment of growth rate in the *absence* of any further inhibition to neutralise that causes hormesis. As a strategy, this adjustment is likely to supplement fitness, whether or not some later inhibition requires neutralising.

This analysis of the toxicologists' concentration–effect curve is part of the necessary process that Calabrese and Baldwin referred to in a paper published in *Nature* in 2005 entitled *Toxicology rethinks its central belief* [6]. An interpretation of hormesis required nothing less than an explanation, first in terms of the dynamic output of a homeodynamic mechanism, and then an explanation within the integrated concentration–effect curve (Figure 11.1).

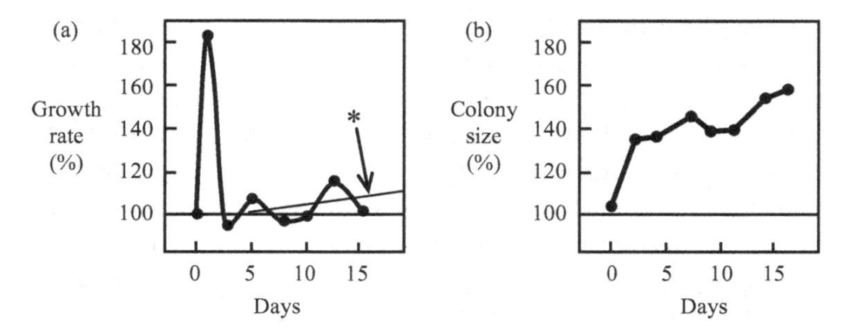

Figure 11.2. Data from an experiment that exposed the hydroid *Laomedea* to a concentration at which goal rate adaptation, leading to hormesis, is clear: (a) a median line (indicated by the *asterisk*) is drawn through the oscillatory data, indicating increased resistance to toxic inhibition; (b) shows the increase in colony size due to, first, the overcorrection at the beginning of exposure and, second, by the gradual increase in the median line about which the rate oscillates. Redrawn from Stebbing [5].

FEEDFORWARD CONTROL

It has been mentioned earlier that two forms of feedforward loop would provide the means by which, first by linking the perturbation with the comparator, a maximal and immediate response could be initiated. A second feedforward loop connecting the perturbation with the goal setting and some integral of the disturbance results in an upward adjustment of goal rate that supplements the capacity to neutralise inhibition (Figure 7.4). Both these responses are apparent in an exemplary output of the specific growth rate ($r\%$). The first loop results in the immediate peak showing a marked overcorrection to inhibition, to a level of inhibition well within the capacity of the control mechanism to neutralise it (Figure 11.2a). The second loop shows a sustained and gradual increase in growth rate over time (Figure 11.2a). This is shown as an inclined increase about which the output oscillates, indicating an increase in the goal setting which amounts to ~10% over 16 days.

While the adjustment in growth rate may be small, it is significant, particularly when considered over time by cumulative growth, as colony size gives a fivefold amplification of the increase in growth rate. This is noteworthy because it is the larger increase in size by which we know and recognise hormesis, yet it is the lesser increase in rate that is the cause and provides the explanation.

Further examination of these two plots also reveals that the initial maximal response to inhibition can be seen in the control output data (Figure 11.2a), but also makes a significant contribution to the increase in colony size at the outset of the experiment (Figure 11.2b). We now see that hormesis is due to two factors. The first is the initial rate overcorrection, while the second is due to the progressive adjustment of growth rate, or goal setting of the control mechanism.

Hormesis is a significant and obvious feature of the β-curve, and it is contrary to expectation that a small change in growth rate could be responsible for such significant increases in size. A return to the travel analogy is helpful in understanding this. Like growth rate, it is the *rate* of travel that is controlled. Large and transient increases in speed have obvious consequences, but what about slower and lesser increases? We may be unable to detect a difference in speed of 1 mph in cars overtaking us at 100 mph. But if this difference in speed were maintained throughout a 24-hour race the two cars would end up 24 miles apart. In the hydroid data we see a 10% increase in rate resulting in a 50% increase in size over 16 days. Integration of the variations in the rate output over time contributes to the overall size of hydroid colonies, whether sustained or transient. We can therefore conclude that hormesis is caused by small growth rate increases in response to low levels of inhibition by toxic agents, which are capable of resulting in large increases in size over time.

ACQUIRED TOLERANCE

It may have been noticed that in the experiment described in the previous section, adjustment of the goal setting was still increasing at the end of the experiment (Figure 11.2a). We can now ask what happens when rate adaptation is allowed to take its full course. Experiments with hydroids lasting for long periods of time showed that pre-exposure for a period of 3 weeks is sufficient to maximise adaptation. It was then possible to compare the effect on hydroids that had been copper-adapted (to a concentration of 10 μg ℓ^{-1}) for 3 weeks with others that had never been exposed to copper. The two groups were later compared for their sensitivity to the same metal over a range of concentrations (0–50 μg ℓ^{-1} copper). It was found that hydroids that had been pre-exposed to toxic agents were more resistant to subsequent exposure (Figure 11.3a). One can consider the data in two ways. If one takes an arbitrary level of inhibition (100% down to 60%) on the vertical axis and reads across to the curves, it can be seen that the pre-exposed group showed an increased

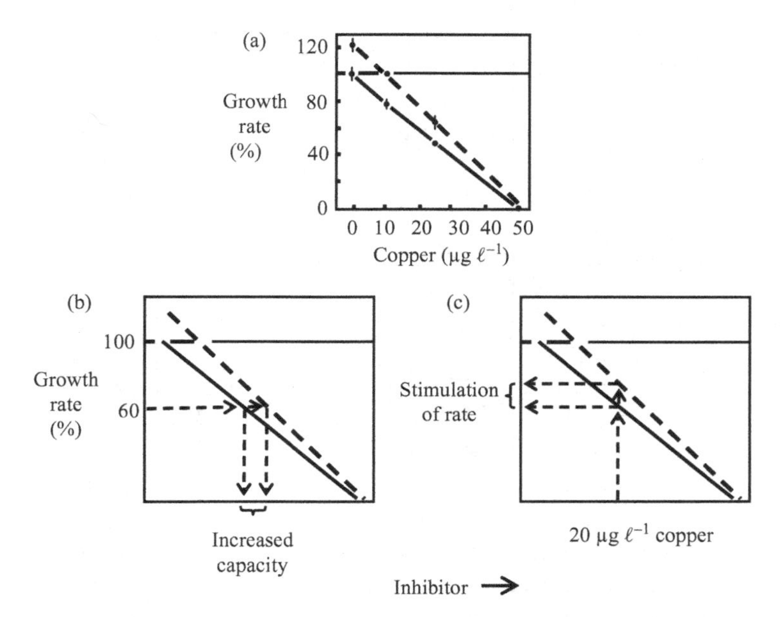

Figure 11.3. Data from an experiment with the hydroid *Laomedea* showing the effects on susceptibility to a metal after 3 weeks' pre-exposure to copper (at 10 µg ℓ$^{-1}$) to subsequent exposure to the same metal (a); diagrams (b) and (c) show two interpretations of the data (see text for discussion).

capacity to resist copper (Figure 11.3b). Conversely, for a given inhibition of growth rate, those hydroids that had been pre-exposed to copper required a higher concentration to cause the same amount of inhibition. Greater resistance to inhibition is clear; pre-exposure to one concentration gives greater resistance to both higher and lower concentrations. On the other hand, for a given concentration of copper, the pre-exposed hydroids are less inhibited (Figure 11.3c); or pre-exposure has the effect that growth is stimulated with respect to those not pre-exposed. Increasing the baseline growth rate by sustained exposure has the effect of increasing the growth rate when later exposed to the metal. We can now see that the linear adaptive increase in growth rate over 16 days that causes hormesis, if continued for 3 weeks, is likely to result in a 20% increase in rate, which would account for the levels of tolerance seen in the acquired tolerance experiment (Figure 11.3a). Over time a 20% increase in growth rate is likely to result in substantial increases in size, and so cause hormesis. While we tend to think of the effect of pre-exposure to a toxic metal as likely to cause resistance

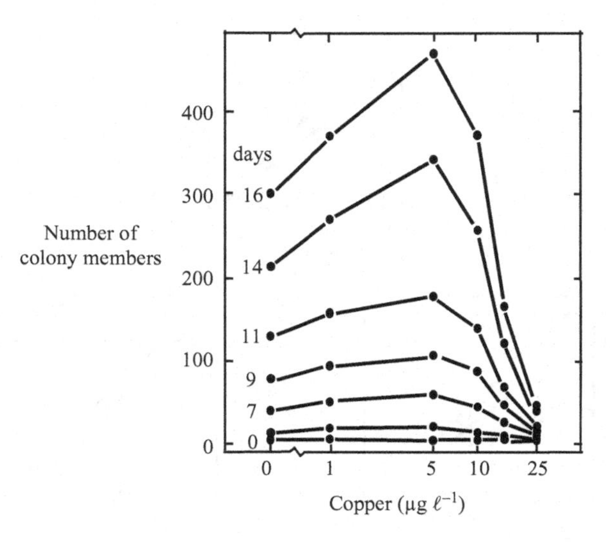

Figure 11.4. Growth of colonies of the hydroid *Laomedea* exposed to a range of metal concentrations (1–25 μg ℓ^{-1} copper) over a period of 16 days. The concentration–effect curves show the development of the β-curve over time. Redrawn from Stebbing [5].

to inhibition, we could equally well refer to such cultures as being stimulated to grow at a faster rate with respect to those cultures that were not pre-exposed.

Finally we return to the β-curve, and once again see the consequences of rate adaptation, which becomes increasingly significant over time. In fact the margins of increase widen as the experiment proceeds, which is due to the exponential growth of hydroid colonies (Figure 11.4). It is also important to note that in the early stages of the experiment, the data adopt an α-curve; it is only later that they exhibit a β-curve.

We now see that hormesis in hydroids is a product of a process of adaptation that results in acquired resistance to toxic agents, due to initial overcorrection and an increase in the baseline growth rate due to sustained exposure. The adaptation confers a greater capacity to resist toxic exposure. But when this increase in growth rate is not devoted to countering toxic inhibition, and there is no inhibition to be neutralised, the increased growth rate results in the stimulation of growth, and thus hormesis.

LINKING THE GENERALITY OF HORMESIS AND HOMEODYNAMICS

The generality of hormesis is beyond doubt; Calabrese and his colleagues have demonstrated that hormesis is a general phenomenon to be found in all groups of organisms used by toxicologists, and can be due to all kinds of toxic agents. For hormesis, as a generalisation, to have meaning and to be accepted as a single phenomenon, then all the examples must share a common mechanism and explanation. Inevitably any such consideration is speculative, as here we only have experiments with one species of organism and a toxic metal to provide a rationale to account for hormesis. Furthermore, the experiments considered above relate only to growth.

Calabrese identified broad categories of biological processes that exhibit hormesis. Examples of hormesis are known from a variety of disciplines including biochemistry, physiology, reproduction and ethology, besides growth and replication. Any process that is self-regulating, self-limiting or self-adjusting requires a feedback mechanism; actually this statement is a tautology because the feedback loop is the means by which 'self' is referred to. As physiological processes are typically controlled, it follows that they are likely to be homeodynamic and to require cybernetic mechanisms. Growth control occurs in relation to rate, density, physiological load and other preferenda found at different levels; such mechanisms permeate all levels of hierarchical organisation (Chapter 9), from the molecular (Chapter 6) through to populations (Chapter 8).

The evolution of homeodynamically controlled processes has been a feature of all organisms, increasing in complexity and functionality throughout evolution from the protozoans to mammals. Homeodynamic control is the most widely distributed physiological principle in living systems. It remains the fundamental concept around which human physiology is understood and taught. We are reminded of the quotation of Lyapunov's that control is the most universal property of life. All living organisms feature feedback mechanisms that can be expected to behave as the hydroid model does in regulating growth. Some form of control is a property of all life as, without it, growth is undirected, uncontrolled and unlimited. For such reasons, any organism can be used as a model to study homeostasis, and any organism is expected to exhibit the adaptive behaviour that has hormesis as a byproduct.

Generalised responses to perturbation add considerable fitness to organisms, as they enable organisms to respond adaptively to any kind of perturbation or inhibition. This is not to overlook the various specific systems that organisms possess to metabolise, excrete or sequester toxic agents foreign to them. Here the importance lies in the generalisation that such control mechanisms respond in the same way to any agent that perturbs the controlled process from its goal setting, as the response is to the perturbation caused rather than to the agent that causes it. However likely the generality of the hydroid interpretation of hormesis may appear, it is now necessary to examine how other organisms and homeodynamically controlled processes respond to toxic agents in perturbation experiments.

CATCH-UP GROWTH AND RELAXATION STIMULATION

There is another growth phenomenon, as interesting as hormesis, for which the Maia hypothesis provides an explanation. 'Catch-up' growth is known mainly from research on humans and agricultural animals. It occurs when growth is held back due to a period of disease, deficiency or malnutrition. When the problem is past, recovery results in spectacular accelerations in growth. The impression has been given by those in the field that growth behaves as if it must catch up by the extent to which it was retarded. In doing so, growth exceeds normal rates, and in some cases growth does actually catch up by the amount it was held back by stress, disease or deficiency.

It was in the 1960s that J.M. Tanner and colleagues' work on the growth of children described a phenomenon which they called 'catch-up growth', and which implied the action of a control mechanism [7,8]. Tanner proposed that it was located in the central nervous system and compared body size to a goal setting or target size for age, by way of a neuroendocrine. The problem with his idea was how the organism might maintain some awareness of what its size should be for a given age, since any target size must be continually changing as the animal grows and becomes older. This hypothetical mechanism was expected to gradually slow growth rates to zero as the final size is approached. Tanner proposed a target-seeking model that involved an inhibitor as a product of growth, with a concentration of the hormone being proportional to the size of the organism. Tanner's hypothesis was that the concentration is monitored and compared with the concentration anticipated, using a time tally. The discrepancy creates an error signal for the release of a growth-stimulating factor. In this way, growth rate would naturally slow with age.

This hypothesis did not survive because the hormone was never found and could not be demonstrated experimentally. The problem was the size-for-age target, which was very complex compared with other biological control mechanisms known at that time. Nevertheless, the hypothetical mechanism was taken seriously until a few years ago [9], when those working on human growth concluded that their data did not support the idea of a time tally mechanism. The original idea is also complicated by the fact that the human body does not grow isometrically; different parts have differential growth rates, creating allometric growth [10].

Mammals are more complex than other organisms, so one approach to understanding catch-up growth would be to use a simpler organism. Tanner's approach involved size control, and no-one appears to have asked whether the cause of catch-up growth could be due to the underlying rate-sensitive control mechanism.

My interest was prompted by seeing that the results of an experiment with hydroids, designed for another purpose, provided an immediate explanation for catch-up growth. The hydroid experiments revealed that the lowest level of growth control in metazoans is the regulation of specific growth rate. Colonies of the hydroid had been exposed to a range of levels of an inhibitor, allowing enough time for equilibration to the different levels of inhibition. At higher concentrations (15 and 25 μg ℓ^{-1}) it was noted that cumulative growth became slightly inhibited; then, after the inhibitor had been removed at day 16, growth recovered to its previous trajectory (Figure 11.5a). The changes appear small because the data are plotted as cumulative data. Upon removal of the inhibitor, relaxation stimulation becomes a spectacular feature of the data; the stimulatory peaks relate to the concentration of inhibitor to which the mechanism had adapted before its removal (Figure 11.5b). The greatest stimulation in growth rate is seen at the highest concentration level of the inhibitor (25 μg ℓ^{-1}), where rates are stimulated to achieve a rate that is more than four times higher than the normal growth rate.

Relaxation peaks last for as long as the lag in the control mechanism, and indicate briefly the naked counter-response, before it recovers control and oscillates to a new equilibrium. The relaxation stimulation provides a valuable indicator of the counter-response that had been necessary to achieve the earlier equilibrium. It is rather like watching two groups of unruly boys pushing against opposite sides of a swing door; the groups are evenly matched, and the door stays closed. But when the boys on one side whisper to each other, and step smartly aside on the count of three, it takes the others by surprise, and they come tumbling through. This is relaxation stimulation, and is perhaps the only way to reveal the forces involved in keeping the door closed.

It will also be seen that, had we relied upon the cumulative data as Tanner did, we would have drawn the same conclusion as he

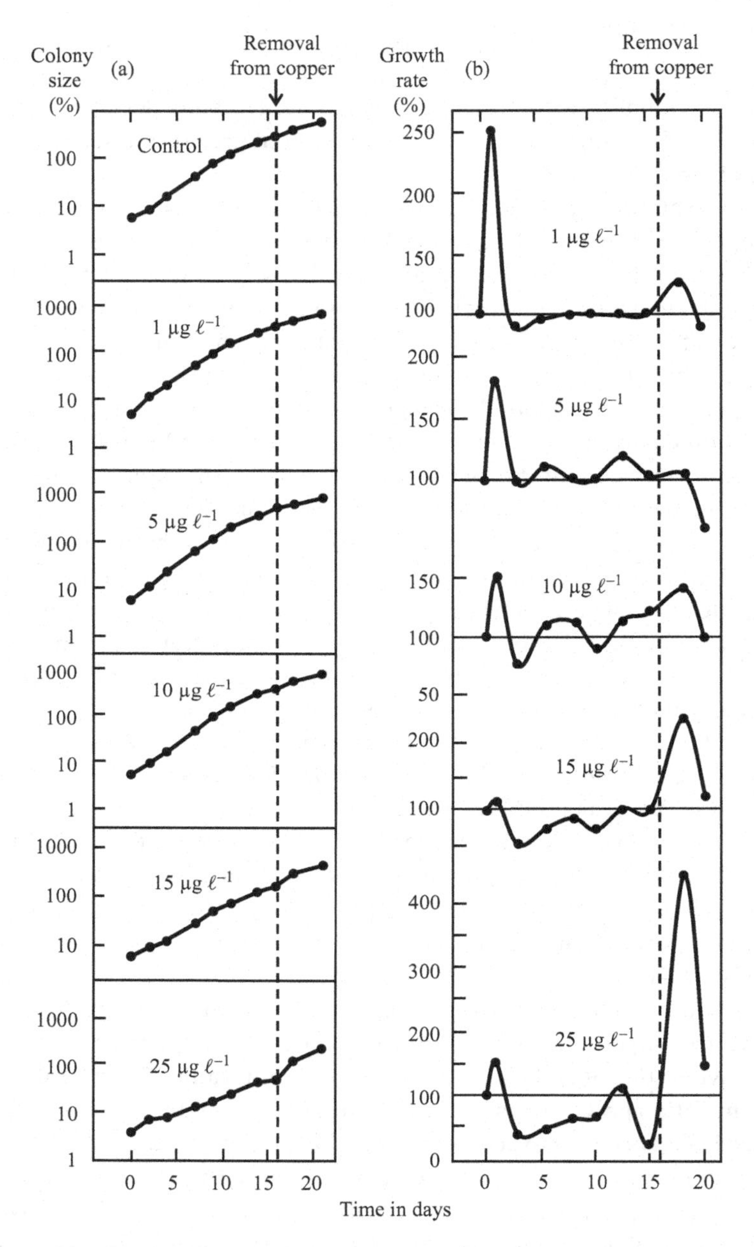

Figure 11.5. Growth of colonies of the hydroid *Laomedea* exposed to a range of metal concentrations (1–25 µg ℓ⁻¹ copper). Exposure was terminated at day 16 to observe relaxation stimulation and its consequences on growth (see text for discussion). Note the small increase in the cumulative growth curve at the highest level of inhibition on removal of the inhibitor (a), which is created by the large relaxation stimulation (b). Redrawn from Stebbing (papers cited in Chapter 5).

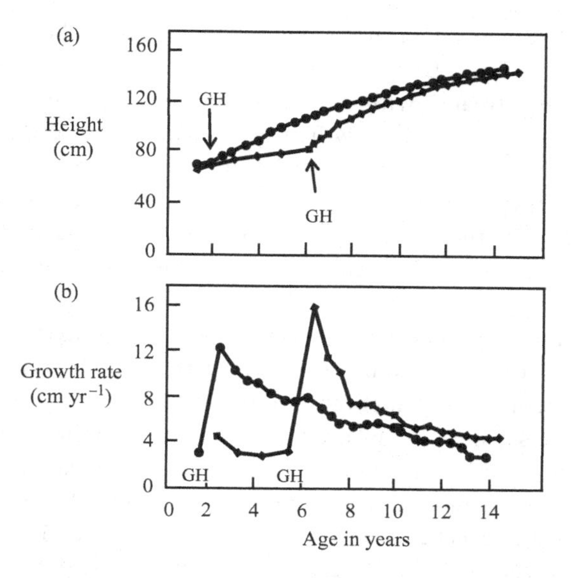

Figure 11.6. Growth data for two boys born with growth hormone deficiency, showing 'catch-up growth' due to the administration of growth hormone. The data are expressed as cumulative heights (a), with *arrows* to indicate treatment with growth hormone (GH). Growth rates in height are given, showing sharp increases in growth rate following the onset of treatment (b). Redrawn from Tanner [8].

did from the trajectory of growth curves at the highest concentrations. When growth is held back and the obstacle removed, growth rate increases to restore the deficit. But we had already established in hydroids that rate control is fundamental and that inhibition is countered by a stimulatory response to create an equilibrium with the inhibitor. When the inhibitor is removed, it is the counteractive response that is seen briefly, which reveals itself as a transient relaxation stimulation upon removal of load. Such relaxation stimulation can be observed in the behaviour of all control mechanisms, and the concept is well known in cybernetics. A control interpretation is that the peak heights indicate the magnitude of the counter-response that had been necessary to neutralise inhibition and achieve an equilibrium, so that the peak heights are positively correlated with the concentration of inhibitor. Relaxation stimulation briefly exposes the naked counter-response, for as long as the time delay in the system. Then, when the mechanism senses that there is no load, it restores growth rate to the goal setting and returns to the baseline.

One set of Tanner's data shows the height of two brothers who had suffered from birth from growth hormone deficiency (Figure 11.6a) [8]. One was treated with growth hormone from the age of 2 years and showed a linear growth trajectory and nearly normal height for age. This Tanner refers to as a 'classical complete catch-up'. The other brother was not treated with growth hormone until the age of 6 years, and up to that age showed a retarded and linear growth rate until treatment began. Growth rate then accelerated sharply and appears to converge on the same trajectory as his brother's in his late teens. Both brothers ultimately achieved heights only a little less that might have been expected from the height of their parents. In his book, only in passing does Tanner refer to the rate data for the growth of the two brothers (Figure 11.6b), even though they are actually more instructive than the cumulative height measurements, because they reflect more closely growth as it is controlled. Even though the data are not presented as specific rates, they show the sharp acceleration in growth rate as the deficiency in growth hormone is satisfied at the onset of treatment for each boy at ages 2 and 6 years. Here the peaks are interpreted as relaxation stimulations, as the cause of retarded growth is removed, and then over the following years, growth rates steadily decline to a normal rate for age in their early teens. The interpretation suggested by the hydroid data, and the Maia hypothesis, is that the removal of the deficiency, which was responsible for arresting growth, initiated steep relaxation stimulations of the kind seen in hydroids upon the removal of a toxic inhibitor. At this point, with the removal of an inhibitory load, the graphs show an immediate and pronounced acceleration of growth rate, resulting in the stimulation of growth (Figure 11.5b).

Hydroids replicate the same 'catch up' phenomena as Tanner and his colleagues observed, but as hydroids exhibit the same behaviour without having a target size, it is likely that he was observing the output of a rate-sensitive control mechanism. The appearance of catching up is fortuitous; it is simply a mechanistic consequence of removing load from a rate-control mechanism, where an equilibrium had been established between load and response. The greater the response necessary to counter the inhibition, the greater is the stimulation upon removal of load. On removing the inhibition, the response becomes apparent for as long as it takes for a new equilibrium to be established, or for as long as there is a lag time in the system.

Although Tanner was searching for a satisfactory hypothesis based on control theory, it now seems that the Maia hypothesis provides a satisfactory explanation. The fundamental difference between the two approaches is that Tanner assumed that growth is regulated in terms of size, as measured by height, whereas the Maia hypothesis accepts growth rate, as measured by specific growth rate (*r*), as the process used to control growth. The point is that only rate control provides for the effective regulation of processes; measurements of size at intervals in time obscure the process, which is lost in the cumulative products of the process. Nevertheless, size control is also necessary, but is to be found at a higher level. This implies the existence of a dual loop control mechanism of the kind already put forward for Maia, with an inner control loop regulating rate, while an outer control loop limits size.

Some mention is required of the fact that the growth rates and relaxation stimulations have markedly different time scales between hydroid colonies and humans. The hydroids grow rapidly and have a relatively high frequency of oscillation, which is reflected in the duration of the relaxation stimulation (~3 days). The human example reveals a slow growth rate and a low frequency of oscillation, indicated by a long relaxation stimulation (2.5 years in the older boy). This does not negate the possibility that both data sets indicate the behaviour of a rate-sensitive growth control mechanism operating at much slower rates and proportionately lower frequencies. It simply reflects that growth rates and the frequency of their oscillations are related over a range of rates spanning many orders of magnitude.

Relaxation stimulation is a feature of all feedback mechanisms, so it is to be expected that any growth control mechanism will exhibit catch-up growth as a byproduct of control. Since it is due to relaxation stimulation, and does not have catching up as its purpose, the name should be dropped. Like hormesis it is an epiphenomenon, with an interpretation that lies elsewhere than the raw, cumulative growth data.

WHAT IS MAXIMAL GROWTH RATE?

There is another consequence of control phenomena that the reader may have found curious. If there is a maximal counter-response to inhibition that creates an overcorrection, and another overcorrection upon its removal due to relaxation stimulation, then growth will be stimulated following both the addition and the removal of inhibition. This implies that more rapid growth may result from applying and removing low levels of some toxic inhibitor than from maintaining a steady and supposedly optimal state.

Exactly how this might come about can be seen in the experimental data (Figure 11.5b). Both overcorrections can be seen, with the onset of inhibition at day 1, and with its removal at day 16. The overcorrections are much greater than normal growth rates; the initial rate response is 2.3 times higher (at 1 μg ℓ^{-1}), while the relaxation stimulation upon removal of the inhibitor is 4.4 times higher (at 25 μg ℓ^{-1}) than the normal growth rate (= 100%). What is so interesting about these data is that the initial overcorrection is greatest at the lowest concentrations, and the relaxation stimulation is greatest at the highest concentrations. In these data one only sees the extent to which the response *exceeds that necessary* to neutralise it (1–10 μg ℓ^{-1}). The relaxation stimulations are most obvious at higher concentrations (15–25 μg ℓ^{-1}); they increase with concentration because they reveal the counter-response that *had been necessary* to achieve equilibrium.

Linking the responses that cause hormesis and relaxation stimulation leads to some improbable conclusions. The first is that we tend to accept that constant optimal conditions for an organism are those that result in maximal growth. It is clear that the combined effect of an initial maximal correction to inhibition, a slow upward adjustment in the growth rate and a spectacular relaxation stimulation upon removal of inhibition causes a significant enhancement to overall growth. Clearly the capacity for growth to be much greater than that normally observed exists in other organisms, to be mobilised only when they are exposed to varying levels of some inhibitory challenge.

Homeodynamic systems have evolved that cope with change, and without change the capacity to adapt becomes impaired when it is not used. Responses are known to hypertrophy with use and atrophy without, and probably remain healthier when exposed to the changes that they have evolved to neutralise. We therefore need to reconsider what we mean by 'optimal', and the idea that maximal growth occurs under stable and optimal conditions requires re-interpretation. These data show that a constant regime is unlikely to produce maximal growth, so constant conditions should not be used as an indicator of optimality.

Second, the obverse argument is that repeated step changes upon imposing or removing inhibition lead to much more growth than would be possible under the constant and supposedly optimal conditions. Various experimental manipulations are found to increase growth rates. In each case, growth rates exceed those that we would previously have considered possible under optimal conditions.

Growth can be enhanced by changing the inhibitory load upon the homeodynamic system responsible for controlling growth. Given that hormesis alone is capable of stimulating hydroid colony growth by as much as 50% over 16 days, there is scope for a regime that involves a programmed sequence of changes in loads that first cause hormesis, each followed by the removal of load and relaxation stimulation. A cyclical programme of change could be found to maximise the growth of different organisms that are kept under controlled conditions, or in large-scale culture.

In an age when so much depends upon utilising biological growth and productivity, any means of stimulating the growth of organisms will find applications in growing plants and animals for food, in medicine in the culture of cells and tissues, and in accelerating repair processes; and in pharmacology for culturing micro-organisms.

SUMMARY

Due to the comprehensive database created by Calabrese, we now know hormesis to be a truly general phenomenon. It has also been established that concentration–effect (dose–response) curves tend typically to be of the β-form that exhibits the hormetic peak. Feedforward control plays a key role, and hormesis is explained as a product of growth control. It is shown to be due to acquired resistance to sustained exposure to low levels of toxicity, and results in an adjustment that enhances the growth rate and so increases resistance to inhibition. But when such adjustments do not meet further toxicity, the elevated growth rate has the inevitable consequence of increasing growth, which is expressed in hormesis and the β-curve. Relaxation stimulation is due to the removal of load from a control mechanism, resulting in a stimulatory peak that lasts for as long as the lag time. An interpretation of 'catch-up growth' in mammals and humans is provided by relaxation stimulation that follows the end of a period of stress, disease or deficiency. Adjustments to growth rate and relaxation stimulation, both in different circumstances, contribute to hormesis.

12

Cellular growth control and cancer

It is a property of life to grow and multiply, but when cells co-operate in the construction of a complex organism, their growth must be regulated in the interest of the whole organism.

Albert Szent-Györgyi

Cancer is not merely a medical problem: it is a biological phenomenon.

Julian Huxley

It will be time to talk about causes [of cancer] when mechanisms have been worked out.

Nature Editorial (1981)

CANCER: FROM NORMALITY TO CYBERNETIC FAILURE

It was in 1981 that John Cairns, the eminent British physician and molecular biologist, recognised that the study of cancer was not advancing. In an influential review article, *The origin of human cancers*, he made a plea for a return to basic biology [1]. At the time there seemed to be no promising ideas pointing to a breakthrough.

There are approximately 80 trillion cells in the human body, about 2 trillion at birth and 40 times that at maturity. It is estimated that 2% of all cells die each day,[1] and as the total number in adulthood remains fairly constant, an equivalent number of cells must be created. This means that a large number of cells, of the order of 160^{10}, are turned over daily. Billions of cells are lost each day due to wear and tear, so replacement must proceed at a similar rate.

[1] Cells that are turned over slowly are replaced at a rate of 0.1% per day, although typical ranges are of the order of 2.5–3.5% per day. The figure of 2% per day is a median figure.

Cell growth accelerates to close a wound, and stops once healed over. Self-regulating mechanisms are required for each tissue and organ, to ensure that each continues to meet its particular purpose. Skin (epidermis) and bone marrow (the source of replacement blood cells) turn over continuously. The liver and kidneys do not require continuous replacement, though regeneration is rapid following ablation, or injury, while there is enlargement in response to increase in physiological load (see Chapter 4). The balance of these processes requires precise control and lifelong monitoring, with adaptive goals to meet maintenance demands and an ever-ready capability to initiate repairs. Each cell is a member of a community that constitutes a tissue or organ, and collaborates with others in effecting repair and regeneration. Normally, within a tissue, the exchange of signals that prevent cell division ensures the integrity of that tissue.

Cancer is a genetic disease of cell replication, in that the genome of one or a few cells suffers mutations and passes errors on to the descendants of those cells. Most mutations are quickly repaired, and those that remain are often elements of the DNA repair mechanism itself. It is known that the DNA in each cell is likely to suffer 100,000 to 1,000,000 molecular lesions per day. That may appear to be a considerable amount of damage, but it amounts to only one in a thousand of the 6 billion bases that make up the human genome. Such levels of damage are due primarily to cosmic and terrestrial radiation. About 70% of cancers are in principle avoidable, in that they are caused by environmental mutagens, which may be physical, chemical or biological. It must be assumed that this is usual for all organisms, so the rate of repair must typically be of the same order as the level of damage, for otherwise organisms would not survive. It follows that carcinogenesis has as much to do with the capacity of the individual's DNA repair mechanisms, as it is to do with their exposure to mutagens (see Figure 10.5). Nearly all damage to the genome is corrected by the DNA repair mechanism, which operates constantly to repair and check the fidelity of the genome. Studies have shown that between four and six, or even more, sequential mutations are required to cause a carcinoma of the colon. Carcinogenesis is no longer assumed to be due to a single mutation, as was once thought; it is a gradual accumulation of irreparable errors. Other cancers are due to mutations in copying, or an inherited predisposition to cancer. Any cell capable of division can be transformed into a cancer cell, although those cells that replicate frequently are more likely to accumulate genetic errors. At least 85% of cancers occur in epithelial cells, which form the epidermis and the lining of the gut, as they undergo continuous cell division.

Cancer is more likely where there is damage to the genes involved in cell proliferation, particularly those genes that encode signalling molecules. Many tumours are thought to be due to mutations that prevent the synthesis of transforming (or tumour) growth factors – TGFs. The transformation of normal cells into cancer cells may take decades before enough mutations are accumulated for a cell to become free of the controlling influences of the cells that surround it. Carcinogenesis requires the cell(s) to have lost a number of elements of their growth regulatory machinery, before they begin to replicate independently. The cancer cell without its limiting controls tends to behave aggressively, and to grow continuously. It is a mutineer, deaf to the signals and commands that once limited its replication.

Cancer is due to a fundamental failure in the control of cell proliferation. It is a systemic disease of the failure of communication and control. But to begin we must understand normality, and the control of cell populations in the healthy state, before attempting to understand the ways in which the ability to limit proliferation fails. It also needs to be understood as a normal intact system. The feedback mechanism is cryptic, and was for a long time poorly understood. None of the components of biological control systems reflect their roles in any obvious way. Control itself is an emergent property of cells and tissues, and their operation is not apparent in the constituent parts. As there are no obvious components that relate specifically to the business of control, so the mechanism must be studied in the first instance as an intact system. A reductionist approach is unhelpful in attempting to understand invisible systems of process with no recognisable form.

The most informative feature of such systems is their output. We have already established that to observe the behaviour of a control mechanism requires that the controlled process must be disturbed by deviating it from its equilibrium state. It needs to be made to work by having to neutralise some experimental deviation and restore growth to its preferred rate or state. After perturbation, its output then needs to be monitored over time to observe the behaviour that in health and normality restores the system to equilibrium.

It is necessary at this stage to show that the Maia hypothesis applies not only to simple metazoans like hydroids, but also to single-celled organisms like yeasts [2]. Here the model organism is a

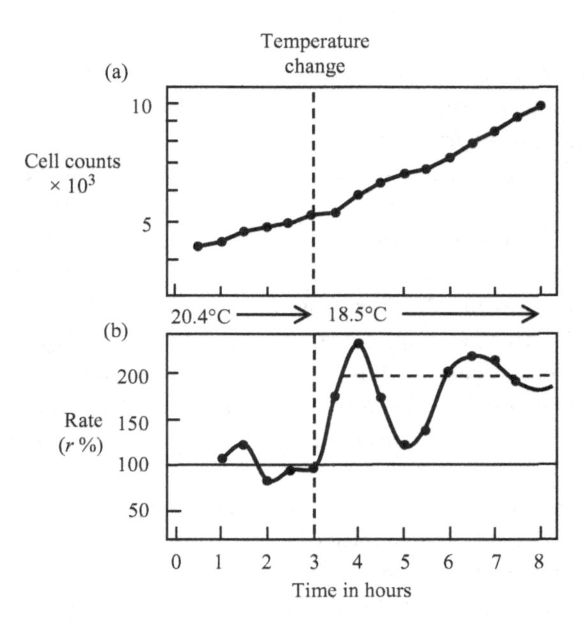

Figure 12.1. Growth of the marine yeast *Rhodotorula rubra* perturbed by temperature change from a high temperature (20.4°C) to one nearer the optimum (18.5°C). Cumulative growth data (a) show a shift in trajectory, but specific growth rates (*r*%) show the oscillatory output typical of a feedback mechanism in the same data (b).

marine yeast, *Rhodotorula rubra*. A link is created with the earlier data for hydroids (Chapter 5) by showing both cumulative growth data and the specific growth rates (*r*%) derived from them (Figure 12.1 a, b). The perturbation here is provided by a temperature change from 20.4°C to 18.5°C; that is, from a temperature that was above optimal to one that is much closer to an optimal temperature. With the change in temperature, growth increases (a), the rate is stimulated and the specific growth rate doubles (b). Control is destabilised by the change in temperature. A relaxation stimulation results, due to removal from a temperature that was suboptimal to a lower temperature that was less so. Normally, when yeast is exposed to toxic inhibition, the initial and normal effect is for growth rate to slow down for the duration of the lag time, before the neutralising response cuts in and growth rate equilibrates at a new and higher rate through a short sequence of oscillations. Here it should be noted that, as for hydroids, the cumulative growth data simply show a change in slope, but when given as specific growth rates, we see the output of the control mechanism regulating biosynthesis showing the typical behaviour of a feedback mechanism.

We now consider the output of the yeast growth control mechanism for a series of experiments in which cultures were exposed to increasing concentrations of a toxic metal (cadmium), and increasing levels of inhibitory load (Figure 12.2). This control mechanism, less complex than the hydroid's, provides simpler responses, more like those described by von Foerster and paraphrased in Chapter 3. On first being exposed to an inhibitor, growth rate falls, creating an error between the growth rate and the goal setting. After a time lag due to the delay required to sense a fall in growth rate, comparison with the goal setting and determination of the sign and size of the error, a full cycle is required before a response can be initiated to begin to neutralise the error.

The hydroid control mechanism has a feedforward loop that initiates the first observed consequence of a toxic inhibitor as a maximal and stimulatory counter-response, rather than unopposed inhibition that occurs in yeast (Figure 12.2). The yeast control mechanism is a simpler system, without the refinements of the hydroid, and there is no immediate response, only the dead time of inhibition. The yeast must wait for the inhibition to create an error and complete the cycle around the feedback loop before it can respond to inhibition. This creates a lasting disadvantage in recovery from the impact of the inhibitor. At low concentrations, after the initial inhibition, recovery is quite rapid, with rates soon exceeding the goal setting. Cultures exposed to concentrations in the middle of the range (2.5 and 4.0 mg ℓ^{-1}) apparently struggle to recover from the initial inhibition, but in both cases appear to be recovering by the end of the experiment, oscillating to levels about the baseline where they started. At a higher level (6 mg ℓ^{-1}), the inhibition is too great from the outset for recovery to be possible, but it is apparent that the control mechanism is still operating, so if the load were removed the population would probably recover. At the highest level (8 mg ℓ^{-1}), following a sharp fall in growth rate, the operation of the control mechanism ceases. This and other effects were lethal to the yeast population.

While the reader will have recognised that the yeast response to toxic inhibition is quantitatively similar to the hydroid data, there are other important differences. The growth of yeast is 25 times more rapid than that of the hydroids, and the frequency of the oscillations is proportionately higher. The period of the yeast oscillations is 2.6 hours, while the hydroid's is 6 days. This is to be expected, as

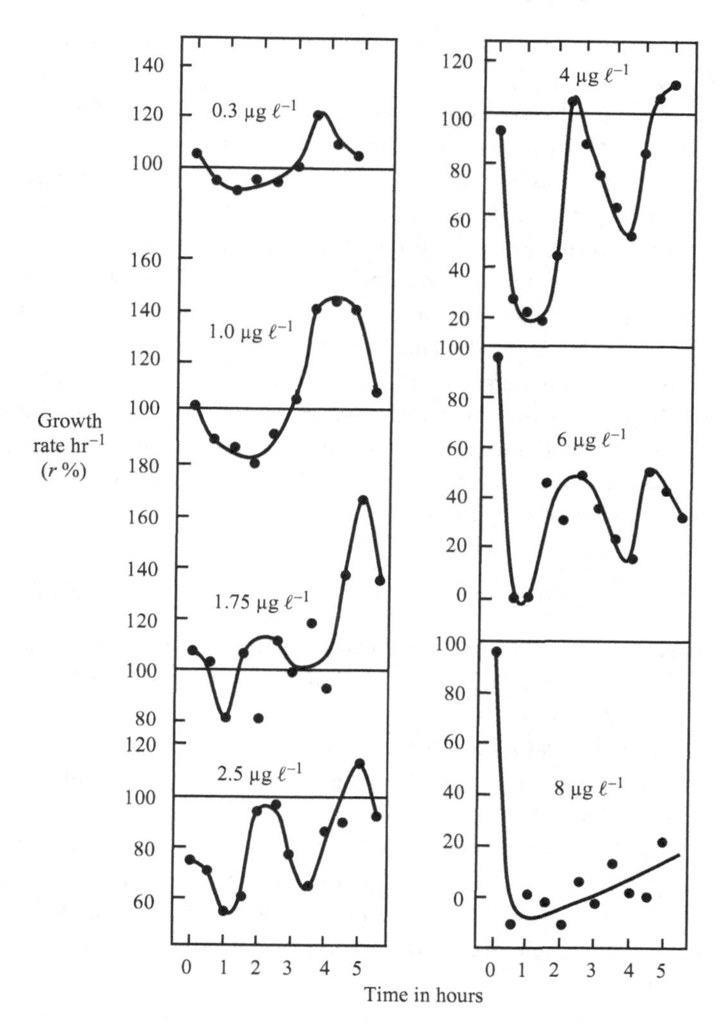

Figure 12.2. The output of the growth rate control mechanism of the marine yeast *Rhodotorula rubra*. Output of the mechanism as specific growth rates (*r*%) shows oscillatory responses to different concentrations of a toxic metal (cadmium), ranging from loads that are easily neutralised to those that cause overload. Redrawn from Stebbing et al. [2].

the specific growth rates (*r*) of all organisms decline with increasing size and complexity (Figure 8.5). It is also true that the delay in control processes at the highest level of organisation is the sum of the delays at all the lower levels. It is to be expected therefore that the growth rates and frequencies of oscillation will be significantly higher in the yeast than the hydroid, which has an additional level of organisation (see Chapter 9).

As for the hydroid growth control mechanism, modelling played an important part in understanding the yeast output as the product of a control mechanism. I had the good fortune to collaborate with the control theorist John P. Norton [3]. Here the dialogue between modeller and experimentalist ensured that levels of loading were optimal for model development, by defining shifts in the form of the control output in relation to load. The series of experiments were adequate to suggest a structure for the control mechanism. A closed loop growth control model was suggested that was linear at any one concentration of inhibitor, but with a response that was amplified by concentration. Here the data and modelling extend and corroborate the Maia hypothesis by suggesting that growth at the cellular level is regulated at a specific rate like that in hydroids.

CAN ANY BIOLOGICAL REPLICATOR BECOME NEOPLASTIC?

Richard Dawkins defined a 'replicator' as anything that makes copies of itself [4]. In particular, biological replicators are those living modules that copy themselves, such as genes, organelles and cells. Organisms too may be replicators when they reproduce asexually and create genetically identical copies that represent a clone. These might include single-celled organisms that replicate by binary fission like an *Amoeba*, or aphids that produce new individuals by parthenogenesis. For multicellular organisms, the exponential increase of cells represents a more acute problem, as any module that multiplies out of control ultimately threatens the whole organism. When this occurs due to the multiplication of cells, it is called neoplastic growth or cancer.

Exponential increase is a positive feedback process in that the products of replication are added cumulatively to the population, and also replicate in an unlimited manner (Chapter 2). Such processes have chaotic consequences, as can be seen from what happened to Mickey Mouse as the Sorcerer's Apprentice in *Fantasia*. His creation of bucket-carrying brooms quickly carried out his task, but multiplied out of control, creating flooding and chaos. Behaving in this way, all replicators are positive feedback mechanisms, so must be constrained by a negative feedback. Such increase may be the driving force of biological growth, the rate-controlled process of biosynthesis that leads to cell division. But a constant specific rate results in exponential increase, so that, however adaptive its products may be in the early stages, ultimately the products of replication become maladaptive. Consequently the evolution of self-limitation that imposes negative feedback is an essential component of any growth control mechanism, if unlimited growth is to be avoided. So, typically in biology, positive feedback systems are not found without some self-limiting capability incorporating negative feedback.

What about other replicators than cells? In principle it would be expected that any replicator could exhibit uncontrolled increase upon the failure of its growth-limiting mechanism; such self-limitation in genes or organelles can be no less fallible than cellular growth control mechanisms. Pseudogenes, or selfish DNA, carry no information needed by the organism and make no contribution to its fitness. Nevertheless, they make many copies of themselves within the genome. The best example is provided by the transposon, which has the ability to move about the genome copying itself into different sites. Such repetitive DNA and its remnants constitute as much as 50% of the human genome. Leslie Orgel and Francis Crick wrote that 'Excessive spreading of functionless replicators may be considered as a cancer of the genome', and speculated that 'the uncontrolled expansion of the genome would ultimately lead to the extinction of the genotype that permits such expansion', implying that some unknown limit to transposon replication is imposed [5].

The theory of endosymbiosis, championed by Lynn Margulis, has accumulated much evidence that organelles of eukaryote cells originated as free-living micro-organisms. Chloroplasts of plant cells are assumed to have a cyanobacterium as their ancestor, while mitochondria are probably derived from a proteobacterium. Free-living bacteria typically replicate by fission and, as organelles, they replicate within their cellular environment. Using the same argument as before, microbial replicators are also assumed to be capable of uncontrolled and exponential growth. The number of mitochondria may vary adaptively between one and several thousand in each cell, depending on their role. It follows that there must be a regulatory mechanism controlling replication which, in failing, could be expected to cause uncontrolled replication.

CONTROL OF ORGAN SIZE BY FUNCTIONAL LOAD

Loss of liver tissue is compensated for by an equivalent amount of growth; such that the more that is lost, the more will grow. This fact is so well known that it is easy to overlook its importance. We now recognise it as the behaviour of a homeodynamic mechanism that regulates organ capacity in relation to load, but the principle has been known since antiquity. Prometheus stole fire from where Zeus had hidden it and brought it to the Earth to give to mankind. As a punishment, Zeus had Prometheus chained to a rock, and sent an eagle to eat his liver. But the amount of liver consumed by the eagle by day grew again by night, so his torment was unending.

In the early twentieth century, the idea of growth in response to functional load took root, with statements like 'the activity of an organ is the cause of its growth', or 'hypertrophy and atrophy are generally the result of use and disuse'. The idea that emerged was the control of growth by functional load. Richard J. Goss, in his book *Adaptive Growth* [6], wrote that functional load is a fundamental concept involving feedback control by which a critical mass of tissue is maintained in the face of disturbing influences. Such growth is of two kinds. The first is 'hyperplasia' and applies to growth due to increase in cell number. It is the predominant kind of growth before birth, as the embryo must acquire the appropriate number of cells for each tissue or organ. The second is called 'hypertrophy' and refers to the increase in the size of cells. Throughout growth into adulthood, it is growth in cell size that accounts for increase in organism size. Of course, there is cell turnover, so hyperplastic growth must therefore continue throughout life, ensuring that the required numbers of cells are maintained in each tissue. Following the removal of part of the liver in rats, mitosis increases to 3–4% of the cells per day, while increasing the functional load on an organ is more likely to initiate hypertrophy. Normally in the adult there is little liver growth, but if three-quarters of the liver of a rat is removed, the lost tissue is regenerated from what remains in 5–7 days. Regeneration following ablation is well known in liver, kidney and endocrine glands, while in other tissues like skeletal muscle, bone, cartilage, sense organs and the adult nervous system this does not happen.

Imposing a greater load on the cell mass has the same effect as reducing the mass, as both impose a greater load on what remains. The signal to grow circulates in the blood, which was found by linking one mouse that was regenerating its liver to the blood supply of another normal mouse. Surprisingly, the liver of the normal mouse grew as well. Growth in response to functional demand is best known in kidneys due to a high-protein diet, and also occurs in the heart due to hypertension and red blood cell formation due to hypoxia.

CONTACT INHIBITION: DIFFERENCES BETWEEN NORMAL AND CANCER CELLS

One kind of growth control that is typical of cells that grow as a single layer is 'contact inhibition'. This simple mechanism is found in cells that grow in monolayers, like epithelial cells. In tissue culture, cells divide every 24 hours and spread out as a monolayer, and those that come in contact with one another cease movement and stop replicating. This process is important in the repair of wounds or breaks in the epidermal layer. Following a cut, while shaving perhaps, the absence of contact between epidermal cells frees them from constraint and they proliferate freely, so the process of repair can begin. Growth continues on either side of the wound until there is contact, and then cell replication stops. This mechanism ensures that there is a continuous layer of epithelial cells that cover the underlying tissues precisely.

This mechanism is not unique; it plays an important role in minimising the effects of competition for space on rock, and algal surfaces on the seashore. Here colonies of bryozoans and other attached organisms compete for space. Where colonies of the same species meet, they stop growing at the point of contact. This implies the existence of rudimentary communication that allows colonies to differentiate 'self' from 'not self', providing the sensory input to a control mechanism that limits overall growth. Such contact inhibition is adaptive, preventing the wasteful consequences of competition and overgrowth. The mechanisms are unknown, but the implication is that something triggers mutual inhibition of growth as the two colonies of the same species come in contact with one another. Where cells touch, adjacent cells form structural links and communication channels that ensure the exchange of information and cell recognition. Contact inhibition between cells ensures that, although they remain capable of multiplying, they do not, because they are touching and adhere to one another as part of a contiguous monolayer. Cells in contact communicate and exchange signals, which provides mutual growth inhibition.

If the contact inhibition experiment is repeated with cancer cells, their behaviour is quite different. Cancer cells do not adhere to one another and are easy to separate. They do not form links and their growth continues, piling up in a heap, because they have not exchanged signals that would normally have halted their growth. Their replication continues without the constraint that limits normal growth. If a disc of normal cells is cut out of the monolayer, it remains clear of cells, but if such a disc is cut out of a culture of cancer cells, the space is soon filled by spreading and multiplying cells. They are competitive and invasive where cooperation and constraint is the norm.

Adhesion between normal cells is achieved by complex devices called desmosomes that are embedded in the cell membrane. Each bears tendrils that radiate out into adjacent cells, ensuring well-rooted connections that securely bind one cell to another. There are other points of adhesion that are called 'tight junctions' which occur between interlocking protein and apparently have the purpose of occluding intercellular space. Most important are 'gap junctions' that incorporate ion channels allowing communication between cells. Some of the junctions are gated, so that channels open and close only to appropriate signals. The channels may be voltage-gated, ligand-gated or mechanically gated, implying that operation of the passage is controlled by charged ions, by the binding of chemical ligands or by molecular size.

Cell-to-cell communication was discovered in the 1960s by W.R. Loewenstein and Y. Kanno [7]. As adhering populations of cells do not replicate or move about, it was assumed they communicate in some way made possible by their being in contact. They made the crucial discovery that in normal liver cells, a current of charged ions from inside one cell passes into adjacent cells, facilitating cell communication over distances of many cells in all directions. Ions move as quickly through ion channels as they do through cellular fluid within the cell. Loewenstein and Kanno were also able to show that cancer cells have no detectable ability to convey such communication, and are unable to communicate in the way that normal cells do, even when they are in contact with one another. Clearly, neoplastic growth could result from the absence of avenues of cell-to-cell communication, apart from the lack of any means of communication.

A crucial requirement for the evolution of multicellular organisms was the development of mechanisms to suppress cell autonomy and their natural tendency to multiply. Without the integrity of their multicellularity, metazoans could not have survived. Normal free-living unicells tend to multiply and behave autonomously, moving about at will. In metazoans, cells communicate and cooperate, acting for the common good by responding to signals to multiply only when required to do so.

AN EARLY FEEDBACK MODEL FOR TISSUE GROWTH

We now need to focus on the means by which cells control their own numbers. This is a complex problem which has long been studied, with erratic progress, over many years. Discoveries in recent years indicate that researchers are closing in on the specific agents that are key to the control mechanisms that, when they fail, cause cancer. In particular, we focus now on the discovery of tissue-specific inhibitory regulators; first for myostatin, which is now recognised as the inhibitory regulator for skeletal muscle. Other statins are now known that are specific to other tissues. It now appears that it may be the failure of a control system, resembling the logistic control mechanism incorporating a specific inhibitor, that might initiate neoplastic growth. It is therefore interesting to trace the idea of specific regulatory inhibitors from their origin over 50 years ago. What follows is an account of how a single line of thought has been pursued by a handful of scientists, each a leader in his field at the time, who strove to take the idea forward. Most of the rest of the chapter is required to follow this single idea and its development to the point where the significance of statins as specific regulators can be recognised.

The first person to attempt to define the laws of biological growth was the Austrian biologist Ludwig von Bertalanffy (1901–1972), best known as one of the founders of general systems theory.[2] He trained and worked in his birthplace, Vienna, later holding university appointments in London, Canada and the USA. His growth equation has become widely used in biological models, and has already been referred to [8]. It is important that he established that growth is the product of two opposing processes. His equation tells us that the rate of growth of an organism is equal to the processes of anabolism, or synthesis, which increases in relation to the 'resorbing surface area' of the organism, minus catabolism or the breakdown of tissue, which is proportional to the weight (W) of the organism. The simplest form of the von Bertalanffy growth equation is given by $dW/dt = Hs - Dw$. The growth rate (dW/dt) is equal to the coefficient of anabolism (H), which is proportional to the resorbing surface area (s) given by Hs, minus the coefficient of catabolism (D), which is proportional to the weight of the organism (w) given by Dw. In this application, the equation relates to the growth of cell populations. We can accept that growth rate in cell numbers, as equivalent packages of biomass, is the difference between cell birth and cell death, given earlier as a rate of turnover of 160^{10} cells per day.

Another Austrian biologist is a more important contributor to our story. Paul A. Weiss (1898–1989) also trained in Vienna in the 1920s. There he met von Bertalanffy,[3] who was three years younger and also a

[2] General systems theory is now seen as a conceptual tool for the study of systems (physical, biological and social) to enable complex and dynamic systems to be understood in broad terms. The preferred tool for studying systems is the computer simulation model. The system as a concept, with key elements represented (often diagrammatically) and assumptions identified, is the necessary preliminary to developing algorithms to transcribe into software for a computer model.

[3] Both were also known to another biologist, Rupert Riedl, whose book *Order in Living Organisms* took a systems approach to the evolution of life.

student, and with whom he discussed biological issues of the day. With hindsight, it now seems that they each should take some credit for founding what became known as the general systems theory [9].

In 1945, Weiss proposed a concept of growth control which was tested in experiments during the 1950s, using the new technique of tissue culture. The results of experiments with heart and kidney tissue demonstrated what Weiss called an 'automatic control principle'. According to Weiss [10], organ-specific growth control depended on two hypothetical agents: 'templates' which promote cell replication, and 'anti-templates' which could slow growth by inactivating the templates. Templates are hypothetical molecules that are characteristic of each cell type, which act as catalysts in promoting growth, such that growth is proportional to their concentration. 'Anti-templates', which carry the identity of the cells they originated from, bind to templates as an inactive complex. They were assumed to be released from the cell into the intercellular spaces, and diffuse through the tissues, acting only upon cells of the same type. Because anti-templates degrade continually, it is essential that they are continually produced. As the concentration of an anti-template in the extracellular space increases, its intracellular density likewise increases, and it begins to inactivate templates. Growth rate declines in all cells bathed in the mixture of templates and anti-templates, such that the rate is measured by the ratio of one to the other, determining the extent by which the growth rate is inhibited. For a growing population, this hypothetical mechanism results in a sigmoid growth curve.

Partial removal of an organ would therefore reduce the production of anti-templates and cause compensatory growth. Implicit in the anti-template hypothesis is that they are short-lived and must be continually replaced at the rate at which they degrade in order to maintain their inhibitory effect. The response time of the system, reacting to provide repair and regeneration, depends on how quickly anti-templates degrade or are inactivated. Weiss estimated the half-life of anti-templates to be about 2 days.

In 1954 the Rockefeller Institute for Medical Research invited Weiss to New York, where he set up a new laboratory specialising in cancer, wound healing and regenerative growth. He embarked on a new programme that was to have great influence on the study of growth control. He appointed a young developmental zoologist, named J. Lee Kavanau, who had not long completed his doctorate. He was to develop a mathematical model based on Weiss's ideas on growth control [11]. Weiss's template hypothesis was already well established before modelling began, and was supported by a considerable body of experimental data. The model depended upon a dual structure: the 'generative mass', which was capable of growth by cell division, and the 'differentiated mass', which consisted of cells incapable of division, that could inhibit the generative mass by negative feedback. Growth is represented by von Bertalanffy's equation such that the net balance of anabolism in relation to catabolism is one of the three differential equations that make up the model.

Not only was Kavanau a trained zoologist, but also an able mathematician and he programmed the model. A decision was made at an early stage that they would code the mathematical model for the new IBM 709 Digital Computer at the Western Data Processing Center of the University of California at Los Angeles. Although the model consisted of only three equations, their initiative was important. In a footnote to one of their papers, Kavanau wrote as a telling aside that 'These calculations would have required many thousands of years to carry out by previously used manual methods'. Clearly, their computer model represented a huge stride forward in technique, and their papers were widely read [12]. There is no doubt in the minds of many that this work, linking experimental data and model building, was a pioneering approach to understanding growth control, and one that was followed in the work summarised here. Their dynamic hypothesis, as a computer model, aimed to replicate the operation of a growth control mechanism. In another sense it was an invitation to experimentalists to produce corroborative data for their simulated output, but it seems that the invitation was ahead of its time.

This must have been one of the first biological models to explicitly incorporate negative feedback. After the series of Macy Conferences on Cybernetics from 1946 to 1953, which attracted a wide range of biologists, cybernetics had such a following that the concept seemed certain to take root in some fundamental area of biology. How appropriate that it should be the co-founder of general systems theory, into which cybernetics was later subsumed, who would recognise the importance of Wiener's concept to biology. Kavanau's later paper provided more instructive model output; both papers attracted great interest and were highly cited.

The impression 50 years later is that Weiss and Kavanau's papers were read more with a sense of wonder at their novelty than in recognition of the direction in which the work pointed.[4] It seems that

[4] I attach as much importance to Kavanau's later paper. Simulation runs were for periods twice as long as those in the first paper, showing more convincing control oscillations, but the basis was laid down in the first joint paper.

no-one was able to generate experimental data that resembled the simulated data, so as to corroborate the model. This would have been difficult at the time, because techniques for determining biomass, or its rate of change, would have required sacrificing the organism, or the culture, to make accurate measurements. Working to a time base was a novel feature of the work, which could not easily be repeated in the laboratory. As can be seen (Figure 12.2), the dynamics of controlled growth are of great interpretive value. Even today there seems to be an aversion to time-course experiments, which had been an important novelty in the simulation. As important was the need to perturb the culture at the outset, but this was apparently overlooked. In the simulation, reduction in the tissue mass, such as would result from ablation, provided the perturbation. The results were presented as a percentage of the unperturbed control, filtering out the effect of perturbation from the underlying growth process as a ratio. Consequently, one only saw the effect of perturbation on growth, revealing fluctuating output that represented the action of the control mechanism as it worked to restore equilibrium.

Seven years later, Goss reviewed the work of Weiss and Kavanau in his book *Adaptive Growth* [6], but did not mention the questions that their work had left hanging in the air. He moved on to consider the specific agents that might be responsible for control, rather than their model output demonstrating control in action. Apparently no-one saw, at that time, that all self-regulating systems embodied cybernetic principles and might be expected to behave in the same way as Weiss and Kavanau's oscillatory model output. It was also not recognised that it was necessary to perturb the system and monitor the consequences over time to visualise the response of the feedback mechanism to perturbation.

Of course, an experiment that required measurements of the same cell population or tissue mass at frequent intervals in time was problematical. The electronic cell counter used in the yeast experiment (Figure 12.2) was not then available to sample and rapidly count suspension cultures. One alternative was to measure growth rate by scoring mitotic figures, but this was extremely laborious, as it was for chalone workers a decade later.

Weiss and Kavanau's model was ahead of its time, but on looking back on their pioneering work, one cannot but think that its principal lessons were lost. Over the intervening years, others have pointed out the paucity of temporal data, in particular, the lack of data showing the growth of cancer cells over time. Such experiments are demanding and inevitably time-consuming, but no-one seemed to appreciate their worth.[5] The only element of Weiss and Kavanau's that was really taken up was the hypothetical anti-template, or regulatory inhibitor.

SPECIFIC INHIBITORS AS UNIVERSAL REGULATORS: AN IDEA IN PARALLEL

In parallel with Weiss's work, a similar idea had been developing for nearly as long in the hands of an experimental biologist, Meryl S. Rose in collaboration with his wife Florence, working together on exocrine-limited population growth in tadpoles. For many summers, at the Marine Biological Laboratory, Woods Hole, Massachusetts, they also worked on regeneration and growth experiments with hydroids. They later retired to Woods Hole, where they enjoyed sailing their sloop, *Mystic*, in Vineyard Sound.

In is impossible to trace the development of Rose's idea, as it involved many organisms, places and collaborators. The papers are too numerous to allow the reconstruction of the growth of the idea, but key papers focus on the hypothesis that emerged from his work. Rose identifies botanical work published in 1905 as the origin of the idea. The central idea, which was supported by much of his work, was that of a 'polarised specific inhibitor' responsible for regulating growth [13]. He meant by 'polarity' that the flow and action of specific inhibitors was directional in relation to the orientation of the organism. The genes that code for orientation in animals have changed little over many millions of years, from the evolution of the fruit fly to the mouse. It follows that the mechanism to maintain it must have existed for as long. Polarity, orientation and symmetry are universal properties of life, for the whole organism and its parts. Rose proposed that the distribution of such regulators is polarised, in that their flux is likely to be in a particular direction, related to the axis and orientation of the organism. He identified a number of such examples in plants, hydroids and cell cultures.

Rose was aware of Weiss's template hypothesis, and wrote that 'the hypotheses of differentiation by polarized specific inhibitions and of growth control by anti-templates are not mutually incompatible. They simply deal with different things'. The search for

[5] For a long time after the corroborative work with yeast and hydroids, I was unaware of any published experiments using the same approach. Even now I am not sure whether perturbation in time-course experiments has been used to access control system output. If so, I would be delighted to hear of such work.

embryological organisers and specific inhibitors, and the development of organs and tissues, turned out to be parts of the same story. Development is determined by inducers and gradients of specific inhibitors in the hydroids, polychaetes and frog embryos studied by Rose, linking to other research by collaborators, on flatworms and chick embryos.

In this way regulation by specific inhibition became known for many and diverse systems. Rose published a major review in which he suggested that it appeared to him that polarised specific inhibition was probably a universal biological phenomenon. Whatever the organism, it was found that the culture fluid from one organ inhibited the growth of the same organ, while not inhibiting the growth of other organs. Rose established that like inhibits like. What was once called 'physiological dominance' by developmental biologists actually found a mechanism in Rose's specific inhibitors. A critical mass of cells of one type continually inhibits cell growth of the same type, so maintaining the tissue or organ at a prescribed size. The concept was found to hold true at the species, individual, organ and tissue levels.

In 1958, Rose took the concept another step forward in a paper entitled *Failure of self-inhibition in tumors* [14]. Rose believed that cells, in the presence of excessive cell death, experienced sustained higher growth rates in achieving tissue homeostasis. Cells in such circumstances, he suggested, fail in self-inhibition, as high cell death prevents limiting concentrations of inhibitor from being reached, so denying them exposure to the normal limiting levels of inhibitor. Cancers not only spread from one kind of tissue to another but can also be transplanted experimentally between animals. However, he found that in transplanting *normal* tissue to another kind of tissue, the explant itself often became cancerous. The implication is that inhibitory regulators are specific to each tissue, such that the transplant is insensitive to the signals that pervade the tissue into which it is introduced.

Rose brought these and other related ideas to the study of cancer, from his wide experience in experimental biology, suggesting 'an untried method of control' of tumour growth. The difference was that, even though Weiss worked with tissue cultures and required the use of animal models, his aim was to understand the human condition, and cancer in particular. Rose, on the other hand, was a biologist who used various invertebrate and vertebrate models, with the aim of establishing the principle of specific inhibition as a universal phenomenon. Convinced that this was so, Rose published a paper in which he proposed that it was the failure of self-limitation that caused a cancer. His idea was overlooked, as is so often the case in the history of science.

A decade later, in 1969, E. Robert Burns devoted a paper to the revival of Rose's 'untried method of control' of tumour growth, and his specific inhibitor hypothesis [15]. Burns summarised the key points. Cells inhibit cells of the same kind by producing cell-specific inhibitors. Self-limitation of size, it was believed, depends on a critical number of cells in a tissue. The phenomenon had been demonstrated in taxonomically diverse systems, so there was reason to believe it had evolved by common descent. Burns not only wished to revive Rose's idea but had also worked with a cancer cell line that he believed would be ideal to test Rose's hypothesis; this was Ehrlich's ascites tumour,[6] a solid, transplantable tumour. The choice of model was a good one, for it appeared that not all vestiges of growth control had been lost. It had a maximum size, suggesting the operation of a self-limiting control mechanism. It was as if the transformation of the cell line to a cancerous state was incomplete. Specific inhibition requires cellular contact; the implication was that as ascites is a solid tumour, contact inhibition could occur, with the specific inhibitor as its active agent, which was to be found in the ascitic fluid. Burns noted that this tumour appeared to regulate its own growth, as normal tissues do. Its dynamics revealed a sigmoid growth curve, with an exponential phase of growth, followed by limitation due to exponential retardation. The form of decline suggests some inhibitory process, related to the size of the mass. The Gompertz function best fitted the curve.

A year later the Danish researcher, Peter Bichel did just what Burns had suggested, and published a series of papers throughout the 1970s. The tumour produced a fluid and was used to inhibit the growth of the same tumour in another organism. Even tumour cells could produce a specific inhibitor. Bichel concluded that 'growth deceleration is probably due to specific feedback inhibition, and, importantly, the inhibition of growth was not due to deficiencies in oxygen, nutrients or the accumulation of metabolites'. Burns' paper was influential because, not only did he show that Rose's idea of specific inhibitors was well founded, but that it was comparable and compatible with Weiss's anti-templates, and also with chalones.

[6] Ehrlich's ascites tumour (EAT) appears to be a form of cancer whose transformation was incomplete. Given its ability to limit mass, the result would be a benign tumour.

THE CHALONE HYPOTHESIS

Like so many truly novel concepts, inhibitory regulators had to be rediscovered several times before the idea took root. In the early 1960s, W.S. Bullough, at the University of London, proposed, on the basis of his experimental work, that the control of cell division lay in the tissues themselves [16]. His group used the ears of live mice and rabbits to study wound healing (using sticky tape to remove epithelial cells) in relation to epidermal mitotic activity. As cells of most tissues are, in principle, capable of indefinite growth, they reasoned that each tissue-specific control mechanism must be anti-mitotic. As about the same time, O.H. Iversen demonstrated the existence of an inhibiting principle in epidermal cells. The oscillatory character of enzyme activity following disturbance pointed to the action of a homeostatic control mechanism. The production of new cells is inhibited by tissue-specific regulators that originated from the cell mass. Bullough called his inhibitory regulators 'chalones' from a Greek root which, loosely translated, means to reduce the sails to slow down a boat.

There appears to have been a general commitment to a cybernetic approach by the chalone group at an early stage; which led to the acceptance of the Weiss and Kavanau model and, more specifically, to anti-templates as analogues of chalones. Rose's hypothesis had more to do with growth during development in a variety of animals, and Weiss's hypothesis to growth following ablation with respect to medical science. But, as we shall see, it was the same growth control mechanism, and the same inhibitory regulator. Despite Burns' efforts, chalone workers, with few exceptions, ignored Rose's work and ideas. In jocular mood, Iversen caught the flavour of the problem facing the epidermal chalone groups. 'Why is the world not filled up with epidermis? Mine just covers my body and there it stops. A beautifully regulated equilibrium between cell loss and cell renewal is constantly maintained.' He proposed that epidermal growth can be measured as:

$$\frac{\text{rate of cell production}}{\text{rate of cell death}} = 1$$

If the ratio remains at <1, the tissue will shrink; at 1 the tissue is in equilibrium; but if the ratio becomes >1 the tissue will grow. If such growth is maintained, it is considered neoplastic.

Other chalones were soon discovered besides the epidermal chalone. Experimental work showed that chalones were tissue- and cell-specific, as chalone extracts from one tissue have no effect upon others. Each tissue was assumed to have its own unique chalone, implying that there are as many chalones as tissues. It was also shown that chalone inhibition is reversible as, in the absence of chalones, proliferation of cells occurs. This was important because it showed that chalones were not toxic to cells. The epithelial chalone in rat, pig and cod had the same inhibitory effect in mouse and human. This implied that a family of chalones for different tissues had been maintained throughout the evolution of the vertebrates.

Epidermal chalone research in the UK and Norway represented the leading edge of chalone research in the early years, followed later by T. Rytömaa's work on granulocyte (a nucleated blood cell) chalone at the University of Helsinki in Finland. The mainly European school of chalone workers grew as the 1960s progressed, holding meetings that fostered communication and collaboration. In 1965 Bullough was certain that mitotic rate was adjusted to meet the requirements of the tissue mass, mediated by the chalone mechanism [17]. Though he paid 'lip service' to cybernetics, Iversen was keen to apply the principles of control theory upon his chalone work, and developed a simple feedback model with an engineering collaborator. At this time Bullough and Rytömaa [18] were convinced that the key to understanding the homeodynamic mechanism that controls mitosis lay within the bounds of cybernetics. They cited a remark by D.W. Smithers in the *Lancet*, which – though overstated – was prophetic. He said 'it is biocybernetics, the science of organization ... which must take over from cancer research'. Bullough and Rytömaa believed that this branch of science must become increasingly prominent in addressing the problems they were grappling with.

In the early years, the chalone hypothesis had been considered heretical, but by the 1970s the idea began to win acceptance, and meetings attracted a large following. The term 'chalone' even found its way into the textbooks. Many new chalones were added to the initial epidermal and granulocyte chalones, including chalones from muscle, liver, kidney, intestine, uterus and testis.

If anyone had been unclear about the aim of the chalone work, all was made clear by the publication in 1968 of three papers in *Nature* under the title *Chalones and Cancer*. Each paper showed that tumour growth could be arrested, regressed or even cured, using chalones. In all cases where chalone content was artificially raised, cell division was inhibited. In vivo experiments showed that when chalones extracted from skin were injected into mice with large tumours, they showed a rapid regression. The hope was expressed that these findings would point the way to treatments of cancer in humans.

Iversen had his doubts and prudently held back, later writing a critical paper. He correctly felt that without knowing the chemical identity of the hypothetical chalones, such claims were not justified. Chalones remained the mitosis-inhibiting principle that they were at the outset. No-one knew what the chemical nature of chalones was, though the range of possibilities has been narrowed to peptides or a glycoprotein; however, this was of little help, as nothing short of a chemical identity would do. Without the chemical identity of a single chalone, the reproducibility of experiments was compromised. They simply represented a hypothetical inhibitory principle that existed within tissue extracts that were likely to contain other chalones or tissue contaminants. Chalones operate at low tissue concentrations, which is an obstacle to their isolation and purification in large enough quantities for chemical analysis.

For chemists, analysis was difficult, as chalones were labile, unstable and short-lived. But these were necessary attributes for an inhibitory regulator, whose rapid disappearance was essential before a positive response could be initiated to repair or regenerate tissues. It was essential to release the brakes on cell division quickly for an adaptive response to be initiated. The response time of different chalone systems varied. Iversen reported a stimulation of mitosis after 3–4 hours, though in many other chalone systems a response took 20 hours. The response time for granulocytic chalone was 14 hours and inhibition ceased after about one cell cycle. Weiss and Kavanau had recognised this problem first, and for their model they had estimated that the life of anti-template molecules was 48 hours. Any adaptive response depended on how quickly the existing chalones in a tissue could degrade or be neutralised, and these times seem rather long. A more rapid reaction may have required the hypothetical antichalone and would be essential for a more rapid response by the chalone control mechanism. We shall see later that the action of the inhibitory regulator myostatin is negated by follistatin, which might be the means of ensuring a rapid stimulatory action to effect repair or regeneration.

Until chalones were chemically identified, no-one could expect to repeat any of the experiments that had such promising results, while the active principle of a chalone extract was unknown. In 1976, work was reported to be under way to extract, purify and identify the epidermal and granulocyte chalones. Clearly, Bullough and his colleagues had moved too soon, and their haste to get into print with their findings that chalones would be an effective anti-cancer agent cost them dearly.

The other great handicap in determining the effect of chalones was the index of the rate of mitosis. The method involved the use of colchicine or colcemid to arrest mitosis at metaphase, so that they could be easily identified as dividing cells. Mitotic cells are prepared as smears or tissue sections on glass slides and then counted under the microscope. To obtain time-course data, thousands of cells for each experimental time interval needed counting. The tedium of making these counts, which required much time spent poring over a microscope, was the principal disincentive from using the technique. A relatively simple experiment would require an untreated control and perhaps more than one treatment. Replicates would have been required, and both might need to be counted at hourly intervals. The person scoring the samples would then have to score many thousands of cells to pick up the small differences in numbers. Finally, a graph of the percentage of cells that were mitotic over time would provide an estimate of the changing growth rates of cells.

The difficulty of this technique was illustrated by an anecdote told by Iversen. It was the story of a drunken man crawling on the pavement in the middle of the night under a street light. A policeman came up and asked him what he was doing. The man replied that he had lost a coin, and the policeman replied: 'I can help you find it, but are you sure you lost it here?' The man replied 'I don't know where I lost it, but this is the only place with light!' Iversen added that the temptation to wander off into the dark was too much for some chalone workers.

In 1975, Bullough wrote another review to assess the chalone work, much admittedly hypothetical [19], and H. Rainer Maurer wrote of the increased effort that had been devoted to the purification and analysis of chalones, and the expectation of 'chemically pure chalones'. Medical workers were becoming interested in chalones, in anticipation of their expected anti-cancer properties. It was evident that several laboratories, including his own, were closing in on a polypeptide (MW ~1,000) from granulocytes that had all the properties of a chalone. In 1976 the status of chalone research was comprehensively summarised and evaluated in a volume edited by John C. Houck [20].

However, chalone research petered out, mainly because the active component could not be identified, and the biochemists did not deliver soon enough for some, given that great things had been promised in 1968. Extravagant claims for chalone had turned influential people against chalone research. Funding agencies had had enough of unfulfilled promises and took their money elsewhere. Researchers moved on, and if they maintained any belief that there was anything in the idea, they kept it to themselves. The work on chalones was discredited because no-one could identify the chemical(s) responsible. Ultimately patience ran out, confidence evaporated and laboratories lost interest, because the long-promised and unidentified chalones were not delivered.

CHALONES FADE FROM THE SCENE

We return to John Cairns' 1981 review on *The origin of human cancers* published in *Nature* [1]. He was an important figure in cancer biology; he held important posts on both sides of the Atlantic, and devoted much of his career to cancer research. The article pointed out the weaknesses of the mechanisms suggested in previous years as plausible explanations of how cancers are caused; human cancer was a problem of biology that was not yet understood. Cairns made no mention of chalones, but his comments must have been written with chalones in mind. His appeal, indicating the importance of cellular homeostasis, alluded to biological cybernetics. Why had cancer research biologists for so long attempted to understand the disease that epitomises failure of control, yet had not looked to the science of control?

The review was accompanied by an anonymous editorial, which went even further. The author had read Cairns' article as a cue to return to basics, such as the physiology of homeostasis, and 'even some long-standing issues of old-fashioned biology'. The point was hammered home: 'The phenomenon of human cancer is a problem in biology that is not yet understood'. The editorial echoed the earlier remark by Julian Huxley quoted at the head of this chapter. It continued, 'It will be time enough to talk about causes when mechanisms have been worked out'. This was a different and important point: it is essential to understand the mechanism of growth control *before* claiming to know how it goes wrong. In cybernetic terms, the knowledge of control mechanisms that fail in cancer was always superficial. The agent of growth control, let alone a mechanism, was not known then, and is still not well enough known to make the kind of claims that were made for chalones. The mistake of promoting the inhibitory principle as a cancer cure helped to bring about the downfall of the chalone paradigm. Looking further back, the idea of specific inhibitory regulators also failed under the guise of chalones because of the premature claims that were made for chalones, but not because the hypothesis itself was flawed. The proverbial baby had been thrown out with the bathwater.

MYOSTATIN AND GROWTH CONTROL OF MUSCLE

During the 1980s there were several claims that the long-lost chalones had been found: these claims were related to somatostatin. It used to be rather clumsily called 'growth hormone-release inhibiting hormone' (GHRIH). It is produced along with growth hormone-releasing hormone (GHRH) in the hypothalamus, and the two hormones regulate the release of growth hormone (GH) from the pituitary. Independently of somatostatin's role is an inhibitor of the growth hormone, researchers have suggested that this, as a growth inhibitor, could act as a universal chalone. This was a false dawn, because so much evidence had been gathered earlier about the specificity of inhibitory regulators for each tissue. This interpretation as a universal chalone was not in accord with Bullough's idea of tissue-specific inhibitory regulators, let alone the specific inhibitors of Weiss and Rose. Yet experimental work with cell cultures showed that it had an anti-proliferative effect, first on cultures of cancer cells, and later even on human tumours. Such a proposal for a specific biochemical as the hypothetical regulator was an idea that could at least be tested by scientists in the field. But a universal inhibitor could not provide tissue-specific control, which the original hypothesis required.

Se-Jin Lee identified a family of genes coding for proteins that he called growth/differentiation factors (GDFs). His insight was that these resembled transforming (or tumour) growth factor β (TGF-β),[7] which was known to display growth-inhibitory properties. But unlike TGF-β, some of Lee's GDFs were tissue-specific. It was in 1997 that A.C. McPherron, A.M. Lawler and Se-Jin Lee, using gene cloning methods, made an important discovery [21]. They found that GDF-8, which coded for a peptide now called myostatin, is a specific regulator of skeletal muscle mass [22]. They realised that cattle breeders had produced hugely muscled breeds such as the Belgian Blue and Piedmontese (Figure 12.3), which had been selected from individuals

[7] Transforming growth factor-β, or activin, prepares cells to respond to other growth factors, stimulates the synthesis of receptors and inhibits the rate at which cells pass through the cell cycle. It may also act as a morphogen.

Figure 12.3. Photo of Belgian Blue bull showing double muscling due to mutations in the myostatin gene. From McPherron and Lee (1997).

with naturally occurring mutations that inactivated the gene coding for myostatin [23]. This had resulted in the growth of abnormally large muscle blocks that would typically have been limited to a genetically prescribed size in normal cattle. Such mutations also occur in some men and women, some of whom have become successful athletes. Various animals have been created that are unable to produce myostatin, because of the commercial implications for agriculture, with the same spectacular results. The original discovery has many implications for applications in agriculture and medicine, let alone sport and bodybuilding.

Links were made to the earlier chalone concept, with the realisation that these growth/differentiation factors (GDFs) met some of the key properties of the hypothetical chalone put forward by Bullough 40 years earlier. 'Chalones are back with a vengeance', one review noted [24]. Myostatin conforms to the criteria for a hypothetical chalone in that it is an inhibitory regulator of muscle size. It is specific to muscle, so must be produced by the same tissue. It is important in development because it limits the number of cells within a muscle, such that the amount of inhibitor increases with the number of cells, which stop dividing at a concentration that indicates the required number of cells has been reached. A regenerative capability is related to the cell number, in that any reduction in number causes a fall in myostatin concentration, such that cell division is initiated and restores the goal number. Myostatin was found to operate generally for skeletal muscle. Later GDF-11 was found to code for another TGF-β-like molecule responsible

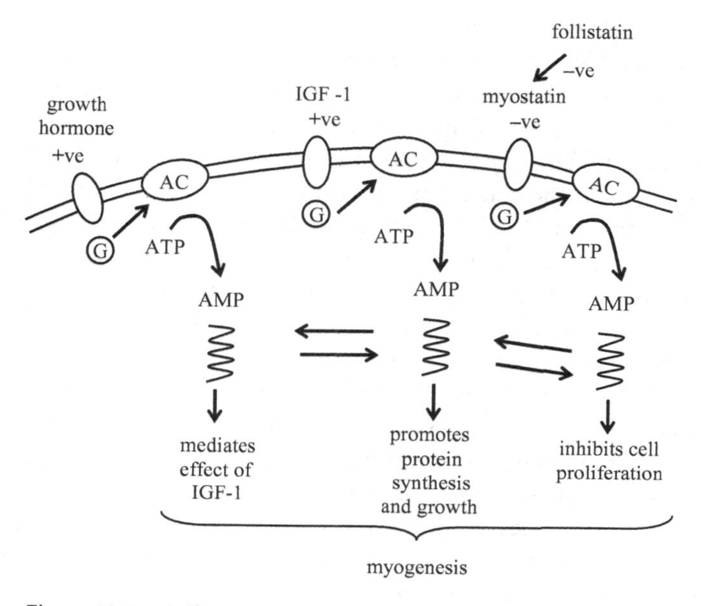

Figure 12.4. A diagram summarising some of the steps in growth control from the first messenger outside the cell to second messengers within the cell. Within this scheme the role of myostatin is shown in steps leading to the production of muscle cells. Abbreviations: G, G-protein; AC, adenylate cyclase; 1GF-1, insulin growth factor 1. Redrawn from various sources.

for controlling the development of nerve tissue. Like myostatin, it was suggested that it acted as an inhibitor within a regulatory loop to regulate differentiation.

Myostatin circulates in the blood and intercellular space, and its concentration is supplemented by other muscle cells. Myostatin binds to an activin receptor in the cell membrane, which allows a signal to pass into the cell, initiating a sequence of kinase reactions (Figure 12.4). The signal is passed though a cascade that amplifies the original signal many times so as to have a greater effect. At the same time, energy is conveyed to perform the required function of inhibiting cell proliferation. Growth hormone is a non-specific growth promoter. Here it binds to the cell, inducing the release of insulin-like growth factor 1 (IGF-1), mediating its effect at the cellular level. It acts as a specific growth promoter by stimulating biosynthesis. Follistatin is also important as an inhibitor of myostatin; the point being that, in blocking myostatin, it in effect acts as a growth promoter, resulting in a more rapid response time. The effect of follistatin in animals that have been genetically modified to create no myostatin is to produce a large muscle mass, appearing just like those animals that naturally cannot produce myostatin.

In normal prenatal animals, myostatin concentrations are low in developing tissues, but rise with the number of cells in a tissue to the point at which the concentration of myostatin, or some other tissue-specific inhibitor, brings growth to a halt. This is similar to the effect represented by the logistic growth curve, which becomes sigmoid as it approaches the carrying capacity (Figure 8.1b). Hyperplasia slows to a level required to balance cell losses with cell gains, while at the same time retaining a capacity to respond to the requirements of stress, exercise and injury. From birth to adulthood, increase in cell mass is due to hypertrophy and increase in cell size. It seems that both processes are under the control of myostatin in muscle tissue, with a shift in emphasis from hyperplasia in the young to hypertrophy in the adult.

The discoveries of McPherron and Lee triggered great interest in statins as growth regulators. There is also considerable interest in their therapeutic use for a wide range of purposes. Here our interest is in their role as analogues of myostatin, as the specificity of myostatin to muscle tissue implies that each tissue must have its own specific regulator. The number of known homologous regulators and control mechanisms has grown, initially as different GDFs were identified. First, regulators for hepatic and nervous tissue were found, and then regulators for heart muscle, pancreatic and endometrial tissues, and lymphocytes. The family of statins also grew - simvastatin, lovastatin, atorvastatin, cerivastatin. New properties were discovered that related to these statins; apart from the inhibition of cell proliferation, statins may also affect the invasiveness or migratory properties of cancer cells and induce cell death by apoptosis.

There remains an added complication that, apart from inhibiting cell proliferation, statins may also cause cancerous growth. At high concentrations growth is inhibited, but at low concentrations they may have the reverse effect and stimulate growth. This is referred to as having a 'biphasic' concentration–effect curve. Although this was unexpected, the stimulatory effects of low concentrations are well known from the field of hormesis (Chapters 10 and 11).

A MINIMAL GROWTH CONTROL MECHANISM

Robin Weiss, an eminent British cancer biologist, long ago realised that 'ecologists'' theories of the regulation of numbers of individual organisms are equally applicable to cell cultures' [25]. There is no obstacle in returning to the logistic equation, due to its use by ecologists to describe population growth. But the logistic equation is used here not simply as a growth equation, but as a control mechanism (Figure 12.5). The reader is referred to Chapter 9 to see how the equation was first recognised as a control model.

Before proceeding, it is important to establish the level of abstraction at which a minimal growth model might be useful. The underlying mechanism within which myostatin operates is much more complex than such a model. The level at which it is represented here is a simplification, with many known biochemicals omitted for clarity (Figure 12.4). The aim here is to identify a model that represents a system in terms of key functional components that have identifiable roles within the control mechanism. It is possible to review the literature surrounding myostatins, and to build up a scheme based on a subset of key components. This is partly because every polypeptide and enzyme seems to be linked to many others, making it difficult to maintain focus on those biochemicals that are essential parts of the control mechanism.

The specificity of signalling is important if overall control is to be achieved, but with local and independent control of lower levels of the tissue or organ. Growth hormone provides non-specific stimulation throughout the organism, while localised tissue-specific control is provided by inhibitory regulators like myostatin. Here then we have the basic activator/inhibitor components of a dual control mechanism whose origins are described in Chapter 5. First messengers, such as growth hormone, necessarily reach the cell at a low concentration so that, once within the cell, the signal must be amplified and energised to be effective. The synthesis of biomass by metabolic pathways within the cell is controlled by regulating the rates of the processes involved. But limitation is in terms of cell numbers in the particular tissue, given by the concentration of myostatin, which is proportional to cell number.

Within the logistic control mechanism (Figure 12.5), the inner loop relates to the control of biosynthetic processes within the cell, while the outer loop relates to the limitation of cell numbers in relation to the requirements of the tissue environment in which the cells exist. We therefore see control over growth rate, creating a positive feedback on cell multiplication, which is constrained by the negative feedback of limitation of cell numbers by myostatin. If natural degradation is not likely to be rapid enough to respond to the requirements

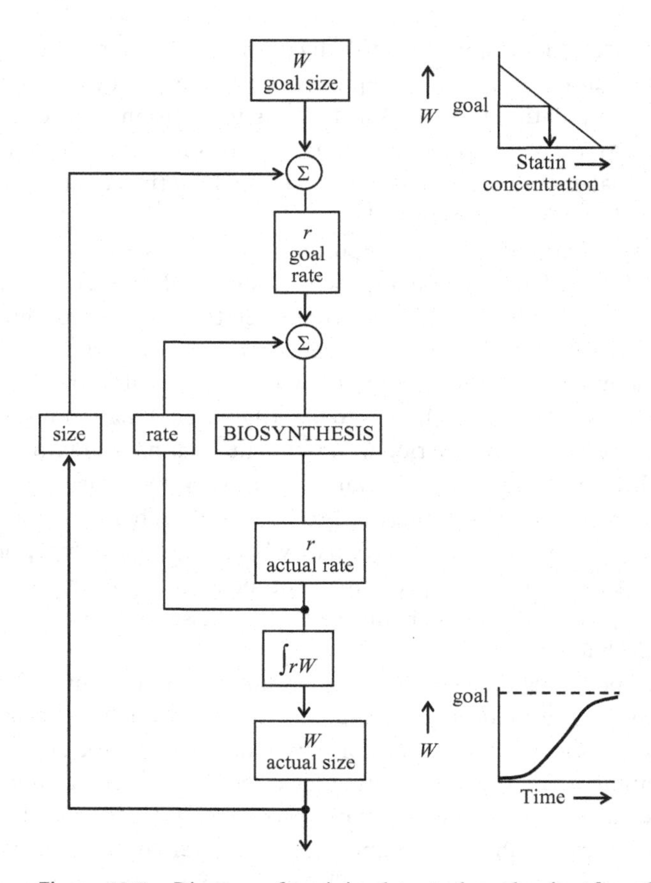

Figure 12.5. Diagram of a minimal control mechanism for cell growth, showing the control of the rate of myogenesis by myostatin, linked to the control of cell numbers that create tissues.

of muscle repair and regeneration, then follistatin neutralises the effect of myostatin (like the hypothetical anti-chalone). In this system of opposites, we see the essential requirements of making a response go faster or slower.

Specificity is a key factor that allows overall growth control from the highest level by growth hormone, but local control is also necessary to provide the semi-autonomous features of control found in different organs and tissues to ensure independent local responses. While conceptually a goal, or preferred setting, is an essential component for any control mechanism, the storage of such information is paradoxical. It is clear from previous chapters that goals are not always fixed, like mammalian core temperature, but can be adjusted to meet adaptive requirements. How goals are maintained is considered in the next section. The aim is to integrate the known and deduced properties of the control mechanism with the system of known biochemicals implicated in the regulation of cell and tissue functions.

We have already established that it is the specific rate (*r*) of this process that is controlled.[8] Perturbation of a controlled process, in order to observe its controlled output, provides a series of oscillations that have a form and duration determined by the strength of inhibition attributable to the concentration of the toxic agent (Figure 12.2). But the baseline about which growth rates oscillate delivers an exponential, or log-linear, increase in numbers. That is to say, the population grows exponentially (Figure 8.1b), even though its rate of growth is held at a more or less constant rate. Rate control of biosynthesis alone results in growth which produces positive feedback in numbers, and creates accelerating rates of increase of cells. The limitation of size of a tissue or organ is necessary, such that in development the tissues reach a prescribed size (*W*) and then stop growing. During adulthood, sizes at these levels are maintained, to the extent that size can adapt to physiological load, losses are regenerated and wounds are healed. The same control mechanisms must operate in both development and adulthood. While the inner loop provides rate control (*r*), the outer loop provides size control (*W*), putting a ceiling on exponential growth (Figure 8.1c). Without control, the system would be unlimited and unsustainable.

Mammalian cells, like protozoans, have the ability to multiply exponentially, but as cells within an organism they do not grow like this for long, as their continued increase would jeopardise the organisation of their host. Protozoans, as independent organisms, produce exocrines that limit their own population density (Chapter 8). Such population control mechanisms were apparently retained in the evolution of metazoans from their single-celled antecedents. The control mechanism was apparently internalised such that exocrines became endocrines, regulating the density and number of cells within tissues and organs.

Thanks to McPherron and Lee and their discovery of myostatin, there is now a chemically defined regulatory inhibitor, which makes experiments reproducible and free of ambiguity in a way that

[8] The specific growth rate (*r*) is the rate of growth per unit biomass per unit time, and is given by $1/W \cdot dW/dt$ or $\log_e W/dt$. It is mathematically the same as the ecologists' intrinsic rate of natural increase for a population (see Chapter 8).

experiments with tissue extracts never could have been. The ceiling to cell growth is provided here by myostatin, which fills the role of the hypothetical regulator, whether Weiss's anti-template, Rose's polarised specific inhibitor or Bullough's chalone. Research on other statins for other tissues seems set to provide a more complete understanding of growth processes at the cellular and tissue levels of organisation.

The legacy of Weiss, Rose and Bullough includes ideas that were parts of their hypotheses for specific inhibitors, which will not necessarily be found in the papers of McPherron and Lee. Nevertheless, as the implications of their findings are explored, some will be relevant. So the reason for this account of the history of the inhibitory growth regulator hypothesis becomes clear. The hypotheses of the worthies summarised here may become useful in inspiring future researchers, helping to make their path less tentative than that of their forbears.

HOW MIGHT A GROWTH CONTROL MECHANISM FAIL?

We ask to what extent might the minimal model mimics the failure of control that accompanies carcinogenesis? If, for whatever reason, the outer loop fails, it is clear that the inner loop would operate on its own, reverting to the primitive condition of rate control alone. Under these circumstances, growth will proceed at a controlled rate which, as we have seen, will result in exponential growth. Failure of the outer loop would allow the inner loop to do what it does naturally, with the result being that the size of a population of cells would grow in an unlimited way, as cancer cells do.

According to recent findings, statins are produced by cells to inhibit their own replication. Cells are stimulated to produce their specific inhibitor by adjacent cells, such that the community of cells contributes to the same pool of inhibitor. When the concentration of a statin in the pool reaches the goal concentration, the inhibition of growth is such that growth rate slows, cell replication ceases and the tissue has reached its goal cell number. The exponential increase of tumour cells would be the inevitable consequence of a cell, or cells, within the tissue community failing to respond to inhibitory statin.

This analysis suggests that tissue-specific inhibitors provide the likely agent that limits cell multiplication. Given a constant specific growth rate, populations increase exponentially as the products of cell division also divide. The principle regulatory requirement therefore is to limit exponential increase. The hypothesis that emerges is that cells have a goal size, or cell number, which provides a limit that is imposed by a tissue-specific inhibitor, by limiting growth rate.

The discovery of myostatin and other members of a family of tissue-specific statins in recent years provides the evidence needed to corroborate this hypothesis, which dates back many years. The logistic control mechanism (Figure 12.5), as the hypothetical mechanism within which statins would play their part, indicates that cells which do not detect an inhibitor specific to their tissue type will not be limited, and so neoplastic growth will follow. It is to be expected, therefore, that metastatic cells which spread to other tissues will continue to grow without stopping. There a different specific inhibitor keeps cells within their goal limit, but this is ineffective for invasive cells from other tissues, and so they multiply without constraint. This interpretation is corroborated by some reports that normal cells introduced experimentally into other tissues may grow in an unlimited fashion to form tumours.

The concept of contact inhibition discussed earlier identifies one way in which a break in the flow of information in the outer loop might come about. It is characteristic of cancer cells that they lack cell-to-cell communication, reflected in the break of information in the outer loop. Cell-to-cell communication and ion channels are essential to ensure the passage of the regulator between cells in order to effect inhibition. A mutation that prevents cells from responding to inhibitory limitation of cell numbers will cause neoplastic growth. This would cause control of cell growth to revert to the more primitive condition provided by the inner loop. Such cells then have an accelerator, but no brakes. The brakes are removed from a system whose natural tendency now becomes the positive feedback of rate control and unlimited exponential growth. Not that the rate is particularly rapid, and the growth rate of cancer cells is often slower that of normal cells; it is just that for such cells the ceiling to growth is removed.

Cancer cells are deaf to the inhibitory limitation, and continue to divide and multiply exponentially, despite signals not to do so. Failure of cells to detect the inhibitor is due to mutations in transformed and neoplastic cells, which may be responsible for the absence of receptors for the inhibitor, the lack of cell-to-cell adhesion, absence of cell-to-cell ion channels and/or the absence of any element in the chain of intracellular agents necessary to effect inhibitory regulation. In some cancer cells (retinoblastoma and colon cancer cells), this is because they do not have receptor molecules to receive signals to prevent cell division.

It is reasonable to assume that, as the route to carcinogenesis requires multiple steps, this route will not be followed by a population of cells. Therefore, there will always be production of the cell inhibitor or statin. Carcinogenesis must be due to one of the various ways in which cells fail to receive the inhibitory signal that keeps their capacity to replicate latent.

This reasoning points to failure of the outer loop of the logistic control mechanism, which is responsible for limiting the size of a cell population or tissue. Carcinogenesis is a result of cells becoming disabled in their ability to receive the inhibitory regulator for their specific tissue.

MEMBRANE-INDUCED CANCER

Over 50 years ago a phenomenon was discovered that now has relevance to the idea of specific regulatory inhibitors. It is raised because we are left with Rose's claim for the polarity of inhibitory regulators. He noted that in some cases regulators pass from cell to cell in one direction only. He cited examples from plants and hydroids, and even cultured fibroblast cells, adding that disruption of polarity can be sufficient to induce tumour growth. If regulators have a directional flux, the seemingly forgotten phenomenon of membrane-induced neoplasia may be relevant.

As part of a series of tests carried out in 1936 on carcinogenic hydrocarbons, pieces of a new material called bakelite were implanted into rats. The result was the induction of tumours. Later it was found that many membranes or foils of glass, metal, polyurethane, cellophane or cellulose could cause tumours to grow in small mammals [26,27]. But when the membranes were ground into a powder and then embedded into laboratory animals, there was no such effect. Clearly the barrier within the tissue was more important than the constituents of the membrane.

It is obvious that a single cell cannot produce so much inhibitor that it would prevent its own cell division. There must therefore be a critical mass of cells to ensure the overall mutual suppression of growth. Early calculations suggest this would be 10^4–10^6 cells. The bounding membrane (a serous membrane or capsule that encloses the organ or tissue) would need to be impermeable to the regulator and must ensure that peripheral cells are not exposed to lower concentrations than others in the tissue. A membrane would also prevent the leakage of the specific regulator from influencing adjacent tissues and organs.

Using various laboratory mammals, it was found that the incidence of cancers could be induced by discs which halved in number when the discs were cut in half. Other workers found that the frequency of cancers due to membranes was related to the area of the cellophane pieces implanted. Other experiments used polycarbonate membrane filters which had the advantage that filter discs with different pore sizes could be used. The idea was for the pore size to control the diffusion rate of an agent that the researchers assumed was implicated. Over a range of pore sizes, the frequency of cancers was inversely related to pore size, so it was recognised that the flow of the agent, which they believed was a specific nutrient or metabolite, was an important factor in inducing tumours. It now seems probable that membranes provide a barrier to the diffusion of a regulatory inhibitor. If the barrier is small, diffusion around it ensures control; if the barrier is perforated by small holes, they have to be large enough to allow sufficient regulator to pass through, in order to prevent tumour cells from being created by the absence of the regulatory inhibitor.

Estimates of the range of inhibitory molecules diffusing through a population of cells is of the order of 100 cell diameters, which – diffusing in all directions – would reach a population of 10^6 cells. But if communication is from cell to cell, each reinforcing the signal that is received, a tissue will receive a concentration of specific inhibitor adequate to suppress cell division. It seems that the membranes simply created a barrier to the passage of information through the tissue. The cells closest to the membrane would receive a weaker signal than that required to inhibit growth, because it came from only one direction. If the signal is directional, some cells in the shadow of the membrane would not receive the signal at all, causing proliferation.

The idea that emerges is that membranes may induce tumours by blocking the flow of inhibitory regulators. Any cells adjacent to the membrane would inevitably receive lower concentrations of the regulator; if the concentration detected does not exceed that necessary to prevent cell division, the cells will begin to multiply. Part of the inhibitory regulator hypothesis is that the inhibitor is produced by the cells within the tissues they control. Each cell produces regulator, neighbouring cells are induced to produce more, and so on, creating a chain reaction within the tissue. If the flux of regulator, or statin, which induces cells to synthesise and release more is polarised, any barrier will create an area of 'shadow' where cells will receive less statin. Where the concentration is less than the threshold concentration necessary to inhibit replication, inevitably cell proliferation will be initiated. It also seems that once cells begin to behave like cancer cells, they continue to do so.

A re-evaluation of membrane-induced cancer is in agreement with the concept of inhibitory regulators such as somatostatin. This provides an explanation for tumour induction by membranes which, at the time it was investigated intensively, appeared to have no satisfactory explanation.

SUMMARY

Cybernetics has made little contribution to our understanding of the control of cell proliferation, or reasons for the failure of growth control in cancerous growth. The approach to growth control in metazoans taken throughout this book is applied to cell suspensions of yeast cells to reveal evidence of feedback control. It is recognised from recent discoveries that inhibitory regulators are crucial in growth regulation at the cellular level. This idea is traced from its origins in the post-war era to the present time, by way of hypothetical anti-templates, polarised specific inhibitors and chalones. Myostatins, as the inhibitory regulator for muscle tissues, opened up the possibility of specific analogues for other tissues. Neoplastic growth results from cells failing to respond to the inhibitor. The logistic control mechanism is used once more as a first approximation. This leads to the conclusion that the failure of control that causes cancer is not the failure of rate control as a whole, but lies in the failure of the outer loop, which is responsible for limiting cell numbers and the size of tissues.

13

Human overpopulation

Population is the problem of our age.

Julian Huxley

Nature always knows where to stop. The great miracle of growth is surpassed in Nature by the miracle of the natural cessation of growth.

E.F. Schumacher

When a single species grows exponentially without regard for carrying capacity, it will suffer an ignoble fate.

Paul Hawken

RECOGNITION OF OVERPOPULATION AS A GLOBAL PROBLEM

The growing human population was seen as a threat by Alexander Carr-Saunders (1886–1966) in his book *The Population Problem* published in 1922 [1]. It inspired his Oxford friend and colleague Julian Huxley (1887–1975), for whom the analysis of populations was to influence his later work and thought. Even in the 1930s, Huxley was lecturing on the importance of family planning and birth control, given what he then recognised as the dangers of overpopulation. In 1931 he wrote the final section on the *Breeding of Mankind* for the *Science of Life*, by H.G. Wells, J.S. Huxley and G.P Wells: a large volume far ahead of its time in advocating birth control. In it he set out the argument that projected overcrowding from growth rates then calculated at 1% per annum.

From 1946 to 1948 Huxley was the first Director General of UNESCO, and in his new role travelled the world, realising for himself the extent of the growing problem of global overpopulation. In 1949 the UN planned a conference on World Resources, and invited Huxley's continued involvement. He pointed out that the conference would lose half its value if it did not also survey the number of people who depended upon those resources. It was not until 1954 that his idea bore fruit with the UN Conference on World Population. As an evolutionary humanist, he must have known what he was doing in taking the conference to Rome. It was the first time that the global population was surveyed as a whole, under the aegis of an international organisation.

Julian Huxley wrote in 1956 of overpopulation as 'the problem of our age' in an article for *Scientific American*. His brother Aldous probably coined the term 'population explosion', and used it in 1959 as the title for an essay based on one of his wide-ranging lectures on the 'human situation', given at Santa Barbara in California [2]. Although the lecture drew on his brother Julian's article, published a year or two earlier, he brought home the power of exponential growth, concluding that the world population was doubling at an increasing rate. The explosion might have been in slow motion, but the metaphor helped to make it seem just as unstoppable. Julian maintained his efforts to bring the issue of overpopulation to the fore with lectures and essays, including an article for *Playboy* (1967), which I read as a student. I well remember his rallying cry. 'If we go on doing nothing', he wrote, 'man will become the cancer of the planet, ruining it and himself with it'. Ronald Clarke, the biographer of the Huxley family, wrote that Julian Huxley 'was largely responsible for pressing the attention of a reluctant world onto a problem that he could see was looming in fearful proportion towards the end of the century'.

The American biologist, Paul R. Erhlich, published his book *The Population Bomb* in 1968, which was an immediate best seller. With his wife, Anne H. Erhlich, he continued to publish books on overpopulation for over 30 years, bringing the problem to the attention of millions. Together they published *The Population Explosion* in 1990 [3].

PRECONDITIONS FOR ACCELERATING POPULATION GROWTH

The development of an agrarian lifestyle, which made larger families possible, started at the beginning of the Mesolithic period around 8,000 BC. It was not until the birth of agriculture that food surpluses could accumulate, removing the burden of constant mobility on family groups. It must have been difficult initially to grow sufficient surplus food to last the winter, but life in a fixed abode became established with the cultivation of crops. Farming needed larger families and more pairs of hands to cultivate crops and tend animals. In time, surpluses made it possible for specialist skills to develop, as food could be exchanged for services. Trades developed, and surplus food was bartered in exchange for tools made by smiths and for the grinding of corn by millers. By the Middle Ages, each town had a market, where demand and supply met, and products could be bartered or sold.

Towns had large and growing populations, where the density of people brought together regularly by markets also permitted the rapid spread of diseases. In 1348 the Black Death arrived in England by way of an infected French sailor who landed at Weymouth. In the following few years the disease spread rapidly throughout England, causing the death of 1.5 million people. The populations of a thousand villages were ravaged by the disease, and some even disappeared. By 1400 the population of Europe was half that of a century earlier, and did not return to its pre-1348 level until 1600. For a long time, disease was the most common cause of death, so few grew old. The brown rat from Asia replaced the black rat indigenous to Europe, and this was infested by the fleas that carried the bacillus responsible for causing the plague. In the seventeenth century, the plague revisited major European cities, spreading to London and causing the Great Plague of 1664–1665 which resulted in 70,000 deaths and spread once more between towns and cities across England. The plague was probably the last pandemic that had a significant effect on global population growth; from this time onward, growth began to proceed at an accelerating rate.

With the Enlightenment of the seventeenth and eighteenth centuries, the power of thought began to improve the lot of mankind. Humanism provided a new perspective from which to interpret life and the world, while reason and experiment begun to challenge divinity and religious dogma. Francis Bacon (1561–1626) established the concept of experiment, and scientism took root, fuelled by the realisation that knowledge can bring power. Science begun to provide understanding for phenomena that had been impenetrable mysteries. It was not until after the Industrial Revolution that the products of civil engineering began to touch the lives of everyone, providing fresh water for all and removing sewage from the streets. The health of people improved; mortality declined and fertility rose. Life expectancy of the population increased sharply, and the mortality of children under five halved.

A further change in the lot of mankind contributed to the population explosion that followed. With the development of the microscope, pathogens could be seen for the first time, and the diseases for which they were responsible became better understood. By experiment Louis Pasteur (1822–1895) and his peers made a succession of major discoveries. They laid the ghost of spontaneous generation, discovering the agents of decay, the role of yeast in fermentation, the germ theory of disease, sterilisation, vaccines and much more. The number of lives saved by such advances is incalculable.

Later Howard Florey and Ernst Chain took Fleming's discovery of penicillin and developed an antibiotic that would save many millions of lives, and provided the knowledge necessary to discover other medicines. Florey reflected in 1967 [4]:

> I suppose we're glad now that it [penicillin] works, but then you've got to see the reverse side of the [Nobel] medal, because I'm now accused of being partly responsible for the population explosion, which is one of the most devastating things that the world has got to face for the rest of the century.

The likelihood of infectious disease increases with the density of humanity, as the focus of the plague on urban areas demonstrated. Now that diseases are largely preventable by improvements in living conditions, we can live safely at much higher densities. Disease and epidemics were prevented by inoculation, and are controlled mainly by abundant clean water, sanitation and good hygiene. Good food, health and the control of infectious diseases also provided the conditions for rapid population growth. The consequence has been increased fertility and reduced mortality, which was the combination that led inevitably to an explosive increase in population numbers.

EXPONENTIAL POPULATION INCREASE

Exponential population increase is described by the Malthusian parameter which is given by the term rN, where the intrinsic rate of natural increase is r, and N is the population size. These two variables determine the intrinsic rate at which a population multiplies and the size of the population that results. The rate at which the human race reproduces is slow compared with mice, for example, but even so the annual increases in the global human population are substantial because the population is already large. Huxley's early estimate of population growth rate was 1% per annum, but this rose to over 2% in the 1960s. The population explosion happens in slow motion, doubling in a generation (~25 years) at its most rapid, unlike the spring phytoplankton blooms during which some species double their number twice a day (Table 2.2). But, as Kingsley Davis (1908–1997) wrote,

> Most discussions of the population crisis lead logically to zero population growth as the ultimate goal, because any growth rate, if continued, will eventually use up the Earth.

These two elements can be teased apart using the relative growth rate, which gives the population growth rate per capita (or per cent). This has the effect of filtering out the positive feedback effect on population size. Within a simple experimental system, such as those described using hydroids and yeast, cultures grow exponentially with a near constant relative growth rate (Chapter 5). Exponential growth is the consequence of the positive feedback, which in other contexts tends to destabilise unless brought under control.

Manfred Eigen and Ruthild Winkler, in their book *Laws of the Game* (1982), provided an accessible analysis of the relationship between the form of the rate functions and their outcome as cumulative curves, which I shall paraphrase here [5]. While the growth rate is constant, the population increases linearly in proportion to time. If the rate grows linearly, in proportion to numbers, then the population grows exponentially with time. This is due to the positive feedback of biological increase, because the progeny of parents become parents themselves, and contribute to the population at large. But if the growth rate increases at a rate that is greater than linear, then growth of the population becomes super-exponential, the rate of population increase itself accelerates and the curve closely resembles the hyperbolic curve (Figure 13.1). Linear growth would reach an infinite size after an infinite time, but an accelerating rate curve will result in a rate that will become infinitely large after a finite time. This is mathematically termed a 'singularity' and reflects the danger of extrapolation, as clearly such growth is a biological impossibility.

For over 50 years, population growth has been perceived to threaten 'overpopulation'. The curve of global population growth is reported to have been super-exponential from 1400 to 1970; not only has growth accelerated over long periods but so has the rate of growth. Its form shows no evidence of control or constraint. Superficially it had all the properties of an explosion, set to grow without limit, and global population growth had a momentum that appeared unstoppable. What is more, it had a delay time between cause and effect of 50 years and more, such that whatever was done to improve the situation had little effect for two generations. The time lags in the system for fertility

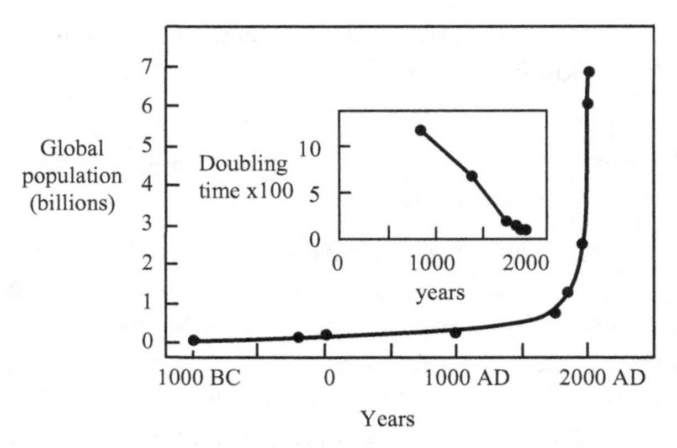

Figure 13.1. Growth curve for the world population from 1000 BC to the present, which has taken a hyperbolic or super-exponential form from about 1400 to 1970. The inset curve shows the decline in doubling time of world population as growth rates have increased.

and mortality make interpretation difficult, as changes in birth rate and mortality rate may require several generations to take effect, and during that time population growth will continue to climb.

Although population growth rate has been slowing since 1964, and is set to plateau by 2100, paradoxically population growth rate is decreasing but the population size is expected to continue to grow until it reaches a plateau at 8.4 billion. This is partly because of the time lag in the system, but also because falling fertility is offset by increasing longevity and falling mortality.

The doubling time is a good index of growth because it provides a measure of growth rate that is independent of population size (Chapter 2). If a population doubles itself at equal time intervals year on year, it is growing at an exponential rate. If growth is at a super-exponential rate, the doubling time of a population is getting shorter. It appears that global population growth has been doing this for centuries (Figure 13.1 inset). It is probably helpful to use the trend of decreasing doubling time as a measure of such growth. The acceleration of population growth rate almost doubled during the nineteenth century, due mainly to reductions in mortality and increased longevity. As a UN spokesman later suggested, 'It's not that people suddenly started breeding like rabbits; it's just that they stopped dying like flies'. In the 1960s the interval between doublings decreased to the point at which the world population doubled with each generation.

Eminent scientists have written sagely, urging action on overpopulation. Carl Sagan's concern was embedded in a fascinating essay on the power of exponentials. He made the important point that 'even if a small proportion of the human community continues for some time to reproduce exponentially the situation is essentially the same – the world population increases exponentially, even if many nations are at Zero Population Growth'. The inference is that overpopulation is a global problem, and must ultimately be solved by international collaboration of all countries.

Heinz von Foerster and colleagues published a paper in *Science* in 1960 entitled *Doomsday: Friday, 13 November, A.D. 2026* as the day on which they calculated that population growth rate would reach infinity [6]. In subsequent correspondence, it was made clear that this point was an esoteric jest: 'such singularities are usually interpreted as an indication of the systems instability in the vicinity of a singularity'. His main purpose was to draw attention to the fact that it was time 'for our species to confront a decision: not limiting our numbers and facing extinction, or limiting our numbers and living forever'. He believed that there would be dire and inevitable consequences '*if* no measures are introduced to keep world population under control'.

It was a mischievous article with a serious aim, to bring attention to the problem at a time when little was being done. There was a feeling that population growth was getting out of control, and further delay just compounded the problems. It was really another call for action, like those of Julian Huxley.

WAYS OF SLOWING POPULATION GROWTH RATES

We now return to the logistic equation, which describes a cumulative sigmoid curve (Figure 13.2). It is symmetrical in that the first half is an exponential growth curve and the second is an exponential decay curve as growth slows. The sigmoid curve is bounded by upper and lower asymptotes on which the curves converge but do not quite meet. Good examples of populations that follow a sigmoid curve came from yeast and *Drosophila* in culture. The curve is followed by cumulative population increase, beginning exponentially from the lower asymptote, before beginning to slow as it approaches an upper asymptote. A bell-shaped curve is derived from it, which provides the trend in rate that creates the sigmoid curve. The sigmoid curve is the cumulative product of the variations in rate over time.

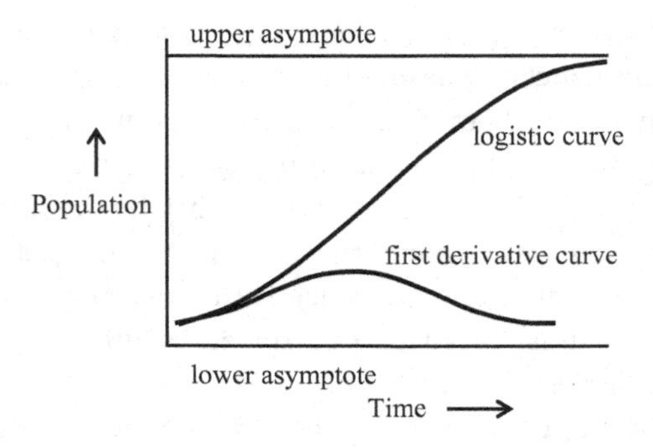

Figure 13.2. The curve described by the logistic equation curve and its first derivative. The logistic curve, as the earliest population model, is often used to represent the cumulative curve of population numbers over time. While the rate curve is the derivative of the first, it reflects the biological rate process that creates the cumulative curve.

While the rate curve is a derivative of the cumulative growth, biological causality is the reverse; so that although interpretation is provided by the rate data, its product is the sigmoid curve. The rate curve, representing changes in the slope of the cumulative curve, rises from zero, reaching a maximum at the point of inflexion mid-way, and declines in a flattened, bell-shaped curve. Traditionally, growth data are plotted as cumulative curves that may approximate the sigmoid curve and tend to obscure the underlying variations in the rate process that is responsible for growth. The rate curve derived from the sigmoid curve more accurately indicates the rate of the process from moment to moment, unobscured by the cumulative products of the process. It is the derivative rate curve that reveals most about population growth as a process, although data are typically collected and represented in a form that resembles the sigmoid curve of the logistic equation.

The global population growth over the last 50 years is represented in Figure 13.3, where graph (a), the population curve, shows a gradual increase in population numbers (in billions) that increases due to feedback of new individuals into the population. Annual increments of 70–80 million people per annum (Figure 13.3b) join the population, and in time become fertile and also breed. The increments are equivalent to creating a new nation on the Earth each year. The new individuals joining the population create a positive feedback effect, and cause the global population to grow at an accelerating rate. Such a mechanism creates exponential increase, which is no different from that found in the re-investment of funds that accrue from a capital

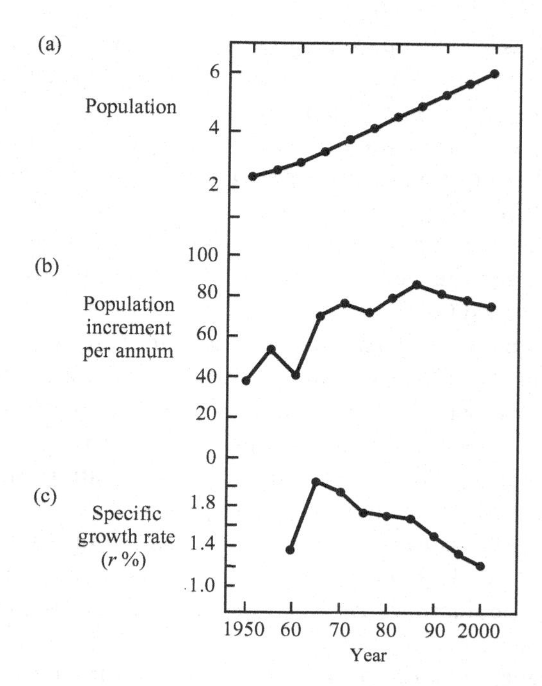

Figure 13.3. World population data over the last 50 years: (a) total world population (billions); (b) annual increments to world population (millions); (c) population growth rate (per capita). Data from the Worldwatch Institute.

investment. A trend of increasing population size can be seen in the cumulative curve until 1965, and then it slowly begins to decline towards 2000.

The relative growth rate is usually given as the percentage growth rate by those who study human populations (Figure 13.3c). This has the effect of removing the population feedback effect, providing a measure that relates to the fertility of individual women. This explains how it can be that the global population will continue to increase while fertility is actually falling. The data run from the highest recorded rate of 2.17% in 1964, which has been falling steadily ever since, despite population numbers that continue to rise (Figure 13.3a). The decline in relative population growth rate is expected to continue to fall to 1% by 2016 (Figure 13.3c).

In 1927, Raymond Pearl published a graph of world population data fitted to the rediscovered logistic curve [7]. As it often represents a good approximation of population growth, the sigmoid curve is frequently used as a first approximation of population growth. There was little justification then to indicate that the increase in global population growth would decline to follow the logistic curve (Figure 13.1). Pearl's bold promotion of his Law of Population Growth attracted much criticism, as his world population data were limited to the middle part of the curve, and did suggest a decline to an asymptote. Confidence that this was so came in part from sigmoid curves for the growth of yeast and the fruit fly *Drosophila*, for which the logistic equation provided a good fit.

Now, with so many more data, and the population growth rate beginning to decline, Pearl's interpretation would not have attracted criticism. Recently Wolfgang Lutz and his co-workers provided a convincing analysis of anticipated world population growth [8]. They forecast, with a probability of 85%, that the global population curve will plateau and growth rate will fall to zero by the end of this century. But the population is expected to reach 8.4 billion in 2100, in accord with an independent forecast of the United Nations. Lutz's prediction shows a population that will reach a plateau by the end of the century.

For the global population to stabilise, average fertility must be limited to no more than 2.1 children per woman. Apart from the need to replace two parents, 0.1 child is allowed for those women who never have any children. The marked fall in fertility over the last 40 years is measured in the sharp reduction in the number of children born to each woman (Figure 13.4). It seems that, with improved education, wealth and awareness, this has led to a decline in birth rates. What has become clear is that the reduction in fertility has in most cases brought the birth rate down to a figure that is lower than the 'replacement' figure of 2.1 children. The global relative growth rate has now been in decline since the 1960s, although population size has continued to grow (Figure 13.3). Longevity has increased spectacularly over the previous century, as living conditions and healthcare have improved. From 1900 to 2002 the expected lifetime of the US male has increased by 30 years (Figure 13.5). Life expectancy is set to continue to increase, varying across different countries from 66 to 97 years by 2100. The United Nations Population Division expects longevity to increase continuously, although at a slowing rate.

It should be added, for the sake of completeness, that the causes of intermittent mortality, even though they may seem large, have little impact on overall population trends. When populations are large, doubling times are so short and annual increments so great that unless

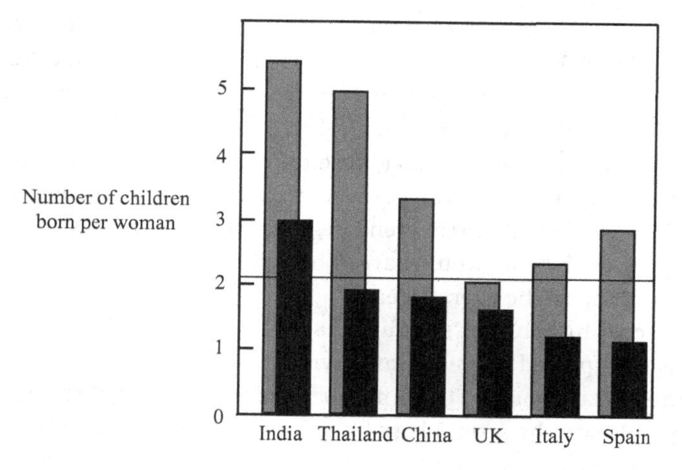

Figure 13.4. The decline in fertility (number of children born per woman) given for various countries from the early 1970s (grey columns) to the early 2000s (black columns). The horizontal line represents the replacement figure of 2.1 children. Data from the United Nations Population Division.

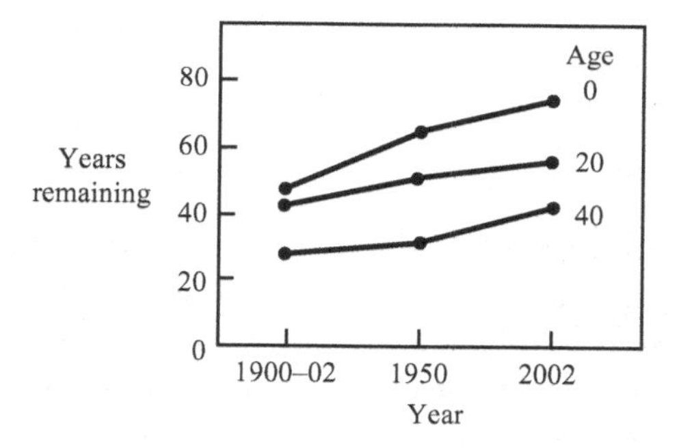

Figure 13.5. The increase in longevity of US men from 1900 to 2002. Data from the US National Center for Health Statistics.

large losses are repeated annually they do not affect the overall trend. The greatest mortalities that come to mind occurred during World War I and the influenza pandemic that followed it. More recently, AIDS has been responsible for the largest number of deaths. At the moment just under 40 million live with the HIV virus that causes AIDS, and in 2006 just under 3 million died of AIDS. The point is that, however significant the loss of life, the annual increment in global population is so great (Figure 13.3), that even the largest losses do not have an impact upon the global pattern of population growth. However variable the increments (Figure 13.3b), the fluctuation of millions is barely detectable in the global increase of billions (Figure 13.3a). Such mortalities are much less than the annual population increments, which over the last 50 years have varied from 41 to 88 million per year. The reason why such mortalities make little difference to global trends now becomes clear. If a major tragedy were to cause the death of 40 million people, it would hold up the trend in global population growth for just 6 months. Such losses would have to be sustained annually for there to be an impact on the overall trend.

According to Lutz's findings [8], the world population growth curve is expected to reach an asymptote, like the logistic curve (Figure 13.2a), around 2100. The current decline in fertility will continue, followed by a time lag of about a generation, and will result in a decline in population size. Actually, such a trend is already happening with the stabilisation of increments (Figure 13.3b), and the more reliable decline in the percentage growth rate (Figure 13.3c) set to fall to 1% or less. Such indices are just beginning to become apparent in a decline in the cumulative curve (Figure 13.3a) which, according to Lutz, will become asymptotic by 2100 at a population of 8.4 billion.

IF ANIMALS CAN CONTROL THEIR OWN DENSITY, WHY CAN'T HUMANS?

While autotrophs do not limit their population size, it is generally accepted that organotrophs do (Chapter 8), though they vary widely in the adaptive means of self-limitation. James T. Tanner, of the University of Tennessee, wrote a review in 1966 of population limitation in animals, which established an important generalisation [9]. He found a negative correlation between population density and growth rate, such that as the density increases, the population falls. This reflects the expected output of the logistic control mechanism discussed earlier (Chapter 9). Tanner reviewed results for 71 species, including animals such as insects, fish, birds and mammals, and concluded that, whether control is intrinsic or extrinsic to the organism, it is clear that there is a limitation of numbers due to their sensitivity to density.

It is known that mammals limit their population numbers by regulating their own density by innate mechanisms. The American ecologist John J. Christian concluded from a long experimental study of populations in mammals [10,11] that:

> The evidence ... supports the existence of endocrine feedback mechanisms which can regulate and limit population growth in response to overall 'social pressure', and which in turn are a function of increased numbers and aggressive behaviour.

So this mechanism is the direct consequence of overcrowding, which reduces fertility and thereby limits population numbers to create a negative feedback loop. But the question is, to what extent is density-dependent limitation the direct effect of density due to mortality, or loss of fertility? or is there sufficient evidence to accept the generality of adaptive self-limitation of numbers at some lesser density in all organotrophs? Certainly there is a strong body of evidence for adaptive self-limitation by exocrines in aquatic organisms (Chapter 8).

Let us consider a little-known data set for what is arguably the longest and largest experiment on the growth of a mammalian population. It lasted a century and involved millions of sheep. Of course, it was not an experiment in the strict sense, rather a serendipitous data set, but the results are important and require further interpretation. In September 1803, 16 ewes, four ewe lambs and three wether lambs (castrated males) were taken ashore at Risdon Cove on Tasmania off the south-east Australian coast. These sheep later crossbred with later introductions of Bengal and Cape breeds. By May 1809, a government census of stock reported 1,091 sheep. A century later there were over a million sheep. Counts were made of the whole population annually.

In 1934 James Davidson, the Australian ecologist, published a paper using these historical data [12]. Davidson plotted the data and fitted a logistic curve. The data were published by Eugene P. Odum and Howard T. Odum in their *Fundamentals of Ecology*. In the 2nd edition, published in 1969, they wrote:

> The numbers of sheep followed the sigmoid curve and the population reached an asymptote about 1850. Since then the population has fluctuated irregularly above and below 1,700,000 sheep, the fluctuations being presumed to be due to variations in climatic factors which limit the number of sheep that could be maintained properly.

What was overlooked at the time is that not only did the population follow a sigmoid curve following the exponential phase, but over the following 80 years the system fluctuated systematically, settling to an equilibrium through a series of oscillations of decaying amplitude at a carrying capacity of 1.7 million sheep (Figure 13.6). On close examination of the data, and comparison with similar control data, it seems that a feedback mechanism is responsible for the observed behaviour.

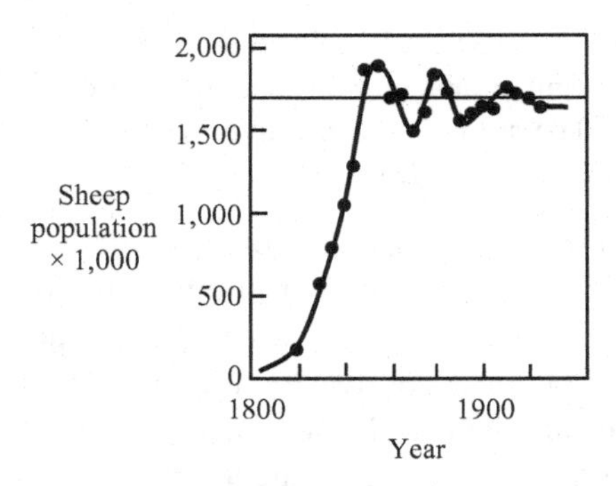

Figure 13.6. Population data for sheep in Tasmania from their introduction in 1803 to 1924. Data from Davidson [12].

The data closely resemble the distinctive behaviour of a stabilising feedback mechanism of the kind anticipated by G.E. Hutchinson in 1948 (Figure 8.1c).

A significant point made by Tanner from his review of density-dependent limitation is that man does not behave as animals do. Human populations grow with a higher rate at greater density. Tanner wrote, 'in the world today the association between [population] growth and density is negative not positive'. Why should our species be an exception to Tanner's generalisation? Animals in general limit numbers in relation to their density, so why should the human species show no evidence of such control?

THE CONTROL OF FERTILITY BY EARLY MAN

It is important to realise that we evolved as hunter-gatherers for over 2 million years before our ancestors adopted an agrarian lifestyle from about 7000 BC onwards. Life as part of a wandering tribe meant that foraging a territory over considerable areas required constant movement to find enough food. Carr-Saunders made the observation that primitive peoples had stable populations and practised some form of population control, leading to an optimal number for the habitat. Wynne-Edwards, whose ideas about population homeostasis had been contentious in the 1960s (see Chapter 8), having read Carr-Saunders, welcomed the ideas as an independent corroboration of his own hypothesis.

Anthropologists have shown that hunter-gatherers maintained populations at a lower level than their environment could support, aware of the need to leave food for the future. Among the Polynesians the need to limit numbers in relation to available food, and the social pressure this creates, were factors in keeping population numbers down. With respect to the !Kung bushmen of the Kalahari Desert, it was an individual decision that was not articulated, though among Australian Aborigines it was, as decisions that affected the band were typically made by the elders. It seems that primitive peoples around the world have always understood the necessity of controlling their own numbers, and carried it out in a reasoned and pragmatic way.

The eminent American anthropologist Joseph Birdsell (1908–1994) is best known for his work with Aboriginal people, leading several expeditions to Australia [13]. His colleague and mentor was N.B. Tindale [14] and, as a result of their studies, the Aborigines became the best-understood hunter-gatherer peoples. Tindale and Birdsell made a study of 183 tribes spread over a large area of the Australian interior. The life of the Aborigines required great knowledge and skill of both the men as hunters, and the women in gathering roots, berries and small animals. Each band had a territory and walked it continuously, moving camp every day or two. Nevertheless, a large area was required for even a small band to eke out a subsistence living; constant mobility was a necessity in order to find enough food. There was also the need to arrive at the right place at the right season to find ripe berries and plump beetle larvae.

It seems that infanticide has long been endemic among Aboriginal tribes. In Victoria, 30%, and in Western Australia 19% of births led to infanticide, and among tribes in South Australia it was customary to kill the firstborn. When mothers died, the child was often buried as well. In southern tribes every second child was eaten, in the belief that the strength of the first child would be doubled as a result. At childbirth a 'granny woman' attended the mother, who knew what the elders required her to do with the newborn. If the mother already had an unweaned child, the baby was not allowed to 'come to life', by preventing it from taking its first breath. The granny held her hand over the baby's face at birth so that it never breathed, freeing her of the guilt of murder. In some groups, as many as half of births ended in this way.

The mobility of a band is determined by the speed of the slowest, so all indirectly had an interest in the women keeping up with the band. With a child under four that required carrying when tired, a newborn at the breast would sap her energy, restricting mobility and her ability to catch and carry food. Infanticide was the practical solution to the imperative of mobility for the band. More boys than girls were kept, which reduced the future fertility of the band.

The !Kung[1] tribesmen of the Kalahari Desert evolved a similar system. Women were sexually abstinent after childbirth, and breastfed their babies for three years, carrying them for up to four years. Once again, the burden of carrying the baby, and the demands of breastfeeding, are a handicap in gathering and carrying food home for the family. According to Richard Leakey and Roger Lewin [15], among the !Kung this results in the birth of a child at intervals of approximately four years, amounting to four or five children in a woman's reproductive lifetime. Child mortality is about 50% so, with the survival of about two children, a population would be stable. If a mother is still nursing when she becomes pregnant, abortion is likely, or the child is killed at birth. The heavier the child as it grows, the less food can be collected and carried. It is estimated that !Kung women walked 3,000 miles per year, and aborigines probably much further. More than one child, together with the need to gather food, is impossible for a mother, but once the child is 4 years old, and can walk with the tribe, life becomes easier.

Birdsell and his colleagues, working with Australian Aborigines, established that for the many tribes they studied, population density was closely correlated with rainfall, even though the average band had a territory of 1,550 km^2 (60 km^2 per head). It is interesting that band sizes of the !Kung and the Australian Aborigines were virtually the same; however, the !Kung had much smaller territories, so their density was greater. The twenty times greater density of the !Kung was made possible by the greater rainfall over the Kalahari than the Australian interior. Antelopes instinctively migrate towards lightning, and the !Kung are reputed to mimic their behaviour, as lightning indicates rainfall, fresh plant growth and game.

Given their distribution over a large area, populations of hunter-gatherers are too sparse for density to be a factor in population control, as wandering bands would rarely meet or see one another. The unit of population control is therefore the band. The limiting factor is the amount of food gathered; and the amount of food gathered can only be increased by greater mobility and spreading the search more widely, which requires walking greater distances.

What is important is that there was no innate mechanism to control population numbers, of the kind typical among animals.

[1] !Kung: the exclamation mark indicates a palatal click, which is a feature of their language.

Population control measures in hunter-gathering tribes fell to the elders and mothers, rather than some innate and automatic mechanism. In this way our species is apparently unique among animals, in that the conscious and reasoned action of individuals is required to limit population numbers.

MODELLING POPULATION GROWTH

When Pierre Verhulst first devised the logistic equation in the nineteenth century (Chapter 8), he realised that some new nations such as the United States of America grew exponentially in their early years, while the growth of mature nations slowed. His equation described the early exponential phase of population growth, which he believed would also slow with time. The logistic equation was rediscovered independently and much later by Raymond Pearl and his colleague Lowell J. Reed in 1920 [16], and put forward as a Law of Population Growth, just as Verhulst had hoped it would be. By adjusting its parameters (r and K), they were able to fit curves to populations, or the growth of organisms from yeast to human.

It was not until 1948 that G.E. Hutchinson realised that the logistic equation was a formal description of a control mechanism. No biological system responds instantaneously; given an appropriate time delay, the output of the logistic equation would oscillate as natural populations do, before finding an equilibrium at the carrying capacity of the system (Figure 8.1c).

It is important to differentiate between a model designed to simulate some data, hypothesis or process and a control model. The distinguishing feature of a control model is that it has one or more target settings for the controlled rate or state, which the system works to maintain at those settings. Such settings define the purpose and meaning of such systems. The logistic control mechanism in this context has two goals: the preferred fertility rate (r) and the preferred population number (K) representing the carrying capacity of the system.

We return now to reconsider the resemblance of such behaviour to the flock of Tasmanian sheep. Davidson used the logistic equation to fit a sigmoid curve to the sheep data, drawing a line through the oscillations that take numbers to an asymptote at the carrying capacity. But it was not until 1948 that Hutchinson interpreted the logistic equation as a control mechanism that required a time lag and caused oscillation.

The expectation is that the flock of sheep should increase at an exponential rate, overshoot the carrying capacity, then, through a series of decaying oscillations, should stabilise at the carrying capacity of the habitat, which is the goal setting for the control mechanism. This was calculated by Davidson as 1.7 million sheep. The period of the oscillations is 28–30 years, and the time delay is approximately half that. The much slower rate of reproduction of sheep adds considerably to the range of population growth rates considered. Over a range of organisms, from yeast to sheep, the frequency of oscillations appears to be positively related to their growth rates. In Figure 13.6, an oscillatory curve has been fitted to the curve by eye, but such an interpretation is new, so a thorough analysis is required to corroborate this interpretation.[2]

Heinz von Foerster's proposal was the creation of a control model for the human population [6]. Von Foerster mused in the concluding paragraphs of his paper that it would be difficult to establish a 'peoplo-stat'[3] to control the world population at some desired level. Since it was obvious that one could not reduce life expectancy, it was clear that any control mechanism must control fertility. This, he continued, would limit families to a little over two children. His principal idea was to propose a population control mechanism:

> whether or not the time has come when man must take control over his fate in this matter and attempt to launch perhaps the most ambitious, the most difficult and most grandiose enterprise in his entire history: the establishment of a global control mechanism, a population servo, which can keep the world's population at a desired level.

The idea was to create a model that could be used as a management tool to control the world's population at a sustainable level. His paper was thought by some to be a joke, and his proposals were not taken seriously.

Jay W. Forrester, as a graduate student at Massachusetts Institute of Technology, first worked under Gordon Brown in his Servomechanism Laboratory, developing feedback control mechanisms for military applications. During the Pacific Theatre of World

[2] It seems that a cybernetic interpretation of these data is new, so a full analysis of the data would be rewarding. It is unlikely that a biological control mechanism is known which stabilises over such a long period.

[3] Control mechanisms were colloquially referred to as 'stats' or, as here, the suffix of 'thermostat' was attached to another word.

War II he went out to the *USS Lexington* to repair a control mechanism on a radar system. In the late 1950s he developed generic methods for simulating dynamic systems, adapting computers to run the models. He devised 'system dynamics', which was a new approach to understanding the behaviour of complex systems. The system reflected Forrester's origins in servo-mechanisms and cybernetics, involving the use of feedback loops of stocks and flows, and time-delayed relationships. The output of such models was often complex and non-linear in ways that could never have been predicted from the components of the system alone. This led to a generic modelling language called DYNAMO (DYNAmic MOdels), which became widely used.

In 1970 Forrester was invited by the Club of Rome to a meeting in Bern, Switzerland to solve what was called by the Club, the 'predicament of mankind'. At that time, there was an expectation of some future global crisis due to the demands being put upon the Earth's carrying capacity by population and economic growth. On the plane returning from the meeting, Forrester drafted in outline the model for the global socio-economic system (*World1*). Following later discussion of his ideas with the Club of Rome about questions of sustainability, *World2* followed, which was incorporated in his book *World Dynamics* (1971) [17]. The book forecast a collapse of the global socio-economic system in the twenty-first century if steps were not taken to reduce the demands on the world's carrying capacity.

Forrester suggested to the Club of Rome that one of his students, Dennis Meadows, should conduct the extended study requested by the Club of Rome, and so the more comprehensive *World3* model was created by Meadows and his colleagues. *The Limits to Growth* was written in a non-technical style by Donella Meadows and others in the Club of Rome group [18]. The book sold 4 million copies and met with both popular acclaim and criticism from social scientists and others. To mark the 20th anniversary of this book, *Beyond the Limits* addressed the criticisms of the earlier book, and used new and recent data to rerun the *World3* model [19].

It may have been at about this time, after the publication of *Limits to Growth*, that Jian Song was a visiting professor at Massachusetts Institute of Technology. As the foremost Chinese control systems theorist, he would have recognised MIT as the home of cybernetics in the West, as well as the institution in which Jay Forrester and his team had developed the *World* models used by the Club of Rome programme. As we shall see, Song became the lynchpin of the Chinese Government's response to the realisation that they too must act to avert a population crisis.

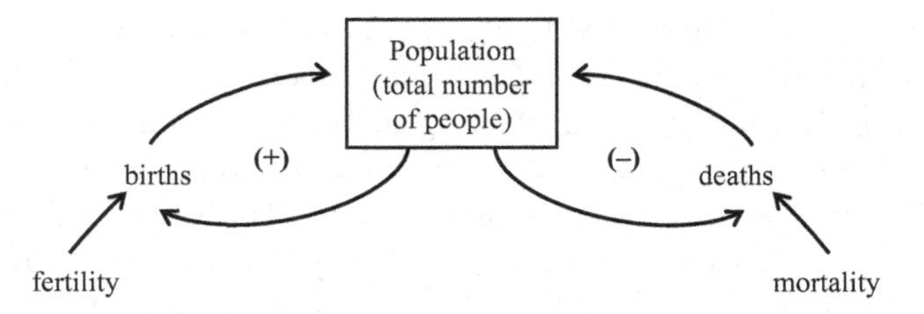

Figure 13.7. Diagram of the population submodel from the *World3* model used in *The Limits to Growth* (1972) and *Beyond the Limits* (1992). Redrawn from Meadows et al. [18].

At a structural level, the population submodel for the *World3* model (Figure 13.7) was simple, and little different in principle from the logistic control mechanism (Figure 8.2). Both models have the same basic features and consist of two antagonistic loops. The inner loop of the logistic model relates to the rate of reproduction. As has already been noted, the growth rate is maintained more or less constant (r), but the integration of growth rates creates what is in effect a positive feedback, creating an exponential increase in growth. The outer loop relates to the limitation of numbers and imposes negative feedback in limiting the increase in population numbers. Thus the control mechanism has the capability of increasing growth, which is ultimately limited when growth reaches a goal setting of the carrying capacity (K), as in Figure 13.6. If left alone, the fertility loop results in exponentially increasing numbers, and negative feedback limitation is essential to provide a constraint on the process.

The conclusions of the second and more recent Club of Rome report *Beyond the Limits* are essentially the same as those published 20 years earlier, but are expressed more strongly. For brevity they are summarised as follows:

1. The rates of use of resources and pollution have exceeded levels that are sustainable. Without reductions in material and energy fluxes, there will be an uncontrolled decline in production, energy use and food output.
2. The decline can be averted by questioning the reasons for sustained economic and population growth, while improving efficiency in the use of energy and materials.
3. We should not allow short-term and less important goals to obscure the long-term and more important goals. Emphasis should be on sufficiency, equity and quality of life, rather than maximising growth and consumption.

The authors were not surprised, after their forecasts in 1971, that by 1991 'many resource and pollution flows had grown beyond their sustainable limits'. This they corroborated with colleagues around the world. We need to focus on securing our long-term future, subordinating much else to the priority of sustainability. Efficiency requires that we live according to our need, rather than our greed, by living a life of sufficiency rather than acquisitiveness. The same requirements for survival lie within the conclusions of *Beyond the Limits*. It seems that such a time must come, as we are currently exploiting the Earth's environmental resources at a level 30% higher than can be sustained. It is forecast that by 2050 two planets will be required to sustain the growing population and its requirements.

CHINA'S EXPERIENCE IN POPULATION CONTROL

Ma Yinchu (1882–1982) was the first Chinese academic to have a doctorate in economics, which he gained in the USA from the University of Columbia. In the early 1950s, as the President of Beijing University, he recognised that China's population was growing rapidly, increasing from 600 million at the first National Census in 1953 to 700 million in 1964. He represented a group of academics who vigorously proposed the adoption of a national family planning policy. In 1957 he published his *New Population Theory* in a national newspaper, forecasting that if the Chinese population grew by 20 births per thousand each year (2% per annum) without limitation, in 50 years numbers would reach 2.6 billion. China would have great difficulty in feeding such great numbers, he warned. Mao Tse-tung's (1893–1976) senior advisors attacked Ma Yinchu for his persistence in promoting family planning, then with the publication of a population theory, his ideas were officially rejected and population studies were virtually banned. He was forced to quit his position at the University. His understanding of the overpopulation problem was ahead of its time, and later there was a saying that 'We lost Ma Yinchu but we gained an extra 300 million people'. By the 1970s the Chinese population had reached 800 million.

Mao Tse-tung remained in power during the Cultural Revolution from 1966 to 1969, and the ideology of the *Little Red Book* remained influential until Mao's death in 1976. A consensus emerged at the highest levels that rapid population growth was a major obstacle to China achieving its modernisation post-Mao. In 1971 population policies were adopted in its planning, with the slogan: 'one couple, one child, and at best, no more than two'. Such planning embraced Ma Yinchu's thinking.

In 1978 China found a new leader in Deng Xiaoping (1904–1997), who re-interpreted Maoism and criticised the policy excesses of the era. Under Deng's influence, a new man came to the fore: this was Dr Song Jian, who had graduated in Moscow in 1960, and went on to receive a doctorate in cybernetics. He became a military scientist and was the first in the East to use optimal control theory in missile control systems design, guiding the development of each new kind of missile, and established the first control theory laboratory in China. He was an eminent control theorist in his own right and worked on a major revision of the standard work in China on *Engineering Cybernetics* by Qian Xuelin, trebling its size in the process.

In the year of his appointment, Deng Xiaoping began reducing investment in military research and development. He was keen to learn from the West to help advance economic reform. Deng learned that China's 'runaway' population growth could become an obstacle to progress, so to avert the anticipated overpopulation crisis, he became a strong advocate for the control of China's burgeoning population,.

It is interesting that 'runaway' is cybernetic jargon, and was wrongly used by Deng as synonymous with overpopulation, implying that he had been talking with cyberneticists. It was also suspected that China's leading cyberneticist Jian Song had Deng's ear. From about this time, Song in his spare time was leading a team researching how to control China's population growth using control theory. The group included two systems specialists and an economist. In taking his work forward, Song travelled widely in the USA, Europe and elsewhere during the 1980s, and visited MIT, Harvard and a number of other Western universities. He returned from one trip with a copy of *Limits to Growth*, and told his group that he had seen the future. He communicated with the Club of Rome, and may well have met Jay Forrester when he was visiting professor at the Massachusetts Institute of Technology.

By the early 1980s, Song's group had developed a simulation model capable of making 100-year projections of China's population, which provided a basis upon which the Party and government could formulate population policy. They used early Chinese computers to simulate the level of fertility required for China to arrive at an optimal population by 2080. The academic Jian Song is said to have 'beguiled' Deng, who had no tertiary education, and Song used computer models 'to create the appearance of a precise forecast' of 'runaway', which would have impressed Deng. Projections involved fertility levels of 1 to 5 children per family measured at 0.5-child intervals, to provide a scientific basis for China's population policy.

This led to the 'one child' policy, which was a more explicit version of the earlier policy dating back to 1971. In 1985 Jian Song, with his colleagues Tuan Chi-Hsien and Yu Jing-Yuan, published *Population Control in China: Theory and Application* [20]. It represented the official Chinese position on the necessity for population control, and was promoted as a model for use by Third-World countries. Jian Song has since been an ambassador for the Chinese approach to population control in India and elsewhere (see Figure 13.4).

Song's study of how to control China's population depended on control theory, systems engineering and computer technology. These achievements provided the scientific basis for the Party and government to formulate its population policy. Over the last 40 years, China's birth rates have fallen from 3.32 children per woman in the early 1970s to 1.9 children per woman in the early 2000s, amounting to a 54% reduction (Figure 13.4). Despite a current population of 1.33 billion, China's population is expected to plateau at 2050: sooner than the rest of the world.

The implementation of the 'one child' policy was harsh, but the consequences of famine were feared to be more so.[4] It involved the sterilisation of one of each couple with two or more children – 12 million in all; and the abortion of unauthorised pregnancies – amounting to 14 million. The outcome was a huge decline in birth rate, although the methods of coercion created immense social suffering among the Chinese people. The 'one child' policy was later relaxed in favour of the more moderate 'later, later' policy, which stipulated later marriage by law (20 years for women and 22 years for men), and later childbearing (22 years for women and 25 years for men).

When his work on population modelling began in the early 1980s, Song had ignored the social scientists working on overpopulation who had fallen from favour under Mao. He later did much to effect the transition of social science 'from a descriptive science to a precision science' by adopting quantitative methods and mathematical modelling. For many, it seemed strange that military scientists became involved in the overpopulation problem, but the origins of cybernetics lay in military technology, and remained so for some time. It was natural, therefore, that those who became involved in population control were

[4] It is important to note that the Club of Rome and *Beyond the Limits* do not advocate coercion in implementing policies on population control, nor is it appropriate to comment here on China's 'one child' policy. The aim here is to give an account of the science that has been applied to these issues.

cyberneticists, and that both Jian Song and Jay Forrester began their careers working with military control mechanisms. Ma Yinchu was reinstated in 1979 and Jian Song became Academician of the Chinese Academy of Sciences in 1991.

WHAT IS THE CARRYING CAPACITY OF THE EARTH FOR HUMANITY?

It is a key property of the logistic control mechanism that the carrying capacity (*K*) for a species represents the maximum population size for a habitat that can be sustained indefinitely. The argument for population self-limitation emerges as a necessity in Chapter 8, and is revealed in examples of specific regulatory inhibitors found especially among aquatic species. Within the logistic control mechanism, that goal setting is represented by *K*, which here represents the maximal carrying capacity for our species on Earth. Any level chosen to represent *K* should not damage the global environment in a way that cannot repair or regenerate itself. The carrying capacity is the goal for the population control mechanism, which defines its purpose in meeting the goal for sustainable numbers.

A comprehensive approach to determining *K* must take into account all the environmental goods and services that mankind uses and which are required for our survival. Naturally this approach is complex and requires the collation of much information. A different approach would be to consider those particular goods and services most likely to become limiting to man's activities. This approach takes into account Liebig's Law of the Minimum, which states that growth is not limited by each of the resources required to make it possible, but by the key essential resources, without which life would become impossible. One obvious example would be potable water; less critical would be the loss of insects that pollinate crop plants and fruit trees.

Estimates of the carrying capacity have been reviewed in recent times by Joel E. Cohen at the Laboratory of Populations at the Rockefeller University in New York [21]. He gives a caution saying that 'the future of the human population, like the future of its economies, environments and cultures, is highly unpredictable'. This is reflected in the spread of estimates of population carrying capacity, which in 1994 ranged from <3 to 44 billion. A tighter range of later estimates of 7.7–12.0 billion found agreement with the United Nations' prediction for 2050 of 7.8–12.5 billion.

It is simpler to begin by asking whether the Earth's carrying capacity is more or less than the current world population. This is complicated by an annual increase of ~70 million per annum which remains certain for some time to come, due to time delays in the system. The answer to this question would immediately tell us whether the current population number needs to shrink, or whether it can grow further. Let us therefore ask whether the carrying capacity should be the present 6.8 billion or less. Implicit in this question is whether the global environment is withstanding the impact of the mass of humanity now. For if it is not, we must assume the carrying capacity would be less than the present 6.8 billion.

Let us consider climate change as a man-made consequence of the population size of humanity and its economic activity. It has been established that the principal causes of climate change are man-made. Carbon dioxide pollution is a byproduct of economic processes causing various effects that are threatening the future of life on Earth, including the human species. Global warming is having more and greater effects that were not imagined when it was first detected; far from simply warming the Earth, it is causing an increase in the frequency of extreme weather. Destabilisation of the weather system is causing more frequent extreme weather events like storms, cyclones, tornados and flooding. Global warming is giving temperate regions earlier springs and longer summers, while it is tropical and subtropical regions that now suffer extreme heat and drought. Such climate change is causing the spread of arid regions and deserts.

Wildlife is on the move in a poleward direction, at a rate of 50–80 km per decade, so life controls its optimal temperature range as the planet warms by moving to cooler climes. Good data now exist that demonstrate that this is true for all animals and plants – for example, butterflies on land and fish in the sea. The greatest temperature increases are to be found at the poles. The frozen soils of the tundra that encircle the polar regions are thawing, releasing methane, which is a much more potent greenhouse gas (×20) than carbon dioxide. Heat is penetrating the deepest oceans and, as it warms, water expands. Add the meltwater from ice on the great Antarctic plateau and Arctic Greenland, and sea levels will rise by at least 0.5 metres this century, and this is accelerating. The melting of Greenland's ice cap alone is capable of raising sea level around the world by 7 metres, and this is now melting faster than at any time in history. Sea level rise was forecast by the IPCC to be about half a metre by 2100. Now that figure is being revised and may need to be doubled.

About 65% of tropical forests have already been cleared, and the land is then used for farming until the natural fertility in the soil is used up. Without trees, the rains they generate cease and the land becomes arid scrub or desert, which is vulnerable to erosion. The capacity to draw down carbon dioxide and to lock up carbon for centuries in the form of wood is decreasing as forests disappear, and the ability to generate oxygen upon which all life depends also decreases.

Carbonic acid brought down from the atmosphere, now rich in anthropogenic carbon dioxide, is increasing the acidity of the oceans. All organisms like molluscs and coral that depend upon calcium will, within a generation or two, become unable to create their skeletons and shells. Less obvious, but more important, are those single-celled organisms with calcareous tests, because in energetic terms they are the giants of the ecosystem (see Chapter 2).

Given that so many man-made global crises have been set in motion, the question must be asked whether any more evidence is required to draw the conclusion that the carrying capacity of the Earth to support the human species has already been exceeded. How could a carrying capacity greater than the present 6.8 billion be contemplated, given the current rate of environmental degradation and the crises we now face? We must therefore identify a figure of less than 6.8 billion as being the optimal number for the Earth to support.

ECOLOGICAL FOOTPRINT ANALYSIS AND CARRYING CAPACITY

The 'ecological footprint' not only provides an index of our individual impact on the environment, but also provides a measure of the Earth's carrying capacity for humanity. While this index is relatively new, it was adopted by the World Wildlife Fund (now known as WWF) in its *Living Planet Report 2002*. In applying it to all nations, the index has been improved, and intercalibration between nations coordinated by the international think tank, the Global Footprint Network, shows that it is a valid approach, and certainly the best integration of data showing mankind's impact on the global environment so far. The synthesis is making it possible to draw new conclusions about human impact upon global ecosystems.

The 'ecological footprint' is a measure of the load that humanity imposes on the planet, expressed in terms of an individual person, a nation or all humanity. The load is divided between those goods and services on which we all depend.[5] They include not just the food we eat and the energy we use, but the demands we put on the environment to assimilate waste or recycle nutrients, and much else. There are trends towards the better use of resources by conservation and substitution, a greater efficiency in their use, and reductions in pollution by processing wastes. But the benefits of such improvements tend to be offset by further economic growth. The reality, the significance of which is lost on many, is that humanity cannot do without the goods and services that the environment provides. This will be discussed in greater depth in the next chapter.

The ecological footprint concept measures our impact upon the environment and helps us recognise the degree of our dependence upon it. Ecological footprint analysis was originated by William Rees of the University of British Columbia, Canada, who for many years taught it to his Environmental and Resource Planning students. In the early 1990s one of these students, Mathis Wackernagel, chose ecological footprint analysis as the subject of his doctoral thesis, which led to the publication of a co-authored book in 1996 [22]. Subsequently Wackernagel joined the WWF's team, using the concept as a tool to help meet the goal of sustainable development handed down by the Earth Summit in Rio de Janeiro. Ecological footprint analysis has made it possible to estimate human uses of the planet in relation to its capacity for regeneration. The level of environmental perturbation cannot be allowed to exceed the capacity of the system to regenerate. That is to say, the load induced by the many uses is only sustainable insofar as it does not exceed that capacity, otherwise irreversible degradation follows. The capacity of the environment to assimilate wastes by their degradation, recycling or sequestration depends in part on biological systems, which degenerate with overloading. Their future capacity to assimilate is thereby reduced. The footprint analysis was published in WWF's *Living Planet Report 2002*, and every year since.

The concept is straightforward. The size of the footprint is expressed in terms of numerous small areas of the Earth which each provide the individual with food and resources: coffee, bananas, hardwood timber for furniture, petrol and oil and so on. The footprint also includes those areas that are used to assimilate wastes in rivers, estuaries and the ocean, and the gases that are emitted from our cars to be

[5] 'Goods and services' is a phrase used by economists to mean commodities provided by economic activity. Here they represent the outputs of ecological activity in the form of materials and processes of importance to the economy.

dispersed in the atmosphere. A footprint can be divided into various major categories. These include Farm Land in near and far-flung places that are required to grow food, and similarly Forest Land that provides timber for construction and wood pulp for paper. Energy Land provides fossil energy consumption, amounting to about a quarter of the total. Consumed Land is the loss of ecologically productive land, used up in development and ending up under cities, highways and airports. The totals for each kind of land make up the ecological footprint of the household, divided by the number of occupants to give the individual footprint, or summed to calculate the footprint of a nation.

The reliability of the concept depends on the quality and completeness of the data. Given that our focus here is on overpopulation, it becomes clear that the amount of biologically productive land per person has been shrinking as global population numbers grow, and as the same land area is shared between more people. The consequence is that some small, densely populated countries, like the Netherlands, have a national ecological footprint 15 times larger than the area of their own country. Conversely, poor countries sell their farming produce to wealthy countries, and forfeit their own land to the footprint of others, and tend to have footprints smaller than the size of their own countries. Table 13.1 compares the average size of ecological footprints of people living in different countries.

The ecological footprint concept also provides an objective estimate of the human population that would be appropriate for the capacity of the planet. If we consider a graph of the ecological footprint against time (Figure 13.8), it can be seen that the ecological footprint of humanity has increased as numbers have grown and the standard of living has increased. This is aggravated by the fact that the amount of productive land is shrinking, consumed by brick, concrete and tarmac, by clearance, erosion and the expansion of deserts. It can be seen that humanity's footprint has grown by 80% between 1961 and 1999.

The WWF *Living Planet Report 2008* provides calculations of ecological footprints for all nations, arriving at an average ecological footprint for all the people on the Earth of 2.7 hectares. Multiplying this figure by the total population provides an estimate of how much of the planet is being utilised. As it turns out, the total footprint of humanity now requires the productive area of more than one planet. Currently our requirement is 30% greater than the Earth's productive area, so we already need a planet larger than the one we have, and are running the equivalent of an ecological overdraft by overexploiting the capacity of ecological systems to regenerate themselves.

Table 13.1. *The average size of ecological footprints of people living in different countries (data from the WWF Living Planet Report 2008).*

	Hectares per person
World average	2.7
High-income countries	6.4
Middle-income countries	2.2
Low-income countries	1.0
Afghanistan	0.5
Argentina	2.5
Australia	7.8
Brazil	2.4
China	2.1
Italy	4.8
Mozambique	0.9
Netherlands	4.0
UK	5.3
USA	9.4

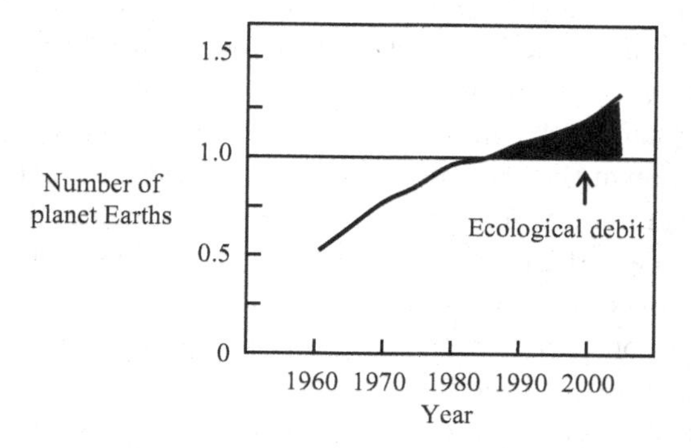

Figure 13.8. Diagram showing the increasing ecological footprint of mankind in relation to the number of planet Earths required, from 1960 to 2005. Redrawn by courtesy of the WWF *Living Planet Report 2008*.

This provides a measure of the extent to which we are living off our capital assets in an unsustainable way. Fisheries are a good example with which to explain the situation. There is a level of exploitation of commercial fisheries, termed the 'maximum sustainable yield', meaning the catch that can be taken from the sea each year without risking the survival of the species, and is the level that can sustain such exploitation. But overfishing has been the problem for many of the world's fisheries. In taking more fish than the maximum sustainable yield, we are taking out the breeding population upon which the future of the fishery depends. The result is that, around the world, fisheries have declined, many are now no longer viable and formerly commercial species are at risk of extinction. Overfishing incurs an ecological debt that we can only repay by allowing the breeding population to recover.

The amount by which the planetary index exceeds unity (Figure 13.8) provides a measure of the extent to which we are building up an ecological debt, due to the way in which the capacity of the planet to sustain human activity is being exceeded. Some effects are already irreversible. We are building up an overdraft of indebtedness to the planet by our overexploitation of capital assets. Coral reefs are one of the wonders of the natural world, with a biodiversity to match the tropical rainforests, yet coral reef ecosystems are expected to die off by 2040, due to ocean acidification. The most obvious irreversible damage is the loss of biodiversity,[6] which has declined by 30% in the last 35 years. WWF has estimated that we will soon require two planets to support humanity, unless a major effort is made to reverse the current trends.

So what is the actual global carrying capacity in terms of the sustainable population size? The intercept of the curve (Figure 13.8) with the 1.0 planet mark in 1986 indicates the time when the global ecological footprint of mankind required the carrying capacity of the whole Earth. The population at the time of the intercept provides us with the figure for the maximum carrying capacity of the Earth. The number of people on the Earth was then about 5 billion. The present population is 6.8 billion and it is expected that this will rise to 8.4 billion before population growth levels off. With a global population expected to grow to a level of 3 billion more than the carrying capacity of 5 billion, is clear that much irreversible destruction is unavoidable before population numbers can be expected to decline. Nations across the globe aspire to the standards enjoyed by the wealthiest countries, but if all humanity enjoyed this standard of living, the Earth would only be able to support 1.75 billion people.

It should now be clear that the global problems we face are not simply due to the presence of large numbers of our species on the Earth. It is our consumption of energy and raw materials that our

[6] Biodiversity is measured here as a proportion of the number of known vertebrate species. Many invertebrate species remain undescribed and unnamed.

technology has made it possible to mobilise. Our species has been reckless, with little concern for those who will inherit the Earth from us. Given that mankind is without an innate mechanism to limit our numbers, we must devise a mechanism for ourselves and implement it. Animals do not overpopulate their habitats because they limit their own density in a way that humanity does not. Early man lost this ability during his time as a hunter-gatherer. The outer loop of the logistic control mechanism that limits numbers was lost rather than failing, as it does in neoplastic growth (Chapter 12). The outer loop has been left to the conscious thought and actions of mankind, but modern morality and ethics restrict the methods that can acceptably be used to achieve population control.

After early pleas from cyberneticists, such as Heinz von Foerster, for the creation of a population regulator for mankind, simulation models have been devised to inform policy-making in an advisory way. The *World3* model was used to simulate the global population growth into the future, and a similar approach has been used to provide a scientific basis for a harsh but effective population management policy in China. The problem with devising a control model is that it requires a carrying capacity for the planet to be used as a goal for the control model. The carrying capacity for man on Earth is about 5 billion, and has already been exceeded by the present population of 6.8 billion. Current trends and forecasts indicate that, even though fertility is declining sharply, the global population is set to grow by a further 1.6 billion to plateau at 8.4 billion by 2100. We are left with the problem of either reducing the number of people on the Earth or reducing the scale of economic activity such that the Earth can continue to accommodate us.

SUMMARY

Human overpopulation has been recognised as a problem for a century or more although, with hindsight, it could have been seen that the path to hyperbolic growth was in place centuries earlier. The growth rate reached a peak in 1964. Even though it has declined steadily since then, the global population is set to continue to increase until 2100. The fundamental problem is that although animal populations are self-limiting, that of mankind is not. Hunter-gatherers consciously limited the size of their tribal groups to maintain their own mobility, but innate control was apparently lost during this period. Jian Song's work on modelling Chinese population growth made it possible to simulate the fall in birth rate necessary to achieve a sustainable population through the Chinese nation's implementation of a 'one child' policy. The global problem is that there is no accepted estimate for the carrying capacity for the Earth. The current environmental crises due to climate change and its multiple consequences can now be thought of as an inevitable result of economic activity, compounded by the sheer number of people. Ecological footprint analysis suggests that a global population of 5 billion is the sustainable maximum, but the population is set to grow to 8.4 billion before a decline in numbers can be expected. The failure in growth control appears to be due to the absence of population limitation by the loss of the outer control loop.

14 Our finite Earth

God has lent us the earth for our life; it is a great entail.[1] It belongs as much to those who are to come after us ... we have no right, by anything that we do or neglect, to involve them in any penalties, or deprive them of benefits which it is in our power to bequeath.

John Ruskin

Wealth is like seawater; the more we drink, the thirstier we become.

A. Schopenhauer

Without a politics of sufficiency there can be neither justice nor peace with nature.

Wolfgang Sachs

If we want everything to remain as it is, it will be necessary for everything to change.

Guiseppe Tomasi di Lampedusa

THE TRAGEDY OF THE COMMONS

Garrett Hardin (1913–2003) was a controversial American ecologist who spent his career at the University of California in Santa Barbara. His main interest was in human overpopulation. He and his wife were members of the Hemlock Society, which promoted the idea that individuals should be able to choose their own time to die. They ended their lives together in 2003. He is best known for his article entitled *The tragedy of the commons*, which became one of *Science* journal's most requested articles. It is an ecological conundrum that occurs when many individuals act independently in their own self-interest; they

[1] Entail: bestowed on man as an inalienable possession.

may ultimately destroy a resource, despite the fact that it is not in anyone's best interest to do so. It was Hardin's belief that the problem had no technical solution, and that only a moral solution could apply. The following re-interpretation suggests that natural selection found a solution to the same problem long ago.

First we must summarise Hardin's metaphor, which gives the article and concept its name [1]. He asked his readers to consider common land used by those who lived nearby to graze their stock. In seeking to maximise his gain, the herdsman asks himself, 'What is the utility to me of adding one more animal to my herd?' As the herd is his own, to add another animal is a benefit to himself. The negative consequence is that adding to the herd contributes to overgrazing, the consequences of which impinge on the other commoners with stock. Since the advantage of adding stock is to himself, and the disadvantage of overgrazing is shared by all the commoners, the best decision in his own self-interest is to add more stock. So he adds one animal and then one more and, following this example, other commoners do the same. Overgrazing is the inevitable consequence and, as Hardin concludes, 'freedom in the commons brings ruin to all'.

The importance of this metaphor is that the problem has many applications. They include natural populations that man exploits, such as the fisheries of the global oceans and the exploitation of the tropical rain forests. But the environmental commons provide negative as well as positive resources, including the use of ecological services. We release smoke and carbon dioxide into the atmosphere, and polluting wastes and effluents are released into rivers and the oceans. We use the environment not only to disperse and dilute effluents, but also to degrade them and recycle nutrients in wastes. These are each global commons from which we not only draw resources but into which we release wastes and effluents.

The vulnerability of the global commons is aggravated by the fact that humanity is now overexploiting many of these ecological services. This is partly because we undervalue their existence and use, as they have no price. In fact nothing has a price until it is bought and sold in the marketplace. So they do not attract monetary value as environmental goods or services, as their cost is only that of extracting resources or depositing waste. Consequently we inevitably undervalue such goods and services, and so the environment becomes overexploited. Of course, there are international agreements and laws with limits and quotas, but on the international commons of the high seas, nothing has been able to save the world's fisheries which, almost without exception, are in decline. The same reasoning can be applied to the seas we pollute, and the atmosphere into which smoke and gases rise from whatever we burn, whether from cars and trucks, ships and planes, or factories and power stations.

There is also a biological interpretation of Hardin's metaphor. Any hypothetical wild herbivorous animal living on Hardin's commons reproduces, and populations grow, quickly reaching an excessive stocking rate, which would overgraze the pastures, causing the animals to die of starvation. In evolution, such behaviour is strongly selected against, and adaptations have evolved that ensure sustainable relationships between an animal and its source of food. Adaptations evolved for self-limitation of the density of organisms, preventing overexploitation that would avoid putting their populations at risk. The way in which nature solved the problem of 'the tragedy of the commons' was, in effect, the evolution of a control mechanism which links reproduction with the carrying capacity of the resource for the species, represented here by the logistic control mechanism (Chapter 8). In this way a balance between grazers and available grazing could be struck, such that grazers survive and the grazing is not overexploited.

Clearly the carrying capacity is not a constant, but varies seasonally as temperature and rainfall cause the growth of pasture to vary. With a high stocking rate and overexploitation, perhaps aggravated by drought, the recovery from grazing is much reduced; circumstances are then reflected in the wasting condition of the stock, which reduces fertility. This is the kind of control that appears to be responsible for self-limitation in the example of a Tasmanian sheep population (Figure 13.6). In biology we now know that organisms have evolved such altruistic behaviour towards their own species, using the mechanism of kin selection first understood by Bill Hamilton, so the evolution of self-limitation is not the problem it was in Wynne-Edwards' time (Chapter 8).

It seems therefore that populations of animals which depend on living food have the ability to overexploit their resource, but do not. Their survival is assured by some form of reproductive constraint, not only to avoid overexploiting the food resources, but to ensure their own survival too. Natural populations apparently limit their own numbers and density in relation to the carrying capacity of their environment, and so ensure pasture for the continued survival of the species. In this way animals solved the problem of overexploiting the commons long ago, as all their resources are 'commons' in Hardin's sense. Our species still needs to find ways of limiting our exploitation of the global commons, if we are not to exceed the carrying capacity of the Earth for the growing numbers of mankind (Chapter 13).

VALUING ECOLOGICAL SERVICES

For mankind, the plight of the commons is aggravated by the undervaluation of environmental resources by the economy, as their use or extraction from the oceans or atmosphere is there for the taking. Humanity has never understood that although the global commons and its natural resources may be 'free', their value is beyond price; yet we continue to overexploit them. The short-term advantage of maximising exploitation is seen as preferable to leaving enough of the resource to grow or reproduce for future use. The problem is similar, but in reverse, to the use of the global commons as a sink for waste and effluents.

An important initiative was taken in 1997 by an international group of environmental economists, who hoped to encourage economists in general to recognise the value of the commons. They were led by Robert Costanza and published an important study [2]. It began, 'The services of ecological systems and the natural capital stocks that produce them are critical to the functioning of the Earth's life support system', recognising the contribution of environmental goods and services to the welfare of man. As many of the assets constitute the global commons, they were vulnerable to overexploitation. Historically it has been just a case of consuming goods, or releasing wastes and effluents into the environment. But now much greater quantities have reduced the ability of these sinks to accept and degrade the inputs of effluents and wastes. Of course, the pre-release processing of effluents has increased on a huge scale, and developing nations allow few untreated effluents to flow into rivers, estuaries and the sea. The group reasoned that if such sinks for wastes are valued appropriately, policies and legislation could be enforced to prevent their overexploitation, and thereby conserve ecological services. Costanza and his co-authors put the importance of their task into perspective, when they wrote:

> The economies of the Earth would grind to a halt without the services of ecological life support systems, so in one sense their total value to the economy is infinite.

The group audited and valued global ecological services, including nutrient recycling, waste assimilation, freshwater supply, food production, genetic resources, gas regulation (CO_2/O_2 balance), climate regulation by Gaian systems, recreation and more. Use of these services is mostly free. As with any resources that are provided for nothing, each is overexploited, so the group estimated the monetary value of ecological services to humanity. Valuations were conservative to ensure the credibility of the final figures. For the entire biosphere, the monetary value in 1997 was in the range of US$16–54 trillion, with an average value of US$33 trillion per year. The global gross national product (GNP) at the time of the study was US$18 trillion per year, so the value of ecological services amounted to nearly twice the global GNP. The authors concluded that their analysis made clear that ecosystem services make an important contribution to human welfare,[2] and therefore should command appropriate weight in policy decisions.

The crucial point is that if these services were not available, they would have to be provided by industrial means, but the global economy would not be adequate to pay for them. Worse, if through overexploitation, these services were destroyed, they would be irreplaceable. But, as has already been pointed out, many of the biological goods and services are being overexploited to the point that their capacity is in decline. This occurs in part because their true worth is not recognised but, more importantly in the long term, their degradation threatens to deny their use to future generations. The initiative of Robert Costanza and his colleagues demonstrated the true value of environmental services to economic growth for the first time. Environmental economists have consolidated these findings since this pioneering work.

In 1998 Hardin made his concept more specific by renaming it 'the tragedy of the *unmanaged* commons' [3]. As the 'dilemma' had no solution, the commons should have been exhausted long ago. As it is, some commons have survived because they are managed. A classic example is the New Forest in southern England,[3] which is still managed in a way that had its origins in the Forest Law that dates back to the Norman kings of the twelfth century. Nowadays commoners have the right to 'common pasturage' for ponies, cattle, donkeys and mules, and are allowed to graze any number upon the common land. Agisters, as officials of the Court of Verderers, have the task of monitoring the stock. If they become lean and out of condition, the

[2] Over 10 years has elapsed since the original study; the figures have changed, but the finding that environmental goods and services were in excess of global GNP has not, so the overall conclusion can still be drawn.

[3] To quote Oliver Rackham, throughout the Middle Ages a forest was a 'place of deer', not necessarily a 'place of trees'.

commoners are required to remove their animals from the common. In this way overgrazing is prevented, at least to the extent that, as stock begins to deteriorate, the numbers are reduced. Clearly, the point at which stock begins to deteriorate occurs when the carrying capacity is exceeded, and the Court of Verderers is able to impose restraint in the stocking level, such that the common pasturage is not overexploited. The system is sustainable, and in such ways the 'tragedy of the commons' is avoided. For other ancient commons, one finds that there are well-established means of avoiding problems by local management of the commons. Nevertheless, the metaphor is a useful one because whatever the current level of legislation and management of the global commons, their depletion is a reality, extinguishing capital assets and overexploiting ecosystem services. What is of interest here is that the ancient commons have been managed and survive to this day under a regime that incorporates the limiting of stock at the first signs of their deterioration. Natural biological systems have their own self-imposed limits that in effect fulfil the same purpose, such that herbivores and predators do not drive their food species to extinction, and destroy the commons.

The point here is that the economy is dependent upon the global environment in ways that are often not appreciated. So the indebtedness of the economy is rarely reflected in a monetary contribution in respect of what is extracted or used, which could help to ensure the future of the commons. There is a need for much better understanding and regulation if the overexploitation and degradation of environmental goods and services is to be avoided. It is to the increasing detriment of all, if this cannot be achieved.

HOW ECONOMIC GROWTH DRIVES CLIMATE CHANGE

It was long thought that limited resources would put an end to continued economic growth, but it is now clear that the limitations are more likely to be due to the overexploitation of ecological services. This is being most clearly demonstrated by global warming and its impact on climate change, sea level rise, ocean acidification, the poleward shift of the biosphere and more.

To demonstrate first that all humanity is responsible for global warming, there is no better example than the emissions of the family car. Although heating the home may involve the release of similar quantities of greenhouse gases, the contribution of our cars relates more closely to the choices we make, particularly by the kind

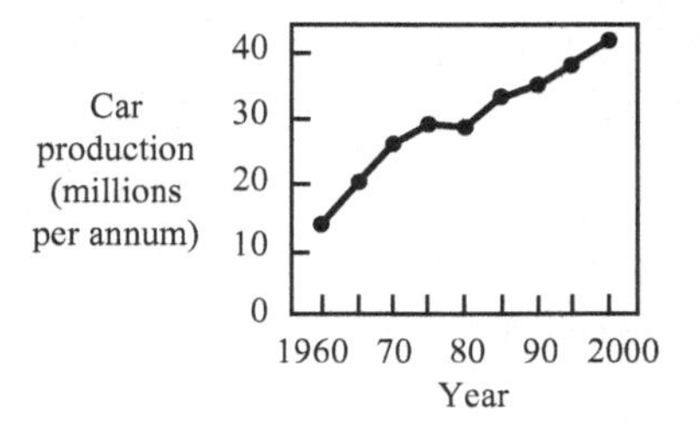

Figure 14.1. Growth in the annual production rate of passenger cars. Data from the Worldwatch Institute, *Vital Signs 2007–2008*.

of vehicle we choose and the distances travelled. As far back as 1967 Julian Huxley noted that the population of automobiles was growing twice as fast as the number of people. Wolfgang Sachs reported that by 1992 the population of cars was growing at a rate four times faster than the population of humans. There are expected to be 1.1 billion cars in the world by 2020, and the global pace of manufacture shows no sign of slowing down (Figure 14.1).

Our way of life now requires the possession of one or more cars for family mobility, employment, recreation and business. They are also the mechanism by which we are individually responsible for releasing carbon dioxide into the atmosphere. The emissions of cars are an important part of our ecological footprints (as emissions per car per year per person). A small family car releases into the atmosphere about 140 grams of carbon dioxide per kilometre travelled. Let us assume that such a car has a lifespan of ~10 years, and accumulates a mileage of about 120,000 (193,000 km); by then it will have emitted about 30 tonnes of carbon dioxide. To adjust such numbers for larger cars, a rule of thumb is that cars emit their own weight in carbon every 6,000 miles (9,700 km). It has become a trend in recent years for people to buy 'sport utility vehicles' (SUVs), or cross-country vehicles with four-wheel drive, for everyday use. These weigh twice as much as small cars, so they are likely to emit twice as much carbon dioxide.[4] Cars make us all contributors to climate change and its various environmental effects due to the combustion of fossil fuels. Our carbon footprint (mostly due to road travel, flying, domestic heating) is over half our ecological footprint, so personal choices make a big difference.

[4] The statistics for car buying show that a higher proportion of small cars are now being bought.

Table 14.1. *Emission of carbon dioxide by different means of travel (grams of CO_2 per tonne carried per kilometre travelled). Data from the UK Department of the Environment.*

Mode of transport	CO_2 emissions (g/t per km)
Ship	30
Rail	41
Road	207
Air	1,206

If we consider the carbon emissions due to travel in terms of grams of CO_2 per tonne (carried) per kilometre (travelled), we can make a realistic comparison of different modes of transport (Table 14.1). It is immediately clear that emissions increase sharply with speed, and that the carbon cost of air travel is six times that of going by car. If we choose to travel by train the cost is five times less, due to low air resistance and the numbers transported more efficiently by a single prime mover.

In 2005, air travellers flew 3.7 trillion passenger kilometres, a figure that continues to increase steadily. In addition, air freight in 2004 amounted to 43 million tons of cargo per day in the USA alone. Much international air freight is for the transport of perishable goods such as flowers, fruit and vegetables to countries where they are out of season, and because of the competitive economic pressure on importers for speed of delivery, the costs of air freight are financially warranted. To put such consumption into perspective, one transatlantic crossing requires 60,000 litres of fuel, which is enough to fuel an average car for 50 years. It is clear that the global economy could not function without a transport system that carries personnel rapidly to satisfy the requirements of business and banking, marketing and sales, let alone the transport of products and goods by road and by air. However normal and everyday such activities may seem, they each make a significant contribution to climate change.

Before going further, we must examine the costs that the economy externalises. The price of goods in the marketplace varies according to the costs of materials and labour incurred by the manufacturer, but does not include the damage costs in extracting resources from the environment, or releasing wastes into the atmosphere. Such costs are not factored in the price so that they could be made available to pay for the regeneration or maintenance of the environment, yet they are real enough. As we have already seen from the work of Robert Costanza and others, their audit and valuation creates a proxy for their true monetary value, to achieve recognition of their *in situ* services to humanity. These values are estimates but, however conservative they might be, they lack the credibility of real prices. The cost implications of the carbon dioxide released into the atmosphere, and its damage in changing the climate and warming the Earth, is almost completely ignored by the economic system. The internalisation of such costs would add much realism to the fractured relationship between the economy and the environment. Nevertheless, load is imposed upon ecological processes, and overload incurs damage that needs to be repaired or given time to regenerate, as the service has a perpetual value to society and the economy.

Herman Daly has made it clear that 'the physical scale of our economy must be kept at a level the planet is able to sustain' [4]. But we are already failing in this endeavour, as the rising costs of repair, adaptation and amelioration indicate. Human numbers and economic activity are reflected in our ecological footprint which already exceeds the carrying capacity of the Earth by 30%. In the USA the cost of climate change is estimated by the Intergovernmental Panel on Climate Change at 3.6% of GDP, which falls within a global range of 1–5%. Globally this amounts to economic costs of US$125 million annually, but will rise to US$340 million by 2030. Insurance companies are also being challenged by the growing costs of insurance payouts as the frequency of extreme weather events increases – another effect of climate change. While globally such costs are relatively low, the major undertaking of raising sea defences around the world to counter sea level rise will add considerably to these costs by 2100. The fruits of economic growth will be taken up to counter its damaging consequences.

Despite initiatives to increase the proportion of our energy requirements from renewable sources, the reality remains that the vast majority of our energy (88%) continues to be derived from coal, gas and oil (Figure 14.2a).

To take the example of travel, it is now probable that the limits to economic growth will not be due to the limitation of resources, but to the increasing amounts of greenhouse gases released in providing the energy to feed economic growth. The growth in the use of fossil fuels continues unabated, and ultimately the rising levels of carbon emissions will impose a limit on economic growth, if some other limit is not imposed first.

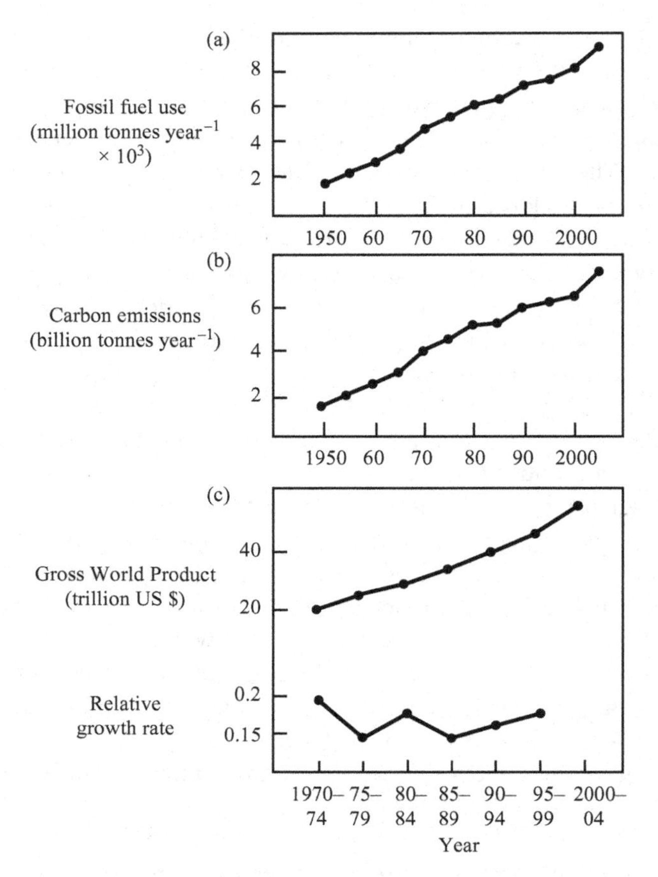

Figure 14.2. The global rates of three linked activities: (a) growth in world fossil fuel consumption (as oil equivalents); (b) carbon emissions from fossil fuel burning; (c) global economic growth as indicated by the Gross World Product, showing the relative growth rate of the global economy. Note that the growth rate of the economy is relatively constant, yet the output of the global economy continues to grow. Data from the Worldwatch Institute, *Vital Signs 2007–2008*.

The juxtaposition of graphs (Figure 14.2) showing the continuing growth in fossil fuel use (a), which results in carbon emissions (b) and generates the energy required for economic growth (c), indicates that they are related. The dependencies between each are understood, implying that the only certain way of reducing emissions is to slow economic growth. Despite the effort that has gone into the reduction of fossil fuel use, it still provides all but one tenth of the energy we use. While wind and solar power production is increasing rapidly, they have not yet resulted in a significant decline in the global use of fossil fuels. Globally only 12% of the electricity generated is derived from nuclear and renewable resources. The predominance of fossil fuel consumption plays its part in increasing the size of our ecological footprint upon the planet, leading to the realisation that we now need more that one planet to satisfy the needs of humanity. The efforts being made to switch to renewable energy are largely being cancelled out by the growth in energy consumption, rather than reducing the reliance on fossil fuels.

The links between manufacturing vehicles, fossil fuel consumption, carbon emissions and economic growth provide a cameo of our global troubles (Figures 14.1 and 14.2). It is not just that they are linked, but that each is growing steadily and showing no sign of slowing down. There is no serious consideration of ways to slow the economy in order to control global warming and climate change.

It is worth looking back to see how long ago it was that climate change and its consequences were first anticipated, as this indicates how long mankind has had knowledge of the problem. Global warming through the accumulation of carbon from fossil fuels in the atmosphere has been a prospect since the Swedish chemist Svante Arrhenius (1859–1927) embarked on a large set of calculations which took him a year.

He completed his calculations in 1894, and published his estimates for the warming of the atmosphere. They revealed that a doubling of carbon dioxide in the atmosphere, on top of natural levels, would cause a temperature rise of 5–6°C. He thought it would take another three millennia for the combustion of fossil fuels to double the concentration of carbon dioxide in the atmosphere. His calculations were close to recent estimates, but he could not have anticipated how fast the rate of use of coal would grow from the global estimate adopted for his 1894 paper of 500 million tons of coal burned annually, to 800 million tons by 1903, and 1,300 million tons by 1920. Arrhenius can be forgiven for failing to foresee the unprecedented growth of industry and the use of fossil fuels. Nevertheless, he identified the mechanism. A rise of 2°C in Europe is now expected by 2050. It is estimated that the amount of fossil fuels currently used each year took a million years to be laid down in the geological past. The multiple effects of climate change are causing environmental change on a large scale, from sea level rise to the gradual shift in the latitudinal ranges of all plants and animals towards the poles. About 90% of the heat from global warming is now in the world's oceans; about half is in the top 300 metres and the rest in deeper waters. The thermohaline circulation distributes this heat around the globe and into the depth of the oceans. The expansion of seawater on warming is contributing significantly to sea level rise, which has been occurring at an average rate of 1.8 millimetres per year for the last century. Recent estimates are significantly higher, at nearly 3 millimetres per year (30 centimetres or nearly one foot per century), as the rate of melting of ice above sea level has increased.

Due to the warming of the globe, all wildlife is on the move towards the poles. This also applies to plants, trees and crop species, although their poleward movement is much slower than that of animals such as butterflies and fish. It is as though two 'travelators' had been installed at the equator to move the biosphere polewards at a rate of 50–80 kilometres each decade. While this is an outrageous oversimplification, it will give the impression of the kind of movements involved. Such shifts in distribution are now well known for many species, including butterflies over land, and fish in the ocean, and will be known for all species in time. The speed of the travelator is likely to be driven by the rate of energy use and the greenhouse effect, as long as fossil fuels remain the principal source of fuel. Our appetite for energy and addiction to economic growth has set the whole of the biosphere on the move, driven by climate change. Around the world the latitudinal shifts in temperature are not only increasing the frequency of extreme events, such as storms and tornados, but in the most severely affected areas the changes are creating local social and economic crises.

In the 50 years from 1950 to 2000, the production of meat per person doubled, and its growth continues to accelerate. Surprisingly, ruminant mammals, including sheep, camels and cows, produce large volumes of methane, which is 21 times more potent as a greenhouse gas than carbon dioxide. The consequence is that ruminants now contribute one fifth of all greenhouse gas emissions that pass into the atmosphere. Meat consumption in developing countries is increasing as more nations become richer and their people acquire similar eating habits.

Another consequence of climate change is less well known. Colleagues at Plymouth Marine Laboratory have been closely involved in the growing concern about the acidification of the ocean, due to carbonic acid falling with rainwater, derived from atmospheric carbon dioxide. They now fear that animals that make calcareous shells, like oysters, cockles and mussels, or that make calcareous skeletons – such as corals, are at risk. Some researchers working on ocean acidification are fearful for the future of coral reefs over the next 50 years. Barbara Brown, who has worked on corals for almost her whole career, points to research in the eastern Pacific, where aragonite[5] saturation is particularly low due to acidification, yet coral reefs are thriving. According to the Intergovernmental Panel on Climate Change (IPCC), there is a lag time of about 40 years between releasing carbon into the atmosphere and it being drawn down again by biological and geochemical processes. Consequently there needs to be a scientific consensus on the risk of acidification to marine life. The lag in the system means that any action taken now (2010) to reduce global warming will not be expected to take effect until 2050.

Deforestation has always been a major factor; it not only remobilises the carbon locked up in wood, but stops living trees from drawing carbon dioxide from the atmosphere in the future. According to the FAO, between 25 and 30% of the greenhouse effect, amounting to 1.6 billion tonnes of carbon dioxide, is caused by deforestation and the burning of wood. Half the world's forests have been cleared and the rate of clearing continues, mostly in the tropics, as if the role of forests in temperature regulation and causing rain were unknown. Once again such environmental services of forests are external to the economy, and so their importance has been underestimated.

Fossil fuel consumption, which provides most of the energy to power the demands of the growing global economy, causes climate change, and this is becoming the most damaging consequence of economic activity on the Earth. It is remarkable that over a century after Arrhenius published his calculations, there is no sign that the steady growth in fossil fuel consumption has started to slow down (Figure 14.2a). This most instructive example of overexploitation of the global commons is due to the release of carbon dioxide into the atmosphere. On the other hand, the capacity of the atmosphere to accommodate carbon dioxide is further reduced by allowing global deforestation to continue. The capacity to accommodate additional carbon dioxide is being exceeded, due to the growing appetite for energy to power the demand for economic growth.

This consideration of climate change, and the metabolic analogy, indicate that the most rapid way to slow climate change may be to slow the rate of economic metabolism. The sustainable rate at which the carbon cycle can accept carbon dioxide without a damaging

[5] Aragonite is a form of calcium carbonate found in molluscs and corals.

greenhouse effect is equivalent to the rates at which carbon dioxide is drawn down from the atmosphere by natural processes. That capacity can be increased by reversing deforestation, and perhaps by stimulating photosynthetic processes in the oceans. A plethora of ideas for reducing the release of carbon dioxide into the atmosphere, or increasing its uptake, are being actively explored, but to date nothing has been effective in slowing the damaging trend of increasing atmospheric carbon dioxide.

POSITIVE FEEDBACK AND EXPONENTIAL INCREASE

We have already seen that exponential increase in biological systems is produced by positive feedback of the products of growth rate to the organism, colony or population. The advantages of exponential growth are used in growth and development, and in *r*-selected species that exploit ephemeral niches, where they grow exponentially. Such systems are known to become unstable and further growth becomes maladaptive, but this is avoided in biological systems by Maia, and exponential growth is brought under control by a negative feedback loop best represented by the logistic control mechanism.

In a 'growth economy', which typically grows at a rate of 3–5%, profits are re-invested to increase future profits, creating exponential increase of the principal over time. In economics too, it is recognised that ultimately such growth becomes destabilising, although there are no inbuilt safeguards to prevent instability.

For reasons that we are familiar with, markets fluctuate, and the economy is very sensitive to whether it is expanding or contracting. Growth for a long time has been the norm, with continued growth as the expectation. A 'bull market' is a stock market in which prices are expected to rise, and in a 'bear market' prices are expected to fall. Life-sized bronze sculptures of a bull and a bear are to be found outside the Frankfurt Stock Exchange, but the raging bull outside the New York Stock Exchange declares a commitment to growth alone.

The economy has the same positive feedback, with the same result as biological growth. Biological and economic growth processes are analogous; 'money breeds money' according to an old proverb, just as 'life begets life'. Biological and monetary systems have these shared properties simply due to the fact that in each case the products of growth also grow; when those products are fed back into the growing population, or back into the principal investment, the growth becomes exponential. The larger a population, or by analogy, the amount of money invested, the faster it grows; simply because it has become larger.

The analogy of the growth of populations and the growth of money by compound interest is mathematically sound and permits close comparison. A population growth rate rN occurs at a constant relative rate r, which is the rate at which new individuals are created. However slow this rate, population size at some time N_t grows faster because the population at N_{t+1} becomes larger.

The relative population growth rate used by ecologists is the same as the specific growth rate of physiologists, and both are given as r. This is also the same as the economist's interest rate, which is given as a percentage, rather than relating to unity. The growth of an investment is given as the interest multiplied by the principal rP, over time t, which is analogous to rN. Then the economic growth rate is $dP/dt = rP \cdot 100$. The term rP has a function r which is the interest rate, or multiplier, that determines the return on money loaned. It is clear that the term for economic growth rate is almost the same as the term for biological increase. The growth of cells and tissues is given by the specific growth rate, and the growth of populations of independent organisms is given by the relative growth rate, which are both given by the term r, expressed as $dN/dt = rN$. The economic and the biological expressions differ only in that the biological measure is expressed in relation to unity or per individual, and the economic term as a percentage or per 100. The two equations share similar properties. In each case a constant rate, however small, will ultimately deliver an exponential growth curve, because the products of biological replication, or the interest, are returned to the population, or to the principal. This positive feedback creates the inevitability of exponential increase over time.

The consequence is that the savings account or the economy will grow, not because r or interest rates are higher, but because of the amounts of money involved. Starting from a small investment and a modest but fixed rate of interest, the principal will grow over time, at a rate that increases continually. The larger the principal becomes, the faster it grows, because it is larger, although the rate of interest has remained constant. Ultimately instability occurs, which makes such growth undesirable.

What is interesting is that the graph shows that the Gross World Product (GWP) is growing at an accelerating rate, yet has a relatively constant growth rate (Figure14.2c). Even with a constant interest rate, the annual product grows at an accelerating rate, apparently because of the growing size of the global economy. This suggests that the acceleration is a consequence of the increasing size of the economy alone.

For such reasons both the growth of capital and populations are seen as the engines of growth, because both the amount of money and the number of people grow faster because, over time, there is more money in the system and more people on the Earth.

There is one significant difference between biological growth and economic growth. Biological growth is limited in relation to the living resources that support it, by a regulator resembling the logistic control mechanism. This ensures that the resources for growth are not exhausted and thereby the animal population is protected from risk. Economic systems have no such mechanism, and so are vulnerable to the risk of exhausting any resources upon which economic growth depends. This includes not only the raw materials for economic activity, but the capacity to dispose of the byproducts of economic activity. These would include not only metals and other raw materials, but also the use of parts of the environment as sinks for the disposal of economic metabolites such as carbon dioxide. Biological systems limit themselves in relation to the resources required, evolving targets for levels of exploitation that can be sustained.

Responsible organisations and central banks monitor the working of the global economic system. As the economy heats up or cools off, control of interest rates and the money supply is used to restore stability. But, unlike biological systems, they have no responsibility for ensuring that the economy is regulated in a manner that conserves the natural systems and resources upon which economic growth depends. To remain sustainable, the economy needs to operate within the carrying capacity of the Earth to sustain the current level of economic activity. But, as we shall see, for much too long, economic growth has been maintained at the expense of ecological capital, degrading irreversibly the natural capital assets of the Earth. For too long, economists have seen no limit to growth, partly because they have not acknowledged its dependence on the products of the Earth, and the sustainability of its environment.

Rather than spend too much time making this point, I have cited some relevant quotations that make these points better than I could.

> The primary freedom of the modern global marketplace is to grow unremittingly, regardless of the consequences to the environment or society (Paul Hawken, businessman).

> Blind faith in economic growth and gain as the be-all, end-all, cure-for-all has been misplaced (Margaret Chan, World Health Organization).

> Our whole philosophy of life is based on the notion that there are no limits to anything at all (Richard Douthwaite, economist).

> Anyone who believes exponential growth can go on forever in a finite world is either a madman or an economist (Kenneth Boulding, economist and interdisciplinary philosopher).

As the economic system continues to serve its commonly perceived function of generating wealth, the global GWP grows at an accelerating rate (Figure 14.2c). The flawed expectation is that poor countries might ultimately enjoy the same standard of living as the rich. It was thought that the resources required for further growth might become limiting, but now it appears that the waste and pollution that are the byproducts of production and consumption will ultimately constrain economic growth. The economy will need to be diverted increasingly to the task of ameliorating and adapting to the environmental consequences of economic activity, while working to regenerate some of the ecosystems that have been destroyed by overexploitation. The American environmental economist Herman Daly suggests that 'economic growth may already be making us poorer rather than richer' by depleting ecological assets. Production and consumption depend upon the capacity of the environment to assimilate their consequences. While wealth creation is a product that is shared by all, humanity seems blind to its environmental consequences of unending growth [5]. Maximising economic growth is sought by those who lead us, urging that we 'grow the economy' at a time when economic growth, unrestrained demand and the overexploitation of environmental goods and services has led us to the point where humanity will require the use of the services of two Earths by 2030. Yet accelerating growth only takes us further beyond the carrying capacity of the Earth to assimilate the byproducts of the metabolism of the economy.

ECOLOGICAL FOOTPRINTS ON THE EARTH

The ecological footprint is defined by the Worldwatch Institute as the amount of land and sea area needed (by an individual, a household, a nation or the world's population) to produce resources, absorb wastes and provide space for infrastructure. The concept was introduced and explained in Chapter 13 (see Table 13.1). Here we are more concerned with the implications for the Earth rather than upon individuals. The size of the average individual ecological footprint increases each year. The Earth's capacity to regenerate and repair ecological systems can no longer keep up with the demands being made upon it, according to WWF [6]. Humans currently use the resources of 1.25 planet Earths. Essentially we are living off the ecological capital, and the demand for ecological goods and services will soon amount to that of two Earths. Costanza and his co-authors write that 'reckless consumption is depleting the world's natural capital to a point where we are endangering our future prosperity', adding that 'our lives depend upon the services provided by the Earth's natural systems' but 'we are consuming the resources that underpin those services much too fast – faster than they can be replenished' [2].

The cycle of economic growth and consumption requires energy in abundance, and the vast majority of it continues to be produced by coal, oil and gas. According to WWF [6], the greatest disparity across the world of the components of the ecological footprint is in the energy used by the rich and the poor nations. People in North America and Europe use 16 times more energy than those in low-income countries. Energy powers economies (Figure 14.2) and the greatest energy use is in those countries that have stable or slowing population growth (Chapter 13). Energy is the fastest growing component of the ecological footprint (2.6% per year), and 88% of that is provided by fossil fuels (Figure 14.2a). Fossil fuels still provide most of our energy, and with the vast reserves of methane gas now being discovered in the Arctic, they will continue to so for the foreseeable future. From the perspective of climate change, nuclear power is the so-called 'least worst' option, as it provides power without carbon dioxide. One might have anticipated an increase in the building of new nuclear power stations, and many are planned, but the number of nuclear power stations has remained virtually unchanged in recent years. Despite the rapid growth of solar and wind power, it is clear that carbon dioxide released to the atmosphere will continue to increase the impacts of climate change. Overpopulation, as such, is not the principal problem, as the richer nations are now controlling their own population growth, while the poorer nations are not. Of course, the world is changing rapidly, but as Fred Pearce, a British writer on global environmental issues, writes: 'Overpopulation is not driving environmental destruction at the global level, overconsumption is' [7].

While the ecological footprints of nations that are growing wealthier catch up with those that presently have the highest incomes, over 60% of ecosystem services are being degraded and are being used in ways that are unsustainable. These problems are being aggravated by reductions in the Earth's stock of land that can provide ecosystem services, as it is being lost in the creation of infrastructure, and so becomes covered in brick and block, tarmac and concrete. As WWF's *Living Planet Report 2008* says, 'Yet our demands continue to escalate, driven by the relentless growth in human population and in individual consumption'. The indications from footprint analysis tell us that our demands upon the Earth are growing, and its capacity to sustain humanity is being exceeded.

SUSTAINABLE DEVELOPMENT

In 1987 one concept emerged from the United Nations World Commission on Environment and Development in Rio de Janeiro led by the Norwegian Prime Minister Gro Harlem Brundtland. The principal issue was to reconcile economic growth and environmental degradation. That concept was 'sustainable development', and was defined as development that 'meets the needs of the present without compromising the ability of future generations to meet their own needs'. Some further explanation of the meaning of the Brundtland Report: *Our Common Future* is needed to make the connection of sustainable development with growth [8].

> Meeting essential need depends in part on achieving full growth potential, and sustainable development clearly requires economic growth in places where such needs are not being met. Elsewhere it can be consistent with economic growth provided the content of growth reflects the broad principles of sustainability and non-exploitation of others.

The concept has been debated thoroughly since then, and new meanings have been given which modify that given above, so that many interpretations now exist. Interestingly, the term 'sustainable development' has been criticised for its apparently contradictory wording, as the first term implies stability and the second implies growth. It is already clear from biological examples that the concepts are conflicting and ultimately incompatible, as they are frequently opposing requirements. The dilemma for the UN meeting in Rio, at its simplest, was to reconcile the polarised views of those who considered economic growth to be essential with those who desired a stable solution that must be enduring. The impossible answer is to try to look in both directions. Tim Jackson, a member of the Sustainable Development Commission, wrote in 2008: 'With the environmental situation reaching crisis point ... it is time to stop pretending that mindlessly chasing economic growth is compatible with sustainability'. We must conclude that the commitment to continuing economic growth is not sustainable.

We are exhorted on radio and television, on billboards as we travel, to spend more and consume more, further feeding the addiction of our society to economic growth and increased consumption. If climate change and its many effects is the greatest threat, to produce less carbon dioxide we must burn less fossil fuel in order to slow that great engine that powers economic growth.

Much that has been written about the issues of economic growth and sustainability includes terms such as 'steady state', 'balancing mechanisms' and 'dynamic equilibrium', which technically all relate to cybernetics. There seems to persist an undercurrent of belief in a role for control mechanisms. With the growth and multiplication of the animal kingdom controlled by Maian mechanisms, humanity has not yet learned the importance of limiting the growth processes, which we have created and drive blindly on. Maia demonstrates that nature's answer has been to grow exponentially until a ceiling is reached, which is determined by an inbuilt limit set at the maximum carrying capacity of the system. Multiplication for animals then slows and stops on reaching a limit, such that population numbers are matched to their habitats. This dynamic equilibrium is sustained automatically. For economic growth, there is a requirement to establish a level that the Earth can accommodate without damaging its ability to provide for our descendents in the same way as it has done for us – indefinitely. For a sustainable outcome, mankind should emulate nature's example.

The reader will begin to recognise a problem that is no different from Hardin's 'tragedy of the commons', as avoiding a tragedy is what sustainable development is about; it is all about growing without overburdening the environment. The exploitation of the commons is limited to the sustainable carrying capacity of the system. Here 'sustainable' means a level of exploitation at a rate no greater than can be regenerated, replaced, recruited or restored as the goods and services are utilised. This is the 'maximum sustainable yield' of fisheries scientists and has a meaning similar to the use of the term 'carrying capacity'; it is helpful because it also makes clear what exploitation must leave behind. The amount of fish remaining must be concomitant with a breeding stock that is required to replace what is taken. The capital asset, or breeding stock, must remain intact; and there must be recruits to the fishery to replace those caught year on year. Similarly, the assimilation of waste and the recycling of nutrients released into the atmosphere are mainly due to the functioning of biological processes. So in harnessing these processes to our needs of waste disposal, we must avoid poisoning the ecosystem with toxic pollutants.

So what is the mechanism that would affect control in these examples? It is necessary to create a link between the growth of exploitation and the carrying capacity, which becomes the goal setting for the control mechanism. As the maximum carrying capacity is reached, the goal is exceeded and population growth is inhibited, thus preventing damage to the environmental asset. But how would it work? It would be imperative that the human use of ecological services should be contained within their capacity to regenerate. Unless this is done, we will lose the services that the Earth provides for all time, demonstrating why control and limitation are required. Control is not as simple as providing a mechanism.[6]

The application of a mechanism to real situations is not straightforward. In the first instance, the worth of such solutions must be accepted politically and embedded in the legislation. Then there is the need to win public support. Research by psychologists is required for such a mechanism to take a socially acceptable form. The only sustainable answer is to become co-owners of such an approach to survivorship. Recent work by Mark van Vugt at the VU University Amsterdam has established that people are more likely to act for the common good if there are institutions involved that instil trust. Incentives may be required that contribute to individuals' self-enhancement and a sense of belonging to a community of like-minded people in a common cause. It is hoped, by doing so, to influence the way in which people behave with respect to decisions and actions for the common good that require personal action and sacrifice.

It is helpful to set the relationship between climate change and the economy into a broader context. There is a long tradition of drawing analogies between ecology and the affairs of humanity that goes back to Karl Marx and his 'social metabolism'. If we consider respiration as a physiological process, we know that the intake of oxygen facilitates the release of energy from organic substances, with carbon dioxide as a byproduct. The faster respiration proceeds, the more energy is made

[6] As Kenneth Boulding implied, but did not say, it is the automation of the control mechanism to regulate social systems that circumvents the moral problems that were an obstacle for Hardin. Such questions need careful consideration, although there are precedents in democracies that are less overtly automated.

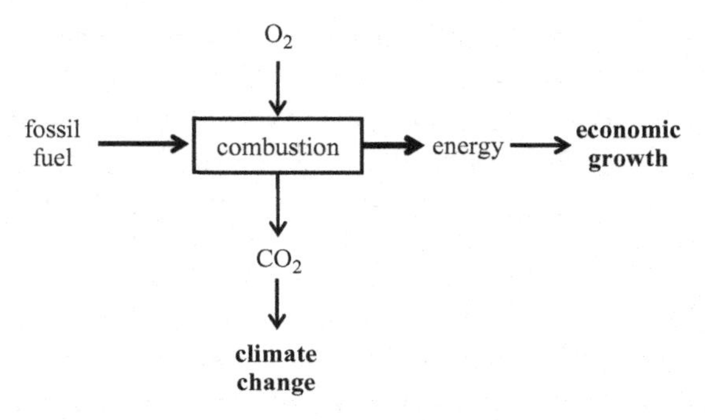

Figure 14.3. A simplified view of economic metabolism that links economic growth to climate change.

available, and the more carbon dioxide is produced. The metabolism of society is analogous, in that the uptake of oxygen is required to combust fossil fuels and release the energy required to power economic growth, with carbon dioxide as the byproduct of energy metabolism that is the principal greenhouse gas driving climate change (Figure 14.3). Climate change and its consequences of warming, ice and snow melting, sea level rise and ocean acidification are due to the increase in the generation of carbon dioxide from the combustion of fossil fuels, and the release of other lesser greenhouse gases like methane. Our addiction to economic growth, as a cure for economic ailments, simply binds us to the wheel of worsening climate change.

Economic growth continues to accelerate, due to the increasing use of fossil fuels (Figure 14.2a). Our addiction to growth as a solution to the economic crisis is deeply mistaken, as new consequences of climate change add to the challenges threatening the Earth.

ECONOMIC CONCEPTS OF QUESTIONABLE WORTH

There is a concern that some of the concepts that are accepted and used within the discipline of economics are no longer valid. With economists asking themselves 'What went wrong with economics?', the title of an editorial in *The Economist* (18–24 July 2009), and many feeling that economic theory is in meltdown, it is timely to question some concepts. Paul Krugman, who in 2008 won the Nobel Prize for economics, takes the view that macroeconomic thinking over the past 30 years has been 'spectacularly useless at best, and positively harmful at worst' (lecture at the London School of Economics, 10 June 2009).

Much doubt has been cast on what some economists thought they knew about economics.

A section has already been devoted to the commitment to economic growth without limit, at a time when it is clear from ecological footprint analysis that we are already exceeding the sustainable limits of the Earth. Here the validity of two more economic concepts is questioned: they relate to *laissez faire* economics and competition. Each, it will be argued, is of questionable worth in reaching out for solutions in finding a sustainable economic system.

Laissez faire economics is a policy which is now clearly seen to be flawed. The economy has long maintained a *laissez faire* 'let do' policy that governments should not intervene in the marketplace. It has long been considered that governments could not improve and should not intervene in the working of a free market which, it was claimed, works best when left to itself. *Laissez faire* economics requires that governments in particular, and society in general, leave the economy alone to do what it does best, that is to generate wealth. The *laissez faire* policy of the market economy results in, and helps to sustain, an inequitable distribution of wealth. The economy provides a global value system that is measured in monetary terms alone, and maintains the objective of creating monetary wealth above all else.

Business at the heart of the wealth-creating machine takes a blinkered view of its role in society. Its purpose is to maximise profits in the interests of its shareholders, to the exclusion of any other consideration, while turning its back on the wider issues of the economy. It is to be hoped that *laissez faire* economics will be remembered for its evils, and that by a greater governmental involvement in the economy, it may be recognised that business and government depend upon, and are part of, the global ecosystem. Society remains blinkered, assuming that economic growth, the generation of wealth and consumerism have a higher priority than issues of sustainability, overpopulation and ecological services.

A second flawed concept is the commitment of the economic system to Darwinian competition. It is a century since Herbert Spencer and his followers persuaded economists that competition could be a creative force in social evolution. The realisation by Charles Darwin of a potential reproductive excess in the numbers of organisms, yet the relative stability of populations, suggested that only the fittest survive in the struggle for existence. Herbert Spencer's famous phrase 'the survival of the fittest' may have captured something of Darwin's ideas, but it is an oversimplification. Competition in that struggle had the effect of partitioning limiting resources, a contest that would be won by those most able to compete successfully. Spencer's interpretation came in the guise of 'social Darwinism' and the similarity between biological evolution and economic development, and gave competition some credibility as a force for good, in that those who survived in competition must be fitter than the others who lost out, justifying the belief that economic life could mimic natural selection. The dogma that competition in the marketplace can function in this creative way is based on the mistaken belief that Darwinian competition is the predominant creative force in evolution.

Why does our species compete so aggressively with others of its own kind, when nature does not? In nature, competition between individuals of the same species takes the form of a conflict that is rarely damaging to those involved. Often there is more facing up to fight than actual conflict, and ritualised skirmishing often helps to maintain a stable stand-off. Biological disputes, whether they relate to territory, food or mates, are typically brief and often more akin to a signalling system than a competitive struggle. Between species, competition is usually avoided and habitats partitioned, with coexistence the typical outcome. In ecology, the competitive exclusion principle tells us that if two apparently competing species coexist in the same habitat, it will be found that they are not competing, but are occupying slightly different niches. Species evolve and partition the habitat, such that competition is avoided by minimising overlap in the multidimensional niche space. Coexistence is achieved in complex communities where niches tend to fit like the irregular shapes of jigsaw pieces with little overlap, and so little competition. Niches tend to remain small and specialised, separating species by their way of life and creating diversity of species and different forms of life. Competition is the means by which resources are divided, and so the effort that goes into competition adds nothing to the value of what is won. It is therefore inefficient, and it is no wonder that organisms have evolved in such a way that the energy devoted to competition is minimised, as ultimately it is the efficient that prevail.

The competition of commerce is different. Competitors do not avoid competition, but rather confront it, and hope ultimately to drive others out of business. Wendell Berry, the American essayist, observed that 'the laws of competition are the laws of war as businesses extinguish one another'. The mechanism is often price warfare in which the largest, with the advantage of economies of scale and resources for loss-leaders, will typically win. The extinction rate of businesses in Europe alone is in excess of 100,000 per year. Such a scenario tends towards an oligopoly, the largest growing to become supermarket chains, with economies of scale and overheads spread between them. Such chains come to dominate the market; the losing competitor is either extinguished or consumed by the larger.

Wasteful duplication of products is a feature of our times, repeated across the market. The range of choices makes 'shopping', as an activity, a necessity if we are not to waste money. The range of models in a car manufacturer's showroom is duplicated in competitors' showrooms nearby. What is so wrong about the commitment to competition is the waste: the duplication of near-identical products, design teams, production lines and factories, not to mention salerooms and advertising. The advantages and benefits of duplication of products cannot be worth the costs incurred across every market sector.

Much has moved forward in our understanding of ecological systems since Herbert Spencer applied Darwinism to society, and economists have failed to update themselves on the advances made in ecology since those Victorian times. In the market, the system tends to an oligopoly with just a few large players, but in nature the climax ecosystem is a large number of species in stable coexistence. When government ministers urge business and the public to be more competitive, the question that springs to mind is, 'But what about Darwinian cooperation?' The actuality is that all life is based upon cooperation, and competition in nature tends to be minimalised or ritualised. Cooperation pervades all organisms, and competition is maladaptive, simply because it is inefficient, and so is typically minimised and avoided.

In recent times there has been growing concern regarding the role of competition in economics. Paul Krugman has written that 'the obsession with competitiveness is not only wrong but dangerous' [9]. Paul Hawken provides an analysis in *The Ecology of Commerce*: 'a large and overwhelming body of evidence that competition in human culture, whether in business or other endeavours, does not improve the species, but is maladaptive and far from being the most intelligent cultural strategy' [10]. Competition for the consumer or between businesses, he concludes, 'is impractical, wasteful, expensive and degrading to all involved'.

Competition commits all involved to greater speed. We hurry to get to a sale before the bargains are snapped up, those on business tend to drive fast, and will often fly wherever possible to save time and beat the competition. To be fast is to be first and win the purchase, the sale, the contract, order or commission. Time urgency is the name of the game, and we are all victims of it. Competition is key in driving the speed of processes and putting us all under stress. Type A behaviour, which leads to heart disease, is typical of the arch-competitor, who is likely to exhibit traits of 'excessive competitive drive, aggressiveness, impatience and a harrying sense of time urgency'. Time, one is often reminded, is money, which transforms life into a never-ending race. The speed of life is driven by competition and, in the process of trying harder to go faster, the quality of life is destroyed.

It can be reasoned from first principles that cooperation is preferable to competition in the more efficient partitioning of resources. Competition is intrinsically wasteful and mostly avoided in nature. In business, competition and duplication add nothing to a product, except the additional costs of competing, such as advertising. Competition is wasteful, as it costs effort, time and money to compete. During evolution, many different evolutionary innovations have been created. It became preferable to pool resources and share skills, rather than for animals to pit themselves against another entity in wasteful competition. Cooperation makes sense as it requires little or no additional effort. All biological life consists of cooperatives; the modules of which living things are composed were once independent and free-living entities. A human being is a community of cells that collaborate and coordinate their activities for the common good.

Kin selection provides the mechanism that makes altruism possible, involving cooperation between parents and offspring, and to a lesser extent among relatives, as the number of shared genes lessens among more distant relatives. Reciprocity theory shows that strategic cooperation is rewarded independently of kin selection. The game theorist Robert Axelrod has shown the robustness of cooperative strategies by giving societal and biological examples. He demonstrates that reciprocity as a strategy occurs between individuals of the same or even different species. Altruism can be established and can survive in a predominantly uncooperative world, which he calls a 'ratchet for evolution', believing it to have implications for private behaviour as well as public policy. These examples lend weight to the interpretation that evolution, and mankind, are much better served by cooperation than by competition.

Simple examples from biology provide insights into far-reaching principles. Seaweeds grow and create new surfaces for colonisation each year, which are quickly covered by sessile animals. There is a race to occupy the space made available, and planktonic larvae of these animals settle and metamorphose. As their larvae settle, growth is directed to where competition for space is least intense, so minimising competition for space. Avoidance is preferable to confrontation. If one examines these crowded surfaces closely, it can be seen that some direct competition between the colonial forms does occur, and sometimes colonies overgrow one another. Some species are adapted to prevent themselves being overgrown; for instance, some grow spines to prevent overgrowth by advancing colonies. There is consistency in the species that overgrow others, or are overgrown: almost a hierarchy of competitiveness. But more important is what happens between colonies of the same or closely related species. Those of the same species stop growing where they meet. It seems that there is mutual inhibition of two colonies where they meet, such as they do not wastefully compete and overgrow colonies of their own kind. Instead growth is diverted in another direction. For competition to be avoided in this way, organisms evolved ways to inhibit the growth of colonies, and this reciprocal inhibition prevents overgrowth between colonies. It is less wasteful to avoid competition, and the necessary adaptations evolved to prevent it. Ultimately it is the most efficient that survive; so competition in nature is avoided rather than confronted.

Biological competition is no longer an accurate example of what happens in the marketplace. The origins of competition lie in social Darwinism, and the strategy of avoidance in biological competition suggests that it may never have been worth emulating in economics. In any case, competition in commerce takes a form that is more confrontational and aggressive than anything we see in biology. Cooperation is intrinsically more efficient than competition, because competition only ever resolves the partitioning of some resource between individuals, costing energy and resources to do so. Competition contributes nothing to the intrinsic value of a product, only adding to its cost.

THE ECONOMICS OF SPACESHIP EARTH

Kenneth Boulding (1910–1993) was born and educated in England, graduating from the University of Oxford with a degree in politics, philosophy and economics. He was granted US citizenship in 1948, becoming a 'magisterial figure' in economics and sociology, contributed to founding general systems theory, and emerged as an interdisciplinary philosopher. In the 1960s, he considered the metaphor of humanity as passengers on a spaceship.

The Greek astronomer Eratosthenes (276–195 BC) first calculated the circumference of the Earth to be 25,000 miles, which is accurate to within 1% of modern measurements. That the Earth is a sphere was confirmed by the circumnavigation of the Chinese admirals sent out in their fleets of large junks by Emperor Zhu Di with orders 'to proceed all the way to the ends of the earth'. They returned in 1421 having circulated the globe. It was previously accepted that this honour fell to the Portuguese explorer Ferdinand Magellan. His depleted fleet returned to Spain in 1522 after a voyage of three years, the loss of four ships and the death of Magellan himself. The Earth is a sphere and therefore a finite entity. In principle, the limits of the Earth have been known since the time of the ancient Greeks, and were demonstrated in the fifteenth and sixteenth centuries by the circumnavigators, but were still not obvious to all until the 1960s, when images of the Earth from space were available for everyone to see.

Images of the blue and white planet slowly spinning in space fascinated a generation, but it became a cliché due to over-familiarity, and the significance of such images was lost. Boulding considered that for mankind to learn that the Earth is finite was 'the most important single fact of our day'. This he wrote in *Earth as a Spaceship*, which was published as a longer article in 1966 [11]. Boulding mused that the human species could have considered the resources of the Earth as effectively unlimited, had it remained 'small in numbers and limited in technology' but, as we know, this has not been the case. Humanity now needs and uses the whole Earth, and its limits to sustain us are becoming something of which we are all aware.

Yet mankind continues to treat the Earth as we always have: as an open infinite system[7] without limits or boundaries. However, the Earth is a 'closed system' in which typically no mass or energy can enter or pass through the system boundaries. The Earth is, strictly speaking, not a completely closed system in that energy from the sun reaches the Earth; some is used and may be transformed in wavelength and then returned into space. This is important in that, however frugal we must become with respect to matter, we will always have solar energy.

It is becoming increasingly apparent that we are reaching the limits of the Earth to accept our polluting wastes and byproducts of economic metabolism. Wastes that do not pass into the atmosphere

[7] There is one other feature of a closed system, which always increases in entropy or disorder. Human activities increase the entropy of the Earth, although the increase in knowledge or information represents an increase in order.

gravitate towards the oceans; we find that large areas of the planet are becoming unfit for life. These include Lake Erie, the northern Adriatic and the Baltic Sea; such water masses have long been accepting pollution loads too great to be assimilated by those systems.

Animals and plants have long 'known' that the Earth is finite, although it would be more accurate to say that they recognise the limits of their habitats as finite. They have always had the power to multiply exponentially, and whether the life form is *E. coli* or an elephant, there has been time enough for all species to reach the limits of their habitats (Chapter 3). Life has evolved ways of circumventing problems that threaten the survival of a species. Maia provides an interpretation. Following a phase of exponential growth in a finite world, life reaches a self-imposed limit to further growth in population numbers; there can only be a turnover of individuals such that those that die are replaced by the newborn. Birth rate and death rate are in a dynamic and varying equilibrium, and there is zero population growth imposed by a ceiling on total numbers. Such systems have evolved in animals, with number or density limited by Maian control mechanisms (Chapter 8). Populations of organisms generally limit their numbers in relation to the carrying capacity of their habitat, with self-limiting properties of a regulator represented here by the logistic control mechanism.

A BIOCENTRIC VIEW OF HUMANITY

It is relevant to ask to what extent our biological past informs the current predicament of mankind, given that overpopulation is becoming more acute and the byproducts of economic growth are causing global changes that are becoming irreversible. The problems we now confront are new to our species, which has never before encountered the limits of the Earth. Jared Diamond's analysis of extinction and survival in our species, under the title *Collapse* is instructive [12], as so many societies have created the circumstances for their demise by their own hand, and did so little to prevent it.

This prompts the broader question, which is to ask how well the human species compares with animals in having life strategies that better equip them for survival. A biocentric approach, allows as ask whether our species has anything to learn from animals in the way strategic problems have been dealt with. These relate to population size limitation, control of growth rates, interactions with own kind, use of resources, and behaviour in conflict situations (Table 14.2). The aim is to see whether, when confronted with the likelihood of unique events of overpopulation and the overexploitation of environmental goods and services, there is something to be learned from the way in which animals evolved.

With this in mind, a comparison is made of the strategies evolved by natural selection, and those that humanity has adopted for itself. The first two comparisons will draw on the Maia hypothesis, and so are already familiar (Table 14.2).

1. It has been shown in developing the Maia hypothesis that the key to biological control and the control of human populations is that they require limitation in relation to the carrying capacity of the habitat. A cybernetic mechanism is required to implement some optimal limit in relation to the carrying capacity. In animals once this point is reached, the limit to further growth constrains absolute numbers, such that birth and death rates become equal. Boulding was clear about the idea of autonomous cybernetic systems providing the means of maintaining a stable population, but in life there is no evidence of an intrinsic mechanism, as hyperbolic growth continues (Figure 13.1). Control models have been applied, or even imposed, as population policies in some countries (Chapter 8). Humanity has no intrinsic self-regulatory mechanism for limiting population numbers, as the human population has grown hyperbolically, with little indication of self-limitation over the last 600 years (Chapter 8). Clearly no innate control mechanisms exist that could work on a global scale, although there may be mechanisms that operate in a contained habitat. It is clear that while animals effectively limit their own numbers, mankind has not yet mastered the problem.

   There is a predisposition of mankind to short-termism that relates to the foreseeable horizons of the individual lifetime, or even human history, rather than to the much longer term of evolutionary processes of which the human species is a product. Super-exponential population growth has taken thousands of years to reach the number of humans who now populate the Earth (Figure 13.1). The overpopulation problem is unique to us, and we have no experience to help us deal with it, other than our own intelligence.
2. The second comparison relates to the growth rate of organisms, for which Maia provides rate control over biosynthetic

metabolism. A feedback mechanism controls the rate of growth, partly because processes require rate control. An advantage is that if each process involved in biosynthesis is set to some fixed goal, other internal processes can relate to it to deliver coupled control, coordinated across numerous processes. This facilitates the evolution of networks of metabolic processes. For humanity, the economy provides the metabolism of society, which is controlled in part by interest rates; although the overall aim is to maintain the economy at some steady rate, as in biological systems, that constant rate inserts an exponential element into the growth of the global economy (Figure 14.2c). Boulding saw the need for automated control to prevent escalation and inflation; although such controls exist in the hands of central banks, growth nevertheless tends to be maximised.

3. The third comparison relates to the interactions between individuals in the animal world which, apart from predator/prey relationships, are predominantly neutral or cooperative. Competition is costly and there are adaptations to minimise or avoid competition, especially within the same species. Adaptations have actually evolved to prevent the waste of competition, and the fittest are those that choose not to compete. As we have seen, cooperation is intrinsically a more efficient strategy, and occurs generally in animals. Mankind seems to be addicted to competition in the marketplace although it is intrinsically inefficient. As the pressure grows to become more efficient in the use of energy and materials, the very process of competing will become recognised as wasteful, not to mention the needless and wasteful duplication of products.
4. In the fourth example of analogous strategies of humans and animals, it is clear that the flow of materials in nature and biological systems is cyclical. One animal's waste is another's resource. Each cycle is in balance and matter of different kinds goes round and round, whether it is the carbon cycle, the nitrogen cycle or the water cycle. In these great cycles, the flux of materials in one direction is matched by their flux in another. The same matter is used and re-used; nothing is wasted and all is recycled. Efficiency is selected for in nature, and those that prevail are those that have adapted to derive more from less. Human systems, by contrast, tend to be linear. We extract a resource from the environment in some raw form, purify the metal or mineral that is needed, manufacture some product, use

it until it is worn out, and then discard it as waste. The level of recycling of the resources we mobilise is a relatively small proportion of the total, but it is increasing. One problem is that manufactured products are not made to be recycled, so the cost of recycling is too great to make it worthwhile from a financial point of view. Recycling costs need to be internalised to become part of the purchase price as an incentive for manufacturers to make recycling less costly. Paul Hawken captured the nub of this problem when he wrote: 'Cyclical biological activity can be the only source of life, because all linear system are, by function and definition, limited and short-lived' [10]. Energy is used in a more profligate way than anything else, but because its use is invisible and once only, such waste is hardly noticed by most people.

5. The fifth strategy, related to the resolution of discord, excludes predator–prey relationships, as these are about nutrition rather than conflict; the predator has evolved to eat the prey. In nature, conflicts between individuals of the same kind tend to be ritualised; a mere skirmish may be all that is required to remind the inferior individual of its place in the hierarchy of dominance. Contests over mates are trials of strength and fitness, and are rarely fatal. Between species that are not predators or prey, peaceful coexistence is likely, and the shared knowledge of strengths and weaknesses enables coexistence without discord. Conflict in nature is costly, so retreat is often preferable to risking all.

The human race would also wish to minimise the cost of war and loss of life involved in conflict, so the progression to war is a sequence of steps that may involve non-cooperation, a trade embargo and cessation of diplomatic relations, before ending up in hostile conflict and war. Nevertheless, a global perspective shows that armed conflicts and wars increased in frequency from the post-war years to the early 1990s. It was Aldous Huxley who long ago realised that the greatest threat to world peace was nationalism. Boulding pointed out that 'there must be a machinery for controlling conflict processes', as conflict is incompatible with a finite Earth [11]. He saw the need for mankind to develop symbiotic relationships in the form of closed cycles, 'with other elements and populations of the world ecological systems'. To survive on planet Earth, loyalty and mutualism must have a higher priority, which requires the recognition that we belong to a single species. Warfare is the most wasteful and least rewarding of all human activities.

Table 14.2. *A comparison of strategies by which key processes are managed in the animal world and by human society.*

	Animal world		Humanity	
	Process	Natural selection	Process	Society
1. Population	Population growth	Self-limitation	Population growth	Unlimited
2. Growth	Growth rate	Optimal control	Economic growth	Maximise
3. Interactions	Intraspecific	Cooperative	Economic	Competitive
4. Resources	Utilisation	Cyclical	Utilisation	Mostly linear
5. Conflict	Discord	Ritualised behaviour	Discord	Warfare

It is worth returning to the question of how much growth is enough, whether in terms of the human population or the size of the global economy. In 1980 Kirkpatrick Sale wrote with respect to other matters, ‘Because we do not know how much is enough, we assume that bigger is better’. No person, nation or international organisation is yet able to say how much growth is enough, or what the carrying capacity of the Earth is for our numbers or the economy. While there is no consensus, and we remain addicted to economic growth and the wealth it creates for us all, why should there be change? In the meantime, globally the great engine of growth through positive feedback drives expansion in population and the economy.

There is an overall conclusion to be drawn from this comparison of biological and human strategies (Table 14.2). In each case biological evolution has created sustainable strategies for life. In one sense this is to be expected, because life on Earth emerged 3.5 billion years ago, and any life strategy that has not been successful in contributing to survivorship has inevitably been lost. The biological strategies we see now are the distillate of billions of years of evolution, and are the proven attributes of animals as survival machines. It is remarkable, therefore, that humanity has repeatedly chosen the opposite strategies to those arrived at by natural selection. From the perspective of the Maia hypothesis, unlimited population growth and the maximisation of economic growth are not sustainable survival strategies. Evolution has come up with strategies that are sustainable, while those chosen by mankind are not. Each biological strategy is fitted and honed for survivorship. Conversely none of the strategies adopted by mankind are optimised for survival, and are all maladaptive for life on a finite planet. The vast majority of all the animals that have ever lived are extinct, so those species that remain are great survivors.

Perhaps Boulding's idea of Earth as a spaceship was too far-fetched even for the 1960s, but now, as the manifold and seemingly unrelated effects of global climate change take hold, there is a greater willingness to accept the Earth as finite. The Intergovernmental Panel on Climate Change has assembled an overwhelming body of evidence, and a unique scientific consensus, so the vulnerability of a finite Earth is becoming clear for all to see. Boulding concluded that 'Man is finally going to have to face the fact that he is a biological system living in an ecological system'.

A STEADY STATE ECONOMY

Mankind is committed and addicted to economic growth because it creates wealth, but in a finite system only a steady state economy is sustainable and can last indefinitely. A finite and sustainable world requires a steady state economy and not a growth economy, which is the only plausible solution in the long term. The antithesis of a steady state economy is the economists' model of modern man, used to illustrate how we behave in the marketplace. This is *Homo economicus*: the 'rational maximiser' of marketing models. *H. economicus* is assumed to be selfish and a natural competitor, who will hunt for a bargain, and who has the expectation of buying more for less. This is the marketer's image of their customer and informs marketing models and advertising. Given that there is so much advertising pressure, it sometimes seems that the psychology of advertising has already created the kind of person they would wish us to be, whose hobby is shopping and for whom a bargain is fulfilment. Modern man is ill-equipped to survive on a finite planet where we can only expect to satisfy needs, and not the appetite for boundless acquisitiveness that advertisers would wish to create. Consumerism of this kind is not compatible with an Earth where our ecological footprint already exceeds that which the planet can sustain.

Herman Daly, as an environmental economist at the World Bank wishes that humanity would grasp Boulding's message that the Earth is finite, and would treat it as such. Instead we continue to do what we always have done, which is to treat the Earth as though its resources are boundless. We are encouraged in this by economists and politicians who wish the economy to grow, apparently blind to the fact that we are currently exceeding the capacity of the Earth to support humanity. Daly tells us that 'As long as our economic system is based on chasing economic growth above all else, we are heading for environmental and economic disaster' [13].

Herman Daly advocates a transition from the growth economy of 200 years' standing, towards a system that is in every respect a steady state economy that relates to the capacity of the planet to contain it. With respect to the way it applies to human numbers, we have already considered essentially the same idea. There needs to be a population that is optimised in relation to the carrying capacity of the planet (Chapter 13). That is to say, there would be zero population growth at a level that would not exceed the sustainable maximum carrying capacity of the Earth. This kind of thinking is already well established, but here Daly takes the natural step towards a steady state economy.

A stable economy would incorporate the same optimising principle applied to each of the sectors of economic activity, from electronics to construction. In each case the scale of that sector would relate not so much to the demand for the products, but would be constrained by the impact of manufacture and its waste products on the environment. Environmental costs would be internalised as part of the total cost of the product; such costs may be invisible to most, yet are crucial in maintaining ecological services. The impact of local timber production, for example, may well be low, but the impact of the electronics industry would be high due to the various metals used in its components, and the limited assimilative capacity of the environment for their toxicity. The cost of tropical fruit and vegetables would include a cost related to travel miles to contribute to the reparative costs of the effects of climate change. All manufactured goods would be long-lasting, with all components replaceable, to maximise durability. Built-in obsolescence in products or components would be outlawed. In an economy of this kind, the environmental impact of every product would be paid for on purchase. The plus side is that, although more expensive, goods would last much longer and all raw materials would be recycled.

Each industry would be of a stable size and limited to the capacity of the environment to assimilate its wastes and effluents. Such an economy would work to replace existing machinery as it wears out, and growth would be allowed to extend only as far as its ecological footprint could be reduced. New and more effective technologies for dealing with its effluents would allow some growth for each sector.

Major household items would be replaced at the end of their life, and the raw materials recycled, as a requirement for their replacement. They would be designed to allow 100% recycling.

As we approach the prospect of economic growth destroying the capacity of the environmental sinks to accept more wastes and effluents, the idea of a transition to a steady state economy will become increasingly attractive. The key aim of a steady state system is that it would be sustainable indefinitely. There would be an awareness of the environmental impact of every product, which would be reflected in the environmental charge in the purchase price. This would create an economy where everyone will have a strong interest in minimising the environmental impact of their products. It would apply particularly to those natural waters receiving noxious effluents, in which living processes must be protected to assimilate and recycle the nutrients in sewage wastes.

Herman Daly identifies three objectives for a transformation to a steady state economy that it will be sustainable over the long term. I quote his plan verbatim [14].

> The economy must be transformed so that it can be sustained over the long run. It must follow three precepts:
>
> 1. Limit use of all resources to rates that ultimately result in levels of waste that can be absorbed by the ecosystem.
> 2. Exploit renewable resources at rates that do not exceed the ability of the ecosystem to regenerate the resources.
> 3. Deplete non-renewable resources at rates that, as far as possible, do not exceed the rate of development of renewable substitutes.

These principles link economic activity, in every sense, to the capacity of the environment to accept or deal with the consequences, and thus will make possible our indefinite existence on a finite planet. Essentially a steady state economy mimics what nature does and what Maia regulates. In every respect, the rates of human activities are determined by the rates at which the environment can respond and deal with the consequences. In principle this is no different from the Maian concept of organisms that control their own numbers in relation to the carrying capacity of their habitat. Their growth and numbers are linked to the limits of the system upon which they depend for life. In principle, their life on Earth is indefinite. In this way they never over-exploit the global commons upon which they depend. I have added little here to what Herman Daly more recently has advocated, informed by Kenneth Boulding's thoughts of 40 years earlier, in attempting to visualise a sustainable future for mankind.

The great engines of growth on the Earth are population growth and economic growth. They incorporate no intrinsic limits to growth, so they grow exponentially. This results in an accelerating convergence of human population and economic growth with the limits of the Earth. There is no place for the exponential increase of humanity on a finite planet, and the only feasible option is a transition to a steady state economy. This would incorporate a limit to the demand imposed on ecological services set by their capacity to regenerate, so avoiding overload and degradation.

THE MAIAN INTERPRETATION OF A SUSTAINABLE EARTH

Evolution has vested much in cybernetic systems, and this interpretation involving control and limitation of growth is called the Maia hypothesis, in part because it is not wholly reality but has become a hypothesis of how reality appears from the perspective of control theory, as the ideas are applied more widely. Maia provides a clear idea of the kinds of control mechanisms that control rate processes (Chapter 5) and limit population numbers (Chapter 8). The assembly of such control modules within a hierarchy suggests a hypothetical organisation for the kind of multi-level control system in organisms that have a number of levels of organisation (Chapter 9).

There is no better example of controlled sustainability than the biosphere, which has evolved steadily over 3.5 billion years in harmony with its environment. If humankind, a relatively new species only 3 million years old, requires a model system by which to guarantee survivorship and stability, we need to consider how the way in which biological systems have survived. The only proven model for survival is to be found in biology, and we can do no better than to emulate biological systems if we wish to do the same. The Maian model developed in Chapter 9 shows that animal populations control their own rate of reproduction internally, and limit the numbers that result externally. A summary of the comparison is set out in Table 14.2 and suggests that the strategies of humanity are opposed to those that organisms have evolved to live on a finite Earth for billions of years.

Boulding saw the need for control as an essential requirement for life on spaceship Earth. He wrote [11]:

> There must be, however, cybernetic or homeostatic mechanisms for preventing the overall variables of the social system from going beyond a certain range. There must, for instance, be machinery for controlling the total numbers of the population; there must be machinery for controlling conflict processes and for preventing perverse social dynamic processes of escalation and inflation. One of the major problems of social science is how to devise institutions which will combine this overall homeostatic control with individual freedom and mobility.

A steady state economy requires cybernetic mechanisms to guide the optimisation of human requirements in relation to environmental services. This is a key principle required to make a steady state economy possible. Maia shows that nature established this principle in the self-regulation of population numbers in relation to the carrying capacity of habitats.

A constant growth rate with exponential growth was a primitive condition and is unsustainable for all but autotrophs, but later in evolution the means of limiting further growth was superimposed in a dual loop mechanism, probably not dissimilar to the logistic control mechanism.

Exponential growth in economic systems has allowed the financial system to grow freely under *laissez faire* policies, allowing wealth to generate more wealth. But biological systems seemingly recognised that more was not better in order to improve fitness, and organisms evolved to limit their own population growth as the overexploitation of shared resources became maladaptive. No one ever considered that for survivorship on a finite planet, it would be prudent not to grow without limit. No economic systems have a built-in limit to growth as biological systems do, so the likelihood is that the limits of the Earth will impose constraints in due course.

Economists have not seen fit to consider the likely need for a steady state economy, and while the system creates wealth so effectively, they have little reason to point out that economic activity is destroying the environment and overloading the Gaian mechanisms that maintain global stability. A convergence of economic and environmental values is required – a single value system that is orientated to sustainability and survivorship. Without this, we are viewing the same problem from different premises, and will not succeed.

For a long time mankind has exploited the Earth, and has been content to trade-off extracted resources with environmental destruction. Human progress is in some senses was epitomised by the growth of economic wealth at the expense of the environment (Figure 14.4). Now the growing economy makes ever-increasing demands upon the

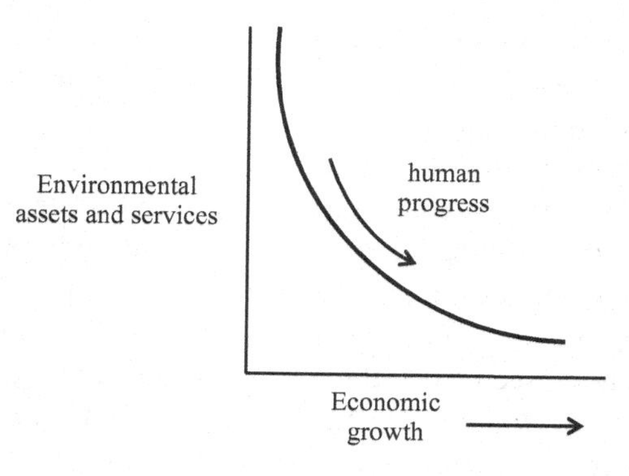

Figure 14.4. Graph indicating the way in which, for most of human history, economic growth has been at the expense of the Earth's environment.

environment, from resource extraction to waste assimilation, creating incompatibility of the economy with the environment. This has been appreciated by many, as the scale of the economy presses harder upon the limits of the global ecosystem. As economic growth accelerates, so does the depletion of the ecological capital.

As the rate at which biodiversity is being lost for all time, there is now an imperative to redress the balance before the damage is too great to regenerate. The direction of human priorities needs to swing towards the importance of survival and environmental regeneration, and away from wealth creation and rampant consumption. In the 1960s, Boulding was appalled by the direction in which society was headed, and wrote, 'that the waste and profligacy of our day will seem too horrible that our descendents will hardly be able to bear to think about us'.

In his study of economic systems, Fritz Schumacher realised the importance of what had evolved long ago, when he observed that 'Nature always know where to stop'. Here the controls and limits to growth that nature discovered for itself are called Maia. For our future, we would do better to emulate what evolution has given us, than to draw on our historical past. The human race has not been far-sighted enough to see the danger of growth processes, either of its own populations or those of its economy.

SUMMARY

The global commons of the Earth, on land, sea and in the air, are being overexploited by mankind, both for their assets and their services. Maia indicates that animal populations avoid overexploiting their 'commons' by limiting numbers in relation to their habitat's capacity to sustain them. The metabolism of society is the economy, which grows exponentially like biological growth, and for the same reasons. So it is that populations and capital are the engines of growth on Earth. But while biological growth is self-limiting, that of the economy is not. The Earth is finite, and the metabolic wastes from economic activity are already stretching the limits of the planet's capacity to assimilate them. This can clearly be seen from the effects of greenhouse gases that are destabilising the climate system. A biocentric view of evolutionary and human strategies shows that we are ill-equipped for survival, compared with the animals that evolved before us. Population growth is unlimited and economic growth is maximised; we have a predominantly linear use of resources and we are committed to wasteful competition in the marketplace. The economy has grown for too long at the expense of the Earth's assets and services, on which we ultimately depend. Biology shows that survival on a finite planet demands that we limit growth processes in relation to the capacity of the Earth to assimilate the byproducts of economic activity. The biological examples once again show that, for mankind, only a steady state system is sustainable indefinitely as its growth is limited to what the Earth can sustain.

15

The Maia hypothesis and anagenesis

Evolutionary progress: increased control over and independence of the environment.

Julian Huxley

Typical of anagenesis: increased independence of the environment and increasing command of environmental factors (progression of autonomy).

Bernhard Rensch

Teleology is a lady without whom no biologist can live. Yet he is ashamed to show herself in public with her.

Ernst Brüke

THE MAIA HYPOTHESIS

Initially the name Maia was given to a cybernetic approach to understanding the control of growth rate in cultured hydroids (Chapter 5). It was reasoned that this may indicate the underlying control of biosynthesis (Chapter 6). The aim was to find a technique to derive the output of the growth control mechanism from traditional cumulative data; and the idea was first applied to the growth of colonies (*Laomedea*) and, later, populations (*Hydra*). A feature of the approach, well known to cyberneticists, is to access the output of control mechanisms using perturbation experiments. The data are characteristically oscillatory, and exhibit cybernetic features such as 'runaway' and relaxation stimulation. The evidence was corroborated with data from a simulation model incorporating feedback, which reproduced the essential features of the hydroid output. This method was later applied to growth output in yeast cell suspensions with similar results (Chapter 12), which showed that the method was portable between taxa. There was reason to believe that the control requirements of other biological replicators might be similar, from the cell to more complex organisms. These experiments provided the inner loop of what proved to be a dual loop feedback mechanism. The products of rate-controlled growth produce a mass of cells or replicators, increasing exponentially due to a positive feedback of individuals to the replicating mass. So the population of replicators grows at an accelerating rate, but eventually such growth becomes maladaptive and requires limitation if the organism or population is to survive. Encompassing the inner loop is an outer loop, which has the function of limiting the further exponential increase in numbers, such that the total is restricted to a genetically prescribed number or mass. Limitation is superimposed upon specific rate control, such that the rate slows and the number stabilises at the goal density or number.

The features of a dual loop control module appear to satisfy the minimal requirements of any replicator, such as an organelle, cell or zooid. But we should reconsider J.B.S. Haldane's idea that endocrines evolved from exocrines, and by implication, population control mechanisms gave rise to growth control mechanisms by endosymbiosis, as internal control mechanisms are assumed to have evolved from external control mechanisms. The simplest single-celled organisms, such as protozoans, have population number or density mechanisms, which limit numbers in relation to the constraints of the habitat.

It was on reading the classical paper of the ecologist G.E. Hutchinson, dating back to the birth of cybernetics, that it became clear that the logistic equation could be interpreted as a control mechanism (Chapter 8). It has proved suitable for describing sigmoid curves giving the growth of populations of organisms from protozoan to human, and from microbial pathogens to neoplastic cells. It appears that the mathematics of biological growth is a stable feature of life.

Exponential growth has great value to biological systems, in the sense that cellular populations of metazoans grow rapidly through development; or populations grow rapidly to fill a vacant niche. In the marine environment, phytoplankton cells multiply rapidly in spring and early summer when conditions become optimal (Chapter 2). These are typically autotrophs that reproduce exponentially, doubling in number 2.8 times each day, to rapidly create huge masses, taking advantage of the seasonal availability of nutrients. It is the degradation of these immense blooms that creates the nutrients that, in being recycled, become available for new growth annually. When autotrophs strip the surface layers of the ocean bare of nutrients to reach astonishing densities of cells, and then crash as nutrients run out, it becomes clear that they have no means of self-limitation. Autotrophs are only able to control their growth rates, and do not limit their numbers. They have a control mechanism with an inner, rate-sensitive loop, which is the more primitive condition.

Organotrophs also have the ability to grow exponentially, but they must not continue in this way. They are consumers and feed mostly on living or dead organisms, and if they were to consume all the available food, they would extinguish the species on which they depend. Self-limitation thus became an essential feature of their lifestyle to avoid driving the organisms on which they depend to extinction. Restraint was essential to ensure that a viable population of their food organisms survived. It can be reasoned that all consumers must behave in this way if they are to avoid causing their own demise by over-consumption of living prey. The argument also implies that self-limitation is essential for the working of food chains and ecosystems which, in effect, pass the same nutrients from one trophic level to the next.

The evolutionary solution was simple, and a second loop encompassed the first, with a goal of limiting numbers of organisms, adding state limitation to rate control. Limitation operates through limiting the growth rate controlled by the inner loop. In this way the two loops are interdependent, but there are two goals. The goal for the inner loop is the specific growth rate for the population, known by ecologists as the 'intrinsic rate of natural increase' or '*r*', which is characteristic of each species. The goal for the outer loop is the maximum sustainable carrying capacity for the habitat, or '*K*'. These terms are well known to ecologists, but not as goal settings for a control mechanism, so re-interpretation of their meaning is required.

Such a mechanism, it seems, would provide the basic minimal regulator for any replicator, from cells (Chapter 12) to populations of individuals (Chapter 8), and so apply to both physiological and ecological contexts. As replicators are also the module-forming agents at each level of organisation, the logistic control mechanism can be adopted as a first approximation of a hypothetical control module for each level of organisation (Chapter 9). Each replicator in a hierarchy is enclosed by those that are larger, and encloses those that are smaller. Control at each level is linked, as no replicators are allowed to replicate autonomously, as this would threaten the stability of the whole hierarchy. Replicators at each level have local rate control that relates to their internal metabolism, but limitation relates to a higher level and, within a hierarchy, is subject to control from above.

So a homeodynamic interpretation of the control hierarchy suggests an adaptive role for hierarchy, independently of its organisational role. By such integration, new properties of the hierarchy emerge, which are mainly unseen adaptations of higher organisms. With respect to hierarchical control, this suggests a coupled and coordinated cooperative that operates for the benefit of the whole. In single-celled organisms such as yeast, it is known that rate-control mechanisms counter the effects of environmental inhibitors. But when such organisms act as modules occupying a level within a hierarchy, control mechanisms must work in concert, each playing their part in minimising the effects of perturbation. The combined resistance of the hierarchy to instability becomes more than the sum of the resistance of its separate control modules, providing a 'stratified stability', as Jacob Bronowski called it.[1] The consequence is that with more levels of organisation and greater capacity to neutralise perturbation, organisms become increasingly complex with each new level [1], and adapted by each level to allow the organism to be more independent of environmental change. This evolutionary adaptation, already alluded to, will be considered in greater depth towards the end of the chapter.

Like other homeodynamic mechanisms, Maia is adaptive such that the feedback mechanism neutralises perturbation to the controlled process. Growth is optimised at a rate that leaves the system with a residual capacity to neutralise negative or positive perturbations of the growth process. This is achieved by stimulatory or inhibitory responses that neutralise each kind of perturbation. It follows that normal growth can never be maximal, as any process operating at its maximum rate would have no residual capacity to counteract perturbation. Growth control mechanisms possess a measurable capacity to neutralise toxic inhibition, which can be defined in terms of the concentration of an inhibitor that can be withstood before overload occurs (Chapter 7). It is necessary that feedback mechanisms have goal settings at rates which are optimal, allowing for a residual capacity to perform its regulatory role. During relaxation stimulation, growth

[1] Bronowski was interested in the way that, during evolution, new levels of organisation were superimposed upon lower levels, adding to the complexity of organisms. This he saw as a central process in evolution; that is, the addition of new levels, each of which is self-stabilising, contributing to what he termed a 'stratified stability' in organisms with multiple levels of organisation.

rates have been shown to fluctuate following perturbation, reaching rates for brief periods that are two to four times higher than their goal settings. It is often assumed that maximal growth indicates optimal conditions for an organism, but this cannot be true. Optimal control requires that whatever the goal growth rate, there must remain a capacity to inhibit or stimulate growth in order to counter stimulatory or inhibitory deviations from the goal setting.

The growth control mechanism is also adaptive to perturbation in the individual, so that, with sustained exposure, its capacity to resist further perturbation increases. We see this in the interpretation of acquired tolerance of organisms exposed to low levels of toxic agents. Adaptation comes about due to adjustments to the goal settings of the control mechanism (Chapter 7). Adjustments to growth rate are made counter to the direction of perturbation; so a stimulatory adjustment to a higher level in response to toxic inhibition has the effect of conferring greater resistance to inhibition.

When such adaptation is not followed by further exposure to toxic inhibition, the effect is the stimulation of growth. This is the phenomenon of hormesis that can be due to any kind of toxic inhibitor (Chapter 10). Maian adaptation of growth rate confers greater resistance to later inhibition such that, when this later inhibition does not occur, the adjustment to rate causes the stimulation of growth, and hormesis. As a result, cumulative stimulation of growth can exceed the normal colony size of hydroids by as much as 50%. Such observations confirm that growth under constant conditions is submaximal, but low levels of any inhibitor to which the control mechanism has adapted are capable of resulting in the stimulation of growth (Chapter 11). Hormesis is an epiphenomenon, and does not appear to have any adaptive value in itself, although limited growth stimulation can hardly be a disadvantage.

The stimulation of growth is also induced by catch-up growth. If growth is suppressed by an inhibitor, the growth control mechanism works to neutralise its effect by means of a stimulatory counter-response, and then restores the equilibrium. This represents a balance of opposing forces, such that the sudden removal of the inhibitor causes relaxation stimulation. It spectacularly reveals the counter-response that had been necessary to achieve an equilibrium, as a stimulatory peak which lasts for as long as the time lag in the control mechanism. It seems probable that this interpretation accounts for catch-up growth, as it is known in humans and other mammals (Chapter 11). It seems likely that the taxonomic generality of the interpretation of hormesis and catch-up growth, as put forward here, also implies the taxonomic generality of their Maian interpretation.

The logistic control module provides a minimal mechanism for the control requirements of any biological replicators, from molecular biosynthesis to populations of organisms, and is put forward as a hypothetical control module at each level of biological organisation in a control hierarchy (Chapter 8). Each of the levels adopted here[2] is represented by a replicator. The addition of new levels of organisation is the principal means by which organisms have increased in complexity during evolution, which has slowed gradually over the last 400 million years. As Lyapunov intimated, a hierarchy of organisation is analogous to a hierarchy of control mechanisms, represented here as a nested system of control loops (Figure 15.1). Each level represents a single control module, with positive and negative feedback loops of rate and state control. The emphasis shifts to process and function, rather than the organisational structure, as structure alone may not be enough to account for the evolution of hierarchy. Homeodynamic mechanisms resist perturbation, and within a hierarchy of such mechanisms, any disturbance tends to cascade from lower to higher levels of organisation. Sequential responses at each level neutralise such disturbances progressively, and help to prevent their impact from reaching higher levels of organisation. What is relevant here is that, for an organism with multiple levels of organisation, the resistance of a control module at one level is supplemented by other control modules at higher levels. This arrangement has the effect of increasing homeodynamic resistance of the organism to perturbation with each additional level. The system is adaptive in organisms with multiple levels of organisation in providing a stratified stability that resists perturbation more effectively (Chapter 9).

The failure of control mechanisms tells much about their normal behaviour. Cancer occurs in cell populations where the failure to limit the growth and replication of one or a few cells results in neoplastic growth and the creation of tumours. Cells transformed in this way replicate without limit, and then spread to create secondary tumours, often leading to the demise of the organism (Chapter 12). In principle, a reversion to autonomy of any replicator may threaten the whole organism. Failure is in the outer loop of transformed cells, which normally limits cell replication. So cancer is the result of removing inhibitory limitation, rather than the stimulation of cell growth.

[2] It is questionable whether tissues and organs should be considered as a level of organisation, as they do not replicate like the other modules.

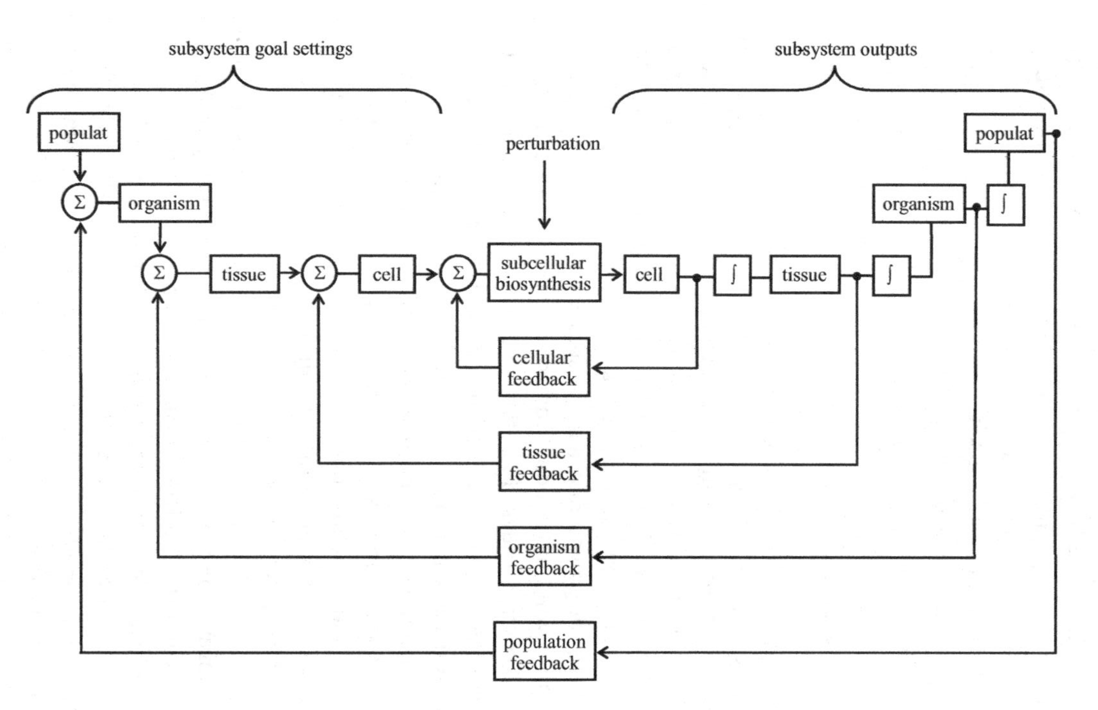

Figure 15.1. The cybernetic diagram of the biosphere as a series of nested feedback mechanisms.

Uncontrolled human population increase is a global problem, which is similarly due in part to the absence of the innate self-limitation typical of other mammals, lost when automatic regulation became superfluous during the evolution of our hunter-gatherer ancestors. From then on limitation has depended upon conscious control rather than innate mechanisms, but with the onset of civilisation, infanticide has incurred moral disapproval. Since the 1400s, world population has increased hyperbolically, and ultimately reached its peak in the doubling of population numbers with each generation during the 1960s. The consequence has been that the human population has now exceeded the carrying capacity of the Earth, due to man's failure to recognise the nature of the problem and to act effectively to curb fertility (Chapter 13).

MAIA AS A META-HYPOTHESIS

The reader will by now be aware that the Maia hypothesis has led to unexpected places far beyond the initial idea. From a simple rate-sensitive control mechanism revealed by perturbation experiments, Maia has become a hypothetical system that is capable of controlling biological growth at all levels of biological organisation, from molecules to mankind. From autotrophs that only possess an ability to control their specific growth rate, a major step was made by the organotrophs with the evolution of a dual control mechanism providing the means of rate-control and self-limitation. Exponential increase represents a threat to all biological systems, as seen in neoplastic growth and human overpopulation. So the evolution of self-limitation was crucial, not simply in controlling growth in terms of its products, but in containing the positive feedback of growth that leads to exponential increase and overexploitation of resources. The threat posed is apparent in the consequences of its failure.

The logistic control mechanism provides a simple system in which the problem of rate control is wedded to state control in a minimal mechanism. However complex specific rate control may become in particular contexts, the logistic control mechanism is likely to represent the nucleus of a control mechanism in all replicators. This has made it possible to consider growth control within a hierarchy, incorporating various levels of biological organisation, adapting the logistic control mechanism as the central control module within each replicator. One can expect the coupling of such modules where complexity is found, creating emergent properties of the hierarchy of biological organisation.

Within this hierarchy, the physiology of growth control comes together with the ecology of population control. At the outset I was happy to embrace the ambiguity of coloniality (see Frontispiece), where its members are zooids that are neither part of an organism or a population, but are both. And in considering the organisational hierarchy, with control at each level, it must be recognised that the biosphere itself is self-controlled. Its latent potential to grow explosively (Chapter 2) is only adaptive as long as it is also self-limiting. Biology could not have evolved as the harmonious profusion of life forms that populate natural habitats without self-limitation. And limitation is not arbitrary, in that species population goals are maximal populations, but are limited in relation to the density of numbers that the habitat can sustain. So it is that life has evolved and existed for billions of years, as self-limitation curbs population growth in relation to resources.

Our own species has failed to recognise this for too long. For centuries the global human population has continued to grow hyperbolically without limit, having apparently lost the innate and inherited mechanism of self-limitation found in other animals. To survive overpopulation, we need to substitute the missing mechanism with another, as von Foerster proposed (Chapter 13). What remains uncertain in this biocentric approach is the goal setting population to which self-limitation must work. That is the carrying capacity of the Earth for the human species, which, according to the recent ecological footprint analyses by WWF, has already been exceeded (Chapter 14).

Just as great a threat, which cannot be separated from overpopulation, is the growth of the global economic system that most would wish to maximise, without considering that its growth must ultimately become limited by the constraints of our planet. Given the scale of economic degradation, and the destabilisation of the global climate system by economic activity, the capacity of the Earth to accommodate its consequences has already been exceeded.

TELEOLOGY AND ANAGENESIS

There is an evolutionary trend that has been neglected for half a century, which relates explicitly to biological cybernetics. It identifies the evolutionary trend that is epitomised by the role of homeodynamic mechanisms, and biological cybernetics in general. The theologian William Paley (1743–1805) provided an explanation for the creation of life that in its time was utterly convincing. To his mind, everything was designed to meet its purpose; he was the champion of teleology. His conviction is reflected in his resounding phrases: 'The marks of design are too strong to be gotten over. Design must have a designer. That designer must have been a person. That person is God'. In 1830, when Darwin was at Cambridge preparing for ordination in the Church of England, Paley was required reading. His aim in *Natural Theology* was to prove that a study of natural history confirmed that God was the divine designer. The well-educated accepted his reasoning as an unassailable link between nature and its Creator. Darwin was examined on his knowledge of Paley for his 'little go';[3] he later wrote, 'I do not think I hardly admired a book more', and later admitted that he knew Paley's arguments by heart. Yet nearly 30 years later he was to create the argument that would overturn Paley's crystal logic. Even then, it must have seemed to Darwin like committing a heresy.

Darwin was keen to link his own concept of evolution by natural selection to the idea of animal-breeding by man so, to help carry the idea, he alluded to agricultural livestock. 'No one objects to agriculturalists speaking of the potent effects of man's selection,' he wrote. But it was a mistake for Darwin to coin the term 'natural selection', and he knew it, because implicit is the idea that in nature a 'selector' is guiding the process. Cast in a positive sense, Darwin feared that 'natural selection' implied an 'active power or Deity', when he knew that natural selection was a process that lacked foresight. Darwin had to concede that 'In the literal sense of the word, no doubt, natural selection is a false term'. But its use continues to this day, and causes some confusion.

Herbert Spencer (1820–1903) came up with the phrase 'The survival of the fittest', which captured Darwin's idea so neatly. He also appreciated that organisms became increasingly complex during evolution, and recognised this as 'progress'. Spencer, with his penetrating insight, realised that, with increasing complexity, organisms also became increasingly independent of their environment. It is this idea that I now wish to examine in these final paragraphs.

[3] 'Little go': the name previously given to the entrance examination for undergraduates at Cambridge.

Julian Huxley explored the idea of evolutionary progress in *Evolution: The Modern Synthesis*, published in 1942 [2]. He associated progress with increasing complexity, although at the time there was no satisfactory way of measuring biological complexity. Huxley equated progress with 'improvement in efficiency of adaptation' and gave temperature regulation as an example. Evolutionary progress, he wrote, was 'a raising of the upper level of biological efficiency, this being defined as increased control over and independence of the environment'. He went further, linking progress with a cybernetic interpretation, alluding to homeostasis and the 'harmony of internal adjustment'. The German evolutionary biologist Bernhard Rensch (1900–1990) had been exploring the idea of progressive evolution and wished to call it 'anagenesis'. He defines anagenesis as the tendency to 'increased independence of the environment', 'increased command of environmental factors' and the 'progression of autonomy' [3]. He also referred to the evolution of homeothermy in higher vertebrates, which thereby became less vulnerable to environmental temperature change. By the time of the translation of his *Evolution Above the Species Level* in 1959, he had discussed this with Huxley, who was convinced that anagenesis could be a product of natural selection.

The evolution of adaptations that confer increasing independence of the environment was given the name 'anagenesis' by Rensch, and was accepted by Huxley as 'progressive'. It can now be seen that they were both referring to the behaviour of homeodynamic mechanisms, which at that time were referred to as homeostasis. Both have the specific function of neutralising the effects of external perturbation, and allow organisms to achieve increasing independence of environmental variation. It is just that at that time no-one had apparently considered it as a form of adaptation that had developed throughout the evolution of life. It is the addition of levels of organisation in organisms during evolution that adds complexity to organisms, but also has the effect of increasing the homeodynamic resistance to environmental change. Escape from the susceptibility of life to environmental change has been a driving force in evolution, which has come about by the increase and elaboration of homeodynamic mechanisms. They are the organs of anagenesis, as they are the means by which the variability of the environment is converted within organisms to near-constancy.

The increasing complexity of organisms has been a trend that has run throughout the course of evolution. So too, it seems, has been the increase in functionality, and the number of homeodynamic systems. With each came more pre-set preferenda and goal settings to provide the optima to which each system corrected its state or rate. Each mechanism has its own purposes, defined by its goals. Collectively the cumulative preferenda represent within an organism its template for life; and meeting cybernetic goals became the purpose of homeodynamic mechanisms during that life. This is not to suggest that the genome became of lesser importance, only that the goals were teleological and adjustable, making adaptation to change possible during the life of the organism. A list of an organism's homeodynamic preferenda, a summation of information embedded in control mechanisms, would allow a reasonable prediction of what its habitat might be.

Information is the currency that makes possible the adaptation of processes at each level of organisation. J.Z. Young stated that 'the whole organism can be seen as a coded representation of the environment', while E.O. Wilson has described that code as a 'template' of its environment. The 'code' is information which is the organism's template that fits it to the habitat in which it lives. It is expressed in different ways at three levels of organisation, each with its own rate of renewal. There is a requirement to continuously update the template. Environmental change is endless, and evolving animals must adapt by following change, without ever being able to catch up. Yet the accumulation of information is crucial to increase functionality in an ever-changing world, and adjustment needs to be made rapidly. Organisms become ever more complex in form and process, reducing vulnerability to environmental uncertainty and increasing in fitness. Despite the inevitable delays to the process of adaptation, the imperative is to track change, following the trail of environmental change, and making adjustments to adapt to change, but never catching up with the actuality of the environment around it. At first each closing step occurred with opportunities for selection or deletion, opportunities that became separated by longer intervals of time for those animals that grow more slowly, become larger and live longer. Later, with secondary cybernetic adaptation, phenotypic adaptation became an important factor in speeding adaptation. But it was not until the evolution of consciousness that responses to change closed in on the heels of change to occur in time present. Adaptation can never be perfect, but during evolution the delay between the level of adaptation and its fit to the changing environment was a drawback; this gap has now been closed.

Evolution has no goal, but it does have direction because it has been driven in the same way for so long by the same maladaptive and sub-optimal environmental factors. As deletion pressure has been driven by the same negative factors for so long, evolution has in effect become progressive. As a consequence, life has become ever more complex and multifunctional in life's escape from unending change. Anagenesis and this trend towards increasing independence of environmental change have been made possible by homeodynamic systems, which have the specific task of automatically achieving internal constancy by countering change. Anagenesis applies to the evolution of homeodynamic mechanisms, and linked adaptations, that work to keep physiological processes constant at some optimal level. Each homeodynamic mechanism contributes to the integrated purpose of the whole: to maintain internal preferenda by neutralising the effect of external change. In the most complex organisms there are thousands of such feedback mechanisms spanning all levels of organisation, driving the physiology of organisms to meet the many homeodynamic preferenda. The cybernetic goals act as a template that map the organism to its niche and none other, and defines the goals that become its life purpose. Driven by the tyranny of endless vacillation, to use Bernard's metaphor, evolution broke the chains of perpetual change by achieving internal constancy. Cybernetic systems allowed higher animals to become 'free and independent' of the environment's unceasing fluctuation.

There still remains a question mark over teleology and the implication that progress implies a destination for evolution. In nature one cannot assume intention of purpose of a mechanistic process that is blind and unknowing. The adaptations that abound in biology so obviously fit animals to the life they lead: Jacques Monod wrote that 'living things without exception are endowed with purpose'. The point is that, in evolution, fitness for purpose is unintended by anyone or anything, as no-one is directing the process. Evolution is not directed by goals, but is driven by pressure that eliminates those individuals that are least fit for life. Evolution is the inevitable consequence of the repeated elimination of the unfit, generation after generation, over many millions of years. It is 'natural deletion' that eliminates the unfit, but for Darwin that would not have tied evolution, as explicitly as he intended, to the everyday experience of improvements in farm stock and domestic species by mankind's selection of breeding stock. Darwin wanted to use the term 'selection', which he hoped would make his idea more acceptable. In the process, he opened the door to teleology and 'intelligent design', which he might not have done had he used the term 'natural deletion', which gives no such directional implication to evolution. Teleology is avoided, and a more realistic view of evolution is provided, if we consider the converse view of natural selection. It seems most likely that natural selection for homeodynamic mechanisms is due to the benefits of escaping the negative effects of environmental variation. There is no evolutionary goal in a 'steady state' as such, rather an escape from the effect of unending change on internal processes made relatively constant by effective control. And if such, unfavourable conditions for life have persisted throughout evolution, then they have provided the selective pressures that have *driven* natural selection, and given evolution its constant direction. The imperative of natural deletion is independence of the ever-changing environment, with internal constancy the result.

Autonomy was never an evolutionary goal, but deletion for those least able to tolerate external changes to internal processes has autonomy as its inevitable outcome. The goal of a carrot hanging in front of a donkey's muzzle could be interpreted as teleological in indicating the donkey's destination, but the application of a switch to its hindquarters would not, although the outcome might be much the same. Evolution is pushed, not pulled to some premeditated objective; the objective is not a goal, but an escape from adverse conditions, and natural deletion provides adaptation yet is not teleological.

The final twist is that, during evolution, mechanisms have evolved that have both purpose and goals. Homeodynamic mechanisms work for the life of the organism to achieve goal preferences embedded within them. Each mechanism has one or more explicit preferenda as goals for the organism, creating a template for a close fit between an organism and its habitat. Goals may even be adjusted rapidly during the life of the individual to provide a closer fit. For the control of growth, homeodynamic mechanisms occur at various hierarchical levels (Chapter 9). Hundreds of other processes at each level are controlled and have analogous roles to maintain prescribed optima. Though the mechanism of evolution is not teleological, physiological mechanisms have evolved that are. Teleological mechanisms have evolved to control processes with the purpose of maintaining physiological goals.

How interesting it is that evolution, that has neither purpose nor goals, and for such reasons could be said to be flawed, has nevertheless produced control mechanisms that possess both purpose and future goals. Apparently evolution has addressed its own limitations. For such reasons, homeodynamic mechanisms are one of the most improbable, yet sensible, products of evolution. In this way one of the weaknesses in evolution has been overcome. Although the term anagenesis was not used until much later, it was Claude Bernard, the great French physiologist, who in 1857 captured the true meaning of the term. He wrote: 'Thus the perpetual changes in the cosmic environment do not touch it; it is not chained by them, but it is free and independent'. The evolutionary trend of anagenesis describes the imperative of life to escape from unfavourable conditions and achieve independence of the ever-changing environment. The principal agent of this trend has been those homeodynamic mechanisms that counter environmental change, which collectively have provided increased freedom from external change, as the number of control mechanisms, and the subtlety of their adaptations, have become more complex. We conclude that anagenesis is the evolutionary trend made possible by cybernetic mechanisms, enabling organisms to become ever more indepedent of changes in their habitats. Anagenesis is the purpose of cybernetic machanisms to life.

SUMMARY

The origins of Maian control mechanisms and their behaviour are summarised as a dual loop control mechanism adapted from the logistic control mechanism. It is suggested that this control module might be used as a first approximation for all biological replicators, as their control requirements are fundamentally similar. This makes it possible to suggest a hierarchical growth control system for metazoans with a number of levels of organisation of coupled control modules. It is proposed that such control has evolved across the animal kingdom, bestowing upon organisms greater independence of their environment as they become more complex. The evolution of Maian and many other control mechanisms reveals a progressive evolutionary trend termed anagenesis, which is coupled with the increase in complexity and refinement of homeodynamic mechanisms during evolution.

Glossary

The aim has been to include those cybernetic terms that biologists might find unfamiliar, and those biological terms that a reader with a cybernetics background might find unfamiliar. Only those terms are included that would not be found in the *Concise Oxford English Dictionary*.

Where possible, all terms have been defined in context, but more difficult terms or jargon used in more than one place are defined in the list below, along with those terms which have been used with a specific meaning in this book.

acclimation	Regulatory compensation enabling increased tolerance to changes in an animal's environment.
aestivation	Torpor during the summer and dormancy during periods of heat or drought.
algorithm	A procedure for solving a problem or, in a genetic sense, the coded procedure for creating an organism.
allometry	Where the growth of parts is unequal, or not in proportion to the whole.
alpha-curve (α-curve)	An L-shaped relationship between concentration and the biological effect of toxicity.
anabolism	The synthesis of larger molecules from simpler ones. *Opp.* catabolism.
anagenesis	A trend in evolution that allows organisms to become increasingly independent of the environment (*sensu* B. Rensch).
asymptotes	The horizontal limits by which a curve is bounded; the curve converges on them without touching.
ATP	Adenosine triphosphate is a carrier of chemical energy in all organisms. Its structural components are linked by covalent bonds which can be hydrolysed to form ADP or AMP, releasing large amounts of energy (*see* cyclic AMP).
atrophy	Reduction in size or function, typically related to decrease in use or work load. *Opp.* hypertrophy.
attractor	A point in phase space which exerts a pull on the system, as if magnetic, such that it is drawn to it.

autocatalysis	Degradation or dissolution of a cell or tissue due to its own secretions.
autonomic nervous system	Parts of the nervous system that provide involuntary control (e.g. viscera).
autotroph	An organism that manufactures its organic constituents from inorganic matter (e.g. photosynthetic plants).
beta-curve (β-curve)	A concentration–effect curve that is biphasic, with a range of concentrations that is stimulatory at subinhibitory levels. *See* alpha-curve.
biodiversity	A general term for the variety of life forms, including various indices of biological variety; sometimes related to a subset of well-known species (e.g. vertebrates).
biosynthesis	The manufacture of biological matter by living organisms.
calcitonin	Hormone produced by the thyroid gland that lowers the concentration of calcium in the blood by inhibiting bone degradation.
capacity	The ability to withstand physiological load, expressed in terms of the agent of loading.
carrying capacity (*K*)	The maximum number of individuals of a species which a particular environment can sustain.
catabolism	The overall effect of enzymatic breakdown processes. *Opp.* anabolism.
catch-up growth	Accelerated growth following the removal of an agent that had been inhibiting growth (e.g. disease, toxic inhibition).
cerebrum	Cerebral hemispheres that evolved due to an extension of the vertebrate forebrain.
cerebellum	The enlarged hindbrain of vertebrates.
chaos	The behaviour of dynamical systems which become unpredictable or random with rate, perturbation or increasing load.
claustrum	A small area of the brain which has two-way connections to most of the cortex. Francis Crick's hypothetical 'seat of consciousness'.
closed system	A system that is sealed from outside interference, so is unresponsive or adaptive to its environment. *Opp.* open system.
coccoliths	Complex calcified scales that cover the cells of unicellular flagellates, called coccolithophores.
competition	Competition occurs between individuals of the same or different species when a shared resource becomes limiting. *Opp.* cooperation.

competitive exclusion principle	The principle states that two or more resource-limited species occupying the same ecological niche, and drawing upon the same resources, cannot coexist for long.
complex adaptive system (cas)	A general term for those systems that are both complex and adaptive. It was created by the Santa Fe Institute in the belief that to identify the general principles would help their wider understanding.
contact inhibition	A long-standing hypothesis for the control of biological growth and population increase by which specific inhibitors are responsible for control by limiting growth.
comparator	A key component of control mechanisms by which the output is compared with the input, which is a goal setting. The mechanism then works to minimise the error or difference between them. *See* error.
contractile vacuole	Spherical vesicle found in protozoans and sponges that is responsible for osmoregulation and excretion.
control hierarchy	A hierarchy of coupled control mechanisms is required where they regulate replicators at each level of organisation.
cooperation	Mutual benefit is achieved in those animals that cooperate. The kinds of cooperation include commensalism, inquilinism, mutualism and symbiosis. *Opp.* competition.
counter-response	Any deviation in one direction is neutralised by a response in the other. Thus a toxic inhibition is counteracted by a stimulatory response.
Cretaceous	The third period of the Mesozoic, approximately 145–65 Ma ago.
cybernetics	Originally defined as 'the science of control and communication in the animal and the machine'.
cyclic AMP (cAMP)	Widespread in animal cells as a second messenger conveying information from hormones to biochemical reactions within the cell. It activates enzymes that catalyse reactions induced by hormones. It is an adenosine phosphate derived from ATP (*see* ATP).
damping	A decrease in the amplitude of oscillation with time, or due to some specific means by which amplitude is reduced.

density-dependence	Refers to regulation of population size in relation to the density of individuals, adjusted by increasing mortality or reducing fertility.
diatoms	Unicellular algae with silica cell walls.
'dose–response' curve	A widely used misnomer for a concentration–effect curve, such as the alpha- and beta-curves. 'Dose' for most purposes is more accurately referred to as 'concentration'. 'Response' is the reaction of the organism, with 'effect' as the overall outcome.
ecological footprint	Measure of the productive surface area of the planet required per person for food, energy and raw materials.
emergence	An outcome which shows that the whole is greater than the sum of the parts or, conversely, a property not apparent or predictable from observations or deduction at a lower level.
endocrine system	A system of glands that secrete hormones into the bloodstream.
endosymbiosis	Symbiotic associations between cells of two or more different species, where larger cells are hosts for smaller cells.
entropy	A measure of the disorder of a system, which increases over time within a closed system.
equilibrium	Stability between some perturbation and the response achieved by a control mechanism with the purpose of equilibration.
error	The difference between the output of a control mechanism and the input or goal setting, which is detected at the comparator.
eukaryote	Any organism consisting of cells in which the genetic material is enclosed in a nucleus. This major division includes all but the simplest organisms. *See* prokaryote.
exocrine	A chemical messenger is an endocrine or hormone that, instead of functioning within an organism, operates externally.
exponential growth	Growth that proceeds at an exponential rate, which is a rate that accelerates linearly according to some exponential function. Populations that grow exponentially have a constant doubling rate.
fecundity	In epidemiology this term refers to the power to reproduce. *See* fertility.
feedback	In a control mechanism information relating to the output is fed back for comparison with goal setting to determine an error, which the mechanism then works to minimise. Self-referencing by feedback is the defining feature of all control mechanisms.
feedforward	Information from earlier in the causal chain of disturbance that is used to give prior warning of disturbance.

fertility — In epidemiology this term refers to the number of offspring produced. *See* fecundity.

gain — The factor by which power is increased; amplification.

genome — The set of genes of an organism in a complete set of chromosomes.

genotype — The genetic composition of an organism. *See* phenotype.

goal setting — The set point, or input to a control mechanism, which is constantly referred to as the mechanism to achieve its purpose.

Globigerina — An important genus of the Foraminifera, an order of mostly marine protozoans with calcareous shells. Individuals are referred to collectively as globigerinae.

granulocyte — A granular leucocyte or white blood cell.

group selection — Classically, natural selection operates on individual organisms, but group selection also occurs between organisms sharing the same genes (kin selection).

growth rates — Two kinds of growth rate are commonly used: the normal growth rate $R = dW/dt$ and the specific or relative growth rate $r = 1/W \cdot dW/dt = dW_e /dt$. The same growth rate ($r$) is used to calculate the 'intrinsic rate of natural increase' used by ecologists in the study of populations.

hierarchy — The vertical organisation of levels in organisms, in which higher levels are dominant over lower levels that are nested within higher levels.

high output failure — This occurs when a control mechanism is counteracting a high level of load, at the onset of overload.

holism — The view that an organism viewed as a whole has properties not apparent in its parts, so all organisms are more than the sum of their parts.

homeodynamic — A general term referring to physiological control mechanisms that encompass homeostasis (control of states) and homeorhesis (control of rates).

homeopathy — An alternative medicine that uses small doses, which in a healthy person would cause symptoms similar to those that allopathic medicine would prescribe to treat.

homeorhesis — The physiological control of rate processes.

homeostasis — The physiological control of states.

homeothermy — The physiological control of internal temperature.

Homo economicus — A term used by economists, mimicking the Linnean name for man, with behavioural traits to incorporate in economic models. These typically include that all human beings are rational, self-interested and will always maximise their gains for the minimum effort.

hormesis — The stimulatory effects of a low level of agents that are toxic at higher concentrations. *See* beta-curve.

hunting — Repeated overshooting, or persistent oscillation, which may also refer to limit cycling.

hydrozoa	Class of the phylum Cnidaria, mostly marine, colonial and sessile organisms. Exhibit alternation of generations between sessile adults and planktonic medusae.
hyperbolic growth	Growth that proceeds at a hyperbolic rate; it is a superexponential in that the rate of increase is in proportion to the square of the population, such that the population increases hyperbolically against time.
hyperplasia	Growth of a tissue resulting from cell division. *See* hypertrophy.
hypertrophy	Growth by increase in volume (cell size) without cell division. An increase in the size of a tissue or organ, typically in response to increased workload. *See* hyperplasia. *Opp.* atrophy.
hypothalamus	Part of the vertebrate brain that projects from its floor and that controls a wide range of physiological processes, particularly homeodynamic processes.
inhibitory regulator	Growth regulation by means of an inhibitory agent.
intrinsic rate of natural increase	The ecological equivalent of the specific growth rate (r).
isometric contraction	Increase in tension during muscle contraction, but without change in length.
isometry	Occurs where growth of a part is proportional to the whole. Organisms that grow in this way retain the same shape throughout life. *See* allometry.
kin selection	Natural selection of those individuals that share the same genes, accounting for the evolution of altruism.
K-selection	K represents the carrying capacity of the environment for species populations that have a sigmoid population growth curve. This implies selection for low birth rates, prolonged development and high survival rates. *See* r-selection.
lactic acid	Produced from pyruvic acid in muscles during strenuous exercise when oxygen is limited.
lag time	The delay in the response of a control mechanism due to various factors, causing the oscillatory behaviour that characterises the output of feedback mechanisms.
leucocyte	White blood cell with a nucleus.
limit cycle	Oscillates, or cycles in phase space continually.

'linear no-threshold' (LNT)	A hypothesis that assumes the linearity of relationships between toxic agents (particularly radiation) and their biological effects.
load	Refers to the burden put upon control mechanisms by external factors that require continuous counteraction. The capacity to withstand load can be expressed in terms of the agent that causes load, and the range of loading that can be withstood indicates the load tolerance.
logistic control mechanism	The control mechanism derived from the logistic equation.
logistic equation	The equation devised by Pierre Verhulst (1838) to describe population growth, which has been much used and modified for different purposes.
logistic map	The map of the logistic equation which first revealed chaos to Robert May, Jim Yorke and Tien-Yien Li.
macrophage	Large white blood cell normally found in tissues that produce blood cells. Able to ingest bacteria and cell debris.
Maia	The Roman goddess of growth and increase.
maladaptive	A conjectural adaptation deemed to be unlikely to confer advantage.
medulla oblongata	The part of the vertebrate brain stem that is continuous with the spinal cord.
medusae	The free-swimming stage in the life cycle of Cnidaria, such as the hydrozoa.
Mesozoic	A geological era that includes the Triassic, Jurassic and Cretaceous periods. It is preceded by the Palaeozoic era and is followed by the Cenozoic, and lasted for 186 million years.
metamorphose	The transformation from larval to adult form in the life cycle of invertebrates and some vertebrates.
metazoan	The Metazoa are a sub-kingdom of animals in which many cells are grouped in tissues and coordinated by a nervous system.
mitosis	The division of one cell to create two, each with a nucleus containing the same genetic make-up as the parent cell. Mitotic activity initiates cell division.
mitotic index	Measure of the rate of cell division.
module	Identical components or multicellular parts that, in a biological context, are genetically identical to others.
morphogenesis	The process of growth and development of organisms that ends with the attainment of adult form.

mutualism	Mutual dependence for social well-being; mutually beneficial symbiosis.
myostatin	A peptide and specific regulator of skeletal muscle mass. This is the growth/differentiation factor GDF-8.
nauplius	The earliest life cycle stage of some crustaceans.
negative feedback	A defining feature of control mechanisms, as the linking of output to the input of the goal setting by feedback enables the error between them to be negated or minimised.
neoplasm	A tumour or cancerous growth, hence neoplastic growth.
niche	The position of an organism in its environment. Elton's helpful analogy was to liken 'niche' to the role of tradesmen in a town.
nucleotide	Bases that are the molecular building blocks of DNA. There are four bases: adenine pairs with thymine, and cytosine with guanine.
open system	Such systems are inextricably embedded, and are part of their environment, exchanging fluxes of energy and materials. *Opp.* closed system.
optimal control	The term given to modern control theory which became established in the 1960s and adopted computer-intensive methods and the laws of control.
optimisation	Maintenance at the most favourable conditions (e.g. for growth or reproduction); best compromise between opposing tendencies; preferendum or goal setting.
organotroph	An animal that depends upon some external source of organic compounds; includes all animals and fungi. *See* autotroph.
oscillation	Cyclic fluctuations about an equilibrium or goal setting.
osmoregulation	The control of osmotic pressure within organisms.
osmosis	The flow of water through a semi-permeable membrane that will permit the passage of the solvent but not the substances dissolved in it.
osteoblast	A cell that forms bone.
osteoclast	A cell that breaks down bony tissue.
osteoporosis	Brittleness of bone due to porosity, primarily from loss of calcium.
overshoot	Time lags result in delays in responding to errors, such that the controlled process is likely to overshoot its goal, before returning to overshoot once more, in a series of decaying oscillations before equilibrium is restored.
PAH	Polycyclic aromatic hydrocarbons. A major combustion product of organic matter.
parthenogenesis	The creation of new individuals by asexual means.

PCB	Polychlorinated biphenyl. A toxic constituent of electrical equipment, especially transformers. Usage now limited.
perturbation theory	A branch of physics that studies the dynamics of systems by perturbing them, in order to better understand equilibria. Such an approach is occasionally used in physiological studies, but in the study of control systems perturbation is a necessity, yet rarely used.
phase space	The dimensionless space in which control systems can be portrayed to reveal their dynamical behaviour in relation to an attractor(s).
phenotype	The characteristics of an organism. *See* genotype.
pheromone	Chemical messengers that are used by organisms in their environment and which influence the development or behaviour of other individuals of the same species. *See* exocrine.
phytoplankton	Algal plankton including motile and non-motile unicells and colonies.
poikilotherm	Animal without active control over temperature, whose internal temperature is determined by that of the environment.
positive feedback	Control mechanism where errors between the output and goal setting are summed, so errors increase in size and the mechanism becomes increasingly unstable. *See* negative feedback.
prokaryote	This includes all organisms that are not eukaryotes, i.e. bacteria and blue-green algae only. *See* eukaryote.
pseudogene	Repeated copies of redundant genes.
Q_{10} rule	The increase in rates of enzyme reactions with an increase in temperature of 10°C. Within physiological limits the value of Q_{10} for physiological processes is 2–3.
Radiolaria	Protozoans of the marine plankton that feature siliceous spicules which radiate from a chitinous capsule.
real time	A time frame that relates to the present time.
reductionism	The belief that the whole may be represented as a function of its constituent parts. *See* holism.
rein control	Rein control, as in a horse's harness, implies that if one rein is used to direct the horse in one direction, then the other must be used to direct the horse in the other direction. Thus growth control requires a stimulator and an inhibitor, and other processes require a minimum of two regulators.
relaxation stimulation	If a control mechanism is exposed to an inhibitory load, to which it equilibrates, the removal of that

	load will cause a relaxation stimulation, which lasts for a long as the lag in the system.
reverse engineering	With respect to biological control mechanisms, where the mechanism itself is often unknown or inaccessible, reverse engineering involves using the output to deduce the nature of the mechanism responsible.
ringing	Sustained oscillation of the output of a feedback mechanism.
r-selection	Selection that maximises the intrinsic rate of natural increase (*r*) of organisms, such that they rapidly colonise newly formed habitats. *See* *K*-selection.
runaway	Loss of control by a feedback mechanism, such that the output passes through ever-wilder oscillations as the system becomes increasingly unstable.
saturation	Occurs when the capacity of the control mechanism to withstand load is completely utilised.
servo-mechanism	A particular kind of control mechanism in which the motion of the output shaft is made to follow the amplified motion of the input.
sessile	Term to describe animals that live their lives attached to a surface.
set theory	A branch of mathematics that deals with sets of well-defined objects, which may be mathematical or not. Set theory was used by Eric Ashby in his early work on control mechanisms.
sigmoid curve	An S-shaped curve described by the logistic equation and used to fit growth curves (e.g. yeast populations). It is bounded by asymptotes with a point of inflexion at the mid point.
singularity	A point at which a function is not analytic, which is also likely to be a region of instability.
specific growth rate (*r*)	Physiologically the most appropriate measure of growth rate, as it reflects the rate of the underlying growth process by representing the growth rate per unit time per unit biomass.
statin	Best known as a drug that lowers cholesterol; it is also a class of tissue-specific growth regulators, of which myostatin was the first to be discovered by Alexandra McPherron and Se-Jin Lee.
step-function	Used here to represent the onset of some experimental treatment.
stochastic	Implying random variation. Simulation models that incorporate some element of random variation.
synergism	The interaction of two or more agents whose combined effect is greater than the sum of their separate effects.
taxonomy	The classification of organisms into specific groupings (e.g. Phylum, Class, Order).

teleology	The term refers to purposiveness, which is inadmissible in evolutionary biology because natural selection can have no aim, goal or target.
template	The coded characteristics that fit an animal to its environment. Here extragenetic features are included, such as the adaptive goal settings of homeodynamic mechanisms.
thermocouple	An electronic instrument for the measurement of temperature.
transfer function	The mathematical function that defines the operation of a simulation model of a control mechanism.
xenobiotic	Any substance foreign to living systems.
zero population growth (ZPG)	The social aim that epitomises the demographic balance at which the numbers in a population neither grow nor decline.
zootype	A representative type of animal.

Further reading

GENERAL TITLES ON BIOLOGICAL CYBERNETICS

Ashby, W.R. (1952). *Design for a Brain: The Origin of Adaptive Behaviour*. London: Chapman and Hall.

Ashby, W.R. (1957). *An Introduction to Cybernetics*. London: Chapman and Hall.

Åström, K.J. and Murray, R.M. (2008). *Feedback Systems: An Introduction to Scientists and Engineers*. Princeton, NJ: Princeton University Press.

Bayne, B.L., Brown, D.A., Burns, K., Dixon, D.R., Ivanovici, A., Livingstone, D.R., Lowe, D.M., Moore, M.N., Stebbing, A.R.D. and Widdows, J. (1985). *The Effects of Stress and Pollution on Marine Animals*. New York: Praeger.

Bennett, R.J. and Chorley, R.J. (1978). *Environmental Systems: Philosophy Analysis and Control*. London: Methuen.

Calow, P. (1976). *Biological Machines: A Cybernetic Approach to Life*. London: Arnold.

DiStefano, J.J., Stubberud, A.R. and Williams I.J. (1976). *Feedback and Control Systems*. New York: McGraw-Hill.

Evans, W.R. (1954). *Control System Dynamics*. New York: McGraw-Hill.

Hofstadter, D. (2007). *I Am a Strange Loop*. New York: Basic Books.

Holdgate, M.W. (1979). *A Perspective of Environmental Pollution*. Cambridge, UK: Cambridge University Press.

Lovelock, J.E. (1979). *Gaia: A New Look at Life on Earth*. Oxford, UK: Oxford University Press.

Lovelock, J.E. (2006). *The Revenge of Gaia*. London: Allen Lane.

May, R.M. (ed) (1981). *Theoretical Ecology: Principles and Applications*, 2nd edn. Oxford, UK: Blackwell.

McPherron, A.C. and Lee, S.-J. (1997). Double muscling in cattle due to mutations in the myostatin gene. *Proceedings of the National Academy of Science*, 94(23), 12457–12461.

Milsum, J.H. (1966). *Biological Control Systems Analysis*. New York: McGraw-Hill.

Rose, S. (1997). *Lifelines. Biology, Freedom, Determinism*. London: Penguin.

Schmidt-Nielsen, K. (1979). *Animal Physiology: Adaptation and Environment*. Cambridge, UK: Cambridge University Press.

Spencer, H. (1862). *First Principles*. London: Williams and Norgate.

Stanley-Jones, D. and Stanley-Jones, K. (1960). *The Kybernetics of Natural Systems: A Study in Patterns of Control*. Oxford, UK: Pergamon.

Turchin P. (2003). *Complex Population Dynamics: A Theoretical/Empirical Synthesis.* Princeton, NJ: Princeton University Press.

Wiener, N. (1948). *Cybernetics: Or Control and Communication in the Animal and the Machine.* Cambridge, MA: MIT Press.

Worldwatch Institute (2007). *Vital Signs: The Trends that are Shaping our Future, 2007–2008.* London: Earthscan.

References

CHAPTER 1. INTRODUCTION

1. Cannon, W.B. (1945). *The Wisdom of the Body*. New York: Hafner.
2. Langley, L.L. (ed.) (1973). *Homeostasis: Origins of the Concept*. Stroudsberg, PA: Dowden, Hutchinson and Ross.
3. Waddington, C.H. (1977). *Tools for Thought*. St Albans, UK: Paladin [published posthumously].
4. Young, J.Z. (1957). *The Life of Mammals*. Oxford: Oxford University Press.
5. Marieb, E.N. (2001). *Human Anatomy and Physiology*. San Francisco: Benjamin Cummings.
6. Maxwell, J.C. (1868). On governors. *Proceedings of the Royal Society, London*, **16**, 270–283.
7. Medawar, P. (1941). The 'laws' of biological growth. *Nature*, **148**, 772–774.

CHAPTER 2. GROWTH UNLIMITED: BLOOMS, SWARMS AND PLAGUES

1. Huxley, T.H. (1894). On a piece of chalk. In *Discourses: Biological and Geological. Collected Essays, Volume VIII*. London: Macmillan, pp. 1–36.
2. Gamow, G. (1962). *One, Two, Three … Infinity: Facts and Speculations of Science*. London: Macmillan.
3. Pasteur, L. (1873). *Études sur le Vin*, 2nd edn. Paris.
4. Baker, J.R. (1952). *Abraham Trembley: Scientist and Philosopher 1710–1784*. London: Arnold.
5. Stebbing, A.R.D. (1971). Growth of *Flustra*. *Marine Biology*, **9**, 267–272.
6. Elton, C.S. (1963). *The Ecology of Invasions*. London: Methuen.
7. Vernadsky, V. (1986). *The Biosphere*. Oracle, AZ: Synergetic Press [first published in Russian in 1926].
8. Smith, F.E. (1954). Quantitative aspects of population growth. In *Dynamics of Growth Processes*. Princeton, NJ: Princeton University Press, pp. 276–241.
9. Malthus, T.R. (1985). *An Essay on the Principle of Population*. London: Penguin Classics.

10. Darwin, F. (ed.) (1929). *The Autobiography of Charles Darwin*. London: Watts [originally published 1887].

CHAPTER 3. SELF-REGULATING SYSTEMS: FROM MACHINES TO HUMANS

1. Spencer, H. (1862). *First Principles*. London: Williams and Norgate.
2. Bernard, C. (1865). *An Introduction to the Study of Experimental Medicine*. London: Macmillan.
3. Smith, A. (1986). *The Wealth of Nations, Books I–III*. London: Penguin Classics [originally published in 1776].
4. Cannon, W.B. (1965). *The Way of an Investigator: A Scientist's Experiences in Medical Research*. New York: Hafner [originally published in 1945].
5. Monod, J. (1972). *Chance and Necessity: An Essay on the Natural Philosophy of Modern Biology*. London: Collins.
6. Zhang, D.Y., Turberfield, A.J., Yurke, B. and Winfree, E. (2007). Engineering entropy-driven reactions and networks catalyzed by DNA. *Science*, **318**, 1121–1125.
7. Foerster, H. von (1953). *Cybernetics: Circular, Causal and Feedback Mechanisms in Biological and Social Systems*. New York: Josiah Macy Jr Foundation.
8. Foerster, H. von (1958). Basic concepts of homeostasis. *Brookhaven Symposia in Biology*, **10**, 216–242.
9. Hayes, B. (2009). Everything is under control. *American Scientist*, **97**(3), 186–191.

CHAPTER 4. THE WEALTH OF HOMEODYNAMIC RESPONSES

1. Kitching, J.A. (1954). The physiology of contractile vacuoles (IX). *Journal of Experimental Biology*, **31**, 68–75.
2. Ashcroft, F. (2000). *Life at the Extremes*: The Science of Survival. London: Harper Collins.
3. Irving, L. (1966). Adaptation to cold. *Scientific American*, **214**(1), 94–101.
4. Heinrich, B. and Kammar, A.E. (1973). Activation of the fibrillar muscles in the bumblebee during warm-up, stabilization of thoracic temperature and flight. *Journal of Experimental Biology*, **58**, 677–688.
5. Stebbing, A.R.D., Turk, S.M.T., Wheeler A., and Clarke, K.R. (2002). Immigration of southern fish species to south-west England linked to warming of the North Atlantic (1960–2001). *Journal of the Marine Biological Association of the United Kingdom*, **82**, 177–180.
6. Southwick, E.E. and Heldmaier, G. (1987). Temperature control in honey bee colonies. *BioScience*, **37**(6), 395–399.
7. Benzinger, T.H. (1961). The human thermostat. *Scientific American*, **61**(1), 134–146.
8. Heller, C., Crawshaw, L.I. and Hammel, H.T. (1978). The thermostat of vertebrate animals. *Scientific American*, **239**(2), 102–113.
9. Slijper, E.J. (1962). *Whales*. London: Hutchinson.
10. Smith, S. (2009). Calcium and bone metabolism during space flight. *Nutrition*, **18**(10), 849–852.

CHAPTER 5. A CYBERNETIC APPROACH TO GROWTH ANALYSIS

1. Medawar, P.B. (1945). Size, shape and age. In *Essays on Growth and Form*. Oxford: Oxford University Press, pp. 157–185.

2. Medawar, P.B. (1940). The growth, growth energy, and ageing of the chicken's heart. *Proceedings of the Royal Society B*, **129**, 332–355.
3. Stebbing A.R.D. (1981). The effects of reduced salinity on colonial growth and membership in a hydroid. *Journal of Experimental Marine Biology and Ecology*, **55**, 233–241.
4. Stebbing, A.R.D. (1981). The kinetics of growth control in a colonial hydroid. *Journal of the Marine Biological Association of the United Kingdom*, **61**, 35–63.
5. Stebbing, A.R.D. and Hiby, L. (1979). Cyclical fluctuations in the growth rate of stressed hydroid colonies. In *Cyclical Phenomena in Marine Plants and Animals*. Oxford: Pergamon Press, pp. 165–172.

CHAPTER 6. A CONTROL MECHANISM FOR MAIA

1. Lotka, A.J. (1910). Contribution to the theory of periodic reactions. *Journal of Physical Chemistry*, **14**, 271.
2. Winfree, A.T. (2001). *The Geometry of Biological Time*. New York: Springer.
3. Belousov, B.P. (1959). A periodic reaction and its mechanism [in Russian]. *Sbornik Referatov po Radiatsionnoy Meditsine [Collection of Abstracts on Radiation Medicine], Moscow*, **1959**, 145 [in Russian].
4. Ball, P. (1999). *The Self-Made Tapestry: Pattern Formation in Nature*. Oxford: Oxford University Press.
5. Zaikin, A.N. and Zhabotinsky, A.M. (1970). Concentration wave propagation in two-dimensional liquid-phase self-oscillating systems. *Nature*, **225**, 535–537.
6. Prigogine, I. and Stengers, I. (1985). *Order Out of Chaos*. London: Fontana.
7. Turing, A. (1952). The chemical basis of morphogenesis. *Philosophical Transactions of the Royal Society B*, **237**, 37–72.
8. Meinhardt, H. (1999). On pattern and growth. In *On Growth and Form: Spatio-temporal Pattern Formation in Biology*. Chichester, UK: Wiley.
9. Meinhardt, H. and Gierer, A. (1974). Application of a theory of biological pattern formation based on lateral inhibition. *Journal of Cell Science*, **15**, 321–346.
10. Dyson, F. (1999). *Origins of Life*. Cambridge, UK: Cambridge University Press.
11. Waterman T.H. (1968). Systems theory and biology: view of a biologist. In *Systems Theory and Biology: Proceedings of the 3rd Systems Symposium at Case Institute of Technology*, New York: Springer, pp. 1–37.
12. Gilbert, W. (1986). The RNA World (News and Views). *Nature*, **319**, 618.
13. Kauffman, S. (1995). *At Home in the Universe*. Oxford: Oxford University Press.
14. Lee, D.H., Granja, J.R., Martinez, J.A., Severin, K. and Ghadiri, M.R. (1996). A self-replicating peptide. *Nature*, **382**, 525–528.
15. Kauffman, S. (2000). *Investigations*. Oxford: Oxford University Press.

CHAPTER 7. THE THREE LEVELS OF ADAPTATION

1. Schmidt-Nielsen, K. (1964). *Desert Animals: Physiological Problems of Heat and Water.* New York: Oxford University Press.
2. Frenster, J.H. (1962). The magnitude of disease as measured by tolerance tests. *Journal of Theoretical Biology*, **2**, 159–164.
3. Stebbing, A.R. (2009). Interpreting 'dose-response' curves using homeodynamic data: with an improved explanation for hormesis. *Dose-Response*, **7**, 221–233.

4. Yerkes, R.M. and Dodson, J.D. (1908). The relation of strength of stimulus to rapidity of habit-formation. *Journal of Comparative Neurology and Psychology*, **18**, 459–482.
5. Ashby, W.R. (1958). Requisite variety and its implications for the control of complex systems. *Cybernetica*, **1**, 83–99.
6. Jacob, F. (1998). *Of Flies, Mice and Men*. Cambridge, MA: Harvard University Press.
7. Kandel, E.R. (2006). *In Search of Memory: The Emergence of a New Science of the Mind*. New York: Norton.

CHAPTER 8. POPULATION GROWTH AND ITS CONTROL

1. Gause, G.F. (1934). *The Struggle for Existence*. Baltimore, MD: Williams and Wilkins.
2. Hutchinson, G.E. (1948). Circular causal systems in ecology. *Annals of the New York Academy of Sciences*, **50**, 221–246.
3. Stebbing, A.R.D. and Pomroy, A.J. (1978). A sublethal technique for assessing the effects of contaminants using *Hydra littoralis*. *Water Research*, **12**, 631–635.
4. Davis, L.V. (1966). Inhibition of growth and regeneration in *Hydra* by crowded culture water. *Nature*, **212**, 1215–1217.
5. Wynne-Edwards, V.C. (1962). *Animal Dispersion in Relation to Social Behaviour.* Edinburgh: Oliver and Boyd.
6. Wynne-Edwards, V.C. (1965). Self-regulating systems in populations of animals. *Science*, **147**, 1543–1548.
7. Hamilton, W. (1964). The genetical evolution of social behaviour: I and II. *Journal of Theoretical Biology*, **7**, 1–16, 17–52.
8. Taylor, R.A.J. and Taylor, L.R. (1979). A behavioural model for the evolution of spatial dynamics. In *Population Dynamics: Proceedings of the 20th Symposium of the British Ecological Society, London 1978*. Oxford: Blackwell, pp. 1–27.
9. Taylor, R.A.J. (1981). The behavioural basis of redistribution. I. The Δ-model concept. *Journal of Animal Ecology*, **50**, 573–586.
10. Taylor, R.A.J. (1981). The behavioural basis of redistribution. II. Simulations of the Δ-model. *Journal of Animal Ecology*, **50**, 587–604.
11. May, R.M. (2002). The best possible time to be alive: the logistic map. In Farmelo, G. (ed.), *It Must Be Beautiful: Great Equations of Modern Science*. London: Granta, pp. 28–45.
12. May, R.M. (1975). Biological populations obeying difference equations: stable points, stable cycles and chaos. *Journal of Theoretical Biology*, **51**, 511–524.
13. May, R.M. (1976). Simple mathematical models with very complicated dynamics. *Nature*, **261**, 459–467.
14. May, R.M. (ed.) (1981). *Theoretical Ecology: Principles and Applications*. 2nd edn. Oxford: Blackwell.

CHAPTER 9. HIERARCHY: A CONTROLLED HARMONY

1. Stebbing, A.R.D. and Heath G.W. (1984). Is growth controlled by a hierarchical system? *Zoological Journal of the Linnean Society*, **80**, 345–367.
2. Woodger, J.H. (1930). The "concept of organism" and the relation between embryology and genetics. Part II. *Quarterly Review of Biology*, **5**, 438–463.
3. Bertalanffy, L. von (1952). *Problems of Life: An Evaluation of Modern Biological Thought*. London: Watts.

4. Feibleman, J.K. (1954). Theory of integrative levels. *British Journal for the Philosophy of Science*, **5**, 59–66.
5. Simon, H.A. (1962). The architecture of complexity. *Proceedings of the American Philosophical Society*, **106**(6), 467–482.
6. Margulis, L. (1998). *The Symbiotic Planet: A New Look at Evolution*. London: Weidenfeld and Nicholson.
7. Pattee, H.H. (1973). *Hierarchy Theory: The Challenge of Complex Systems*. New York: Braziller.
8. Mesarovic, M.D. (1968). Systems theory and biology: view of a theoretician. In *Systems Theory and Biology*. Berlin: Springer, pp. 59–87.
9. Mesarovic, M., Macko, D. and Takahara, Y. (1970). *Theory of Hierarchical, Multilevel Systems*. New York: Academic Press.
10. Bronowski, J. (1970). New concepts in the evolution of complexity: stratified stability and unbounded plans. *Synthese*, **21**(2), 228–246.
11. Gell-Mann, M. (1994). *The Quark and the Jaguar: Adventures in the Simple and the Complex*. London: Little, Brown.
12. Holland, J.H. (1995). *Hidden Order: How Adaptation Builds Complexity*. New York: Basic Books.
13. Aulin-Ahmavaara, A.Y. (1979). The law of requisite hierarchy. *Kybernetes*, **8**, 259–266.

CHAPTER 10. HISTORY OF HORMESIS AND LINKS TO HOMEOPATHY

1. Stebbing, A.R.D. (1982). Hormesis – the stimulation of growth by low levels of inhibitors. *Science of the Total Environment*, **22**, 213–234.
2. Young, J.Z. (1978). *Programs of the Brain*. Oxford: Oxford University Press.
3. Hueppe, F. (1896). *The Principles of Bacteriology*. Chicago: Open Court.
4. Clarke, A.J. (1937). *General Pharmacology*. Berlin: Springer.
5. Calabrese, E.J. and Baldwin, L.A. (1997). A quantitatively based methodology for the evaluation of hormesis. *Human and Ecological Risk Assessment*, **3**, 545–554.
6. Lorenz, E. (1950). Some biologic effects of long-continued irradiation. *American Journal of Roentgenology*, **63**, 176–185.
7. Luckey, T.D. (1994). Radiation hormesis in cancer mortality. *International Journal for Occupational Medicine and Toxicology*, **3**, 175–191.
8. Cohen, B.J. (1995). Test of the linear-no threshold theory of radiation carcinogenesis for inhaled radon decay products. *Health Physics*, **68**, 157–174.
9. Billen, D. (1990). Spontaneous DNA damage and its significance to the 'negligible dose' controversy in radiation protection. *Radiation Research*, **124**, 242–245.
10. Feinendegen, L.E. (2005). Evidence for beneficial low level radiation effects and radiation hormesis. *British Journal of Radiology*, **78**, 3–7.

CHAPTER 11. MAIAN MECHANISMS FOR HORMESIS AND CATCH-UP GROWTH

1. Calabrese, E.J., McCarthy, M.E. and Kenyon, E. (1987). The occurrence of chemically induced hormesis. *Health Physics*, **52** (5), 531–542.
2. Calabrese, E.J. (2005). Paradigms lost, paradigms found: the re-emergence of hormesis as a fundamental dose–response model in the toxicological sciences. *Environmental Pollution*, **138**, 378–411.
3. Calabrese, E.J. (2008). Hormesis and medicine. *British Journal of Clinical Pharmacology*, **66**(5), 594–617.

4. Stebbing, A.R.D. (1998). A theory for growth hormesis. *Mutation Research*, **403**, 249–258.
5. Stebbing, A.R.D. (2003). Adaptive responses account for the beta-curve: hormesis is linked to acquired tolerance. *Nonlinearity in Biology, Toxicology, and Medicine*, **1**, 493–511.
6. Calabrese, E.J. and Baldwin, L.A. (2003). Toxicology rethinks its central belief. *Nature*, **421**, 691–692.
7. Tanner, J.M. (1963). Regulation of growth in size in mammals. *Nature*, **199**, 845–850.
8. Tanner, J.M. (1978). *Foetus into Man: Physical Growth from Conception to Maturity*. London: Open Books.
9. Boersma, B. and Wit, J. M. (1997). Catch-up growth. *Endocrine Reviews*, **18**(5), 646–661.
10. Barron, J., Klein, K.O., Colli, M.J., Yanovski, J.A., Novosad, J.A., Bacher, J.D. and Cuttler, G.B. (1994). Catch-up growth after glucocorticoid excess: a mechanism intrinsic to the growth plate. *Endocrinology*, **135**, 1367–1371.

CHAPTER 12. CELLULAR GROWTH CONTROL AND CANCER

1. Cairns, J. (1981). The origin of human cancers. *Nature*, **289**, 353–357.
2. Stebbing, A.R.D., Norton, J.P. and Brinsley, M.D. (1984). Dynamics of growth control in a marine yeast subjected to perturbation. *Journal of General Microbiology*, **130**, 1799–1808.
3. Norton, J.P. and Stebbing, A.R.D. (1986). Measurement and interpretation of growth-rate oscillations in a marine yeast. *Biomedical Measurement, Informatics and Control*, **1**(2), 101–105.
4. Dawkins, R. (1982). *The Extended Phenotype: The Long Reach of the Gene*. Oxford: Oxford University Press.
5. Orgel, L.E. and Crick, F.H.C. (1980). Selfish DNA: the ultimate parasite. *Nature*, **284**, 604–607.
6. Goss, R.J. (1964). *Adaptive Growth*. London: Logos.
7. Loewenstein, W.R. and Kanno, Y. (1966). Intercellular communication and the control of tissue growth: lack of communication between cancer cells. *Nature*, **209**, 1248–1249.
8. Bertalanffy, L. von (1949). Problems of organic growth. *Nature*, **163**, 156–158.
9. Drack, M. and Apfalter, W. (2007). Is Paul Weiss' and Ludwig von Bertalanffy's system thinking still valid today? *Systems Research*, **24**, 537–546.
10. Weiss, P. (1952). Self-regulation of organ growth by its own products. *Science*, **115**, 487–488.
11. Weiss, P. and Kavanau, J.L. (1957). A model of growth and growth control in mathematical terms. *Journal of General Physiology*, **41**(1), 1–47.
12. Kavanau, J.L. (1960). A model of growth and growth control in mathematical terms. II. Compensatory organ growth in the adult. *Proceedings of the National Academy of Science*, **46**, 1658–1673.
13. Rose, S.M. (1957). Cellular interaction during differentiation. *Biological Reviews*, **32**, 351–382.
14. Rose, S.M. (1958) Failure of self-inhibition in tumors. *Journal of the National Cancer Institute*, **20**, 653–664.
15. Burns, E.R. (1969). On the failure of self-inhibition of growth in tumors. *Growth*, **33**, 25–45.

16. Bullough, W.S. (1962). The control of mitotic activity in adult mammalian tissues. *Biological Reviews*, **37**, 307–342.
17. Bullough, W.S. (1965). Mitotic and functional homeostasis: a speculative review. *Cancer Research*, **25**(10), 1683–1727.
18. Bullough, W.S. and Rytömaa, T. (1965). Mitotic homeostasis. *Nature*, **205**, 573–578.
19. Bullough, W.S. (1975). Mitotic control in adult mammalian tissues. *Biological Reviews*, **50**, 99–130.
20. Houck, J.C. (1976). *Chalones*. Amsterdam: North Holland.
21. McPherron, A.C., Lawler, A.M. and Lee, S.-J. (1997). Regulation of skeletal muscle mass in mice by a new TGF-beta superfamily member. *Nature*, **387**, 83–90.
22. Lee, S.-J. and McPherron, A.C. (2001). Regulation of myostatin activity and muscle growth. *Proceedings of the National Academy of Sciences*, **98**(16), 9306–9311.
23. Lee, S.-J. (2003). Regulation of muscle mass by myostatin. *Annual Review of Cell Developmental Biology*, **20**, 61–86.
24. Gamer, L.W., Nove, J. and Rosen, V. (2003). Return of the chalones. *Developmental Cell*, **4**(2), 143–144.
25. Weiss, R.A. and Njeuma, D.L. (1971). Growth control between dissimilar cells in culture. In *Growth Control in Cell Cultures*. CIBA Foundation Symposium, London: Churchill Livingstone, pp. 169–186.
26. Brand, K.G., Buoen, L.C. and Brand, I. (1967). Premalignant cells in tumorigenesis induced by plastic film. *Nature*, **213**, 810.
27. Boone, C.W., Takeichi, N., Eaton S del A. and Paranjpe, M. (1979). 'Spontaneous' neoplastic transformation in vitro: a form of foreign body (smooth surface) tumorigenesis. *Science*, **204**, 177–179.

CHAPTER 13. HUMAN OVERPOPULATION

1. Carr-Saunders, A. (1922). *The Population Problem*. Oxford: Clarendon Press.
2. Huxley, A. (1978). *The Human Situation*. London: Chatto and Windus.
3. Erhlich, P.R. and Erhlich, A.H. (1990). *The Population Explosion*. New York: Simon and Schuster.
4. Macfarlane, G. (1978). *Howard Florey: The Making of a Great Scientist*. Oxford: Oxford University Press.
5. Eigen, M. and Winkler, R. (1982). *Laws of the Game: How the Principles of Nature Govern Chance*. Harmondsworth, UK: Allen Lane.
6. Foerster, H. von, Mora, P.M. and Amiot, L.W. (1960). Doomsday: Friday, 13 November, A.D. 2026. *Science*, **132**, 1291–1295.
7. Pearl, R. (1927). The growth of populations. *Quarterly Review of Biology*, **2**(4), 532–548.
8. Lutz, W., Sanderson, W. and Scherbov, S. (2001). The end of population growth. *Nature*, **412**, 543–545.
9. Tanner, J.T. (1966). Effects of population density on growth rates of animal populations. *Science*, **47**(5), 733–745.
10. Christian, J.J. and Davis, D.E. (1964). Endocrines, behavior, and population: social and endocrine factors are integrated in the regulation of growth of mammalian populations. *Science*, **146**, 1550–1560.
11. Christian, J.J. (1970). Social subordination, population density, and mammalian evolution. *Science*, **168**, 84–90.

12. Davidson, J. (1934). On the growth of the sheep population in Tasmania. *Transactions of the Royal Society of South Australia*, **62**, 342–346.
13. Birdsell, J.B. (1978). Spacing mechanisms and adaptive behaviour of Australian Aborigines. In *Population Control by Social Behaviour*. London: Institute of Biology, pp. 213–244.
14. Tindale, N.B. (1974). *The Aboriginal Tribes of Australia*. Berkeley, CA: University of California.
15. Leakey, R. and Lewin, R. (1979). *People of the Lake. Man: His Origins, Nature and Future*. London: Collins.
16. Pearl, R. and Reed, L.J. (1920). On the rate of growth of the population of the United States since 1790 and its mathematical representation. *Proceedings of the National Academy of Sciences*, **6**(6), 275–288.
17. Forrester, J.W. (1973). *World Dynamics*. Cambridge, MA: MIT Press.
18. Meadows, D.H., Meadows, D.L., Randers, J. and Behrens, W.W., III (1972). *The Limits to Growth*. London: Pan Books.
19. Meadows, D.H., Meadows, D.L. and Randers, J. (1992). *Beyond the Limits: Global Collapse or a Sustainable Future*. London: Earthscan.
20. Song, J., Tuan, C.-H. and Yu, J.-Y. (1985). *Population Control in China: Theory and Application*. New York: Praeger.
21. Cohen, J.E. (1995). Population growth and the Earth's human carrying capacity. *Science*, **269**, 341–346.
22. Wackernagel, M. and Rees, W. (1996). *Our Ecological Footprint: Reducing Human Impact on the Earth*. Gabriola Island, BC, Canada: New Society.

CHAPTER 14. OUR FINITE EARTH

1. Hardin, G. (1968). The tragedy of the commons. *Science*, **162**, 1243–1248.
2. Costanza, R., d'Arge, R., de Groot, R., Farber, S., Grasso, M., Hannon, B., Limberg, K., Naeem, S., O'Neill, R.V., Paruelo, J., Raskin, R.G., Sutton, P. and van den Belt, M. (1997). The value of the world's ecosystem services and natural capital. *Nature*, **387**, 253–260.
3. Hardin, G. (1998). Extensions of 'The tragedy of the commons'. *Science*, **280**, 682–683.
4. Daly, H.E. (2005). Economics in a full world. *Scientific American*, **293**, 100–107.
5. Douthwaite, R. (1999). *The Growth Illu$ion: How Economic Growth Has Enriched the Few, Impoverished the Many and Endangered the Planet*. Dartington, UK: Green Books.
6. WWF (2008). *The Living Planet Report 2008*. Gland, Switzerland: WWF International.
7. Pearce, F. (1998). Population bombshell. *New Scientist*, 11 July 1998.
8. Brundtland, G.H. (1987). *Our Common Future: The World Commission of Environment and Development*. Oxford: Oxford University Press.
9. Krugman, P. (1994). Competitiveness: a dangerous obsession. *Foreign Affairs*, **73**(2), 28–44.
10. Hawken, P. (1994). *The Ecology of Commerce: A Declaration of Sustainability*. London: Weidenfeld and Nicholson.
11. Boulding, K. (1994). Economics of the coming spaceship Earth. In Daly, H. and Townsend, K. (eds.), *Valuing the Earth: Economics, Ecology, Ethics*. Cambridge, MA: MIT Press, pp. 297–310.
12. Diamond, J. (2005). *Collapse: How Societies Choose to Fail or Survive*. London: Allen Lane.

13. Daly, H.E. (1999). Five policy recommendations for a sustainable economy. *Feasta Reviews*, **1**, 1–11.
14. Daly, H.E. (2008). *A Steady State Economy*. Lecture given to the Sustainable Development Commission UK, 24 April 2008, pp. 1–10.

CHAPTER 15. THE MAIA HYPOTHESIS AND ANAGENESIS

1. Stebbing, A.R.D. (2006). Genetic parsimony: a factor in the evolution of complexity, order and emergence. *Biological Journal of the Linnean Society*, **88**, 295–308.
2. Huxley, J. (1942). *Evolution: The Modern Synthesis*. London: Allen and Unwin.
3. Rensch, B. (1959). *Evolution Above the Species Level*. London: Methuen.

Index